Studies in Systems, Decision and Control

Volume 630

The series "Studies in Systems, Decision and Control" (SSDC) covers both new developments and advances, as well as the state of the art, in the various areas of broadly perceived systems, decision making and control–quickly, up to date and with a high quality. The intent is to cover the theory, applications, and perspectives on the state of the art and future developments relevant to systems, decision making, control, complex processes and related areas, as embedded in the fields of engineering, computer science, physics, economics, social and life sciences, as well as the paradigms and methodologies behind them. The series contains monographs, textbooks, lecture notes and edited volumes in systems, decision making and control spanning the areas of Cyber-Physical Systems, Autonomous Systems, Sensor Networks, Control Systems, Energy Systems, Automotive Systems, Biological Systems, Vehicular Networking and Connected Vehicles, Aerospace Systems, Automation, Manufacturing, Smart Grids, Nonlinear Systems, Power Systems, Robotics, Social Systems, Economic Systems and other. Of particular value to both the contributors and the readership are the short publication timeframe and the world-wide distribution and exposure which enable both a wide and rapid dissemination of research output.

Indexed by SCOPUS, DBLP, WTI Frankfurt eG, zbMATH, SCImago.

All books published in the series are submitted for consideration in Web of Science.

Marialisa Scatà

The Power of Complex Systems

Unlocking Intelligence in Mobile Communication Networks

Springer

Marialisa Scatà
Dipartimento di Ingegneria Elettrica,
Elettronica e Informatica
University of Catania
Catania, Italy

ISSN 2198-4182 ISSN 2198-4190 (electronic)
Studies in Systems, Decision and Control
ISBN 978-3-032-04242-2 ISBN 978-3-032-04243-9 (eBook)
https://doi.org/10.1007/978-3-032-04243-9

This Springer imprint is published by the registered company Springer Nature Switzerland AG
The registered company address is: Gewerbestrasse 11, 6330 Cham, Switzerland

To my beloved family—my mom, dad, and brother.
To the love of my life, Alessandro, and to the light of my existence, my treasures, Emma and Alberto.
To all those who have believed—and continue to believe—in me.

Preface

In the age of hyperconnectivity, where the boundaries between the digital and physical worlds are increasingly blurred, understanding the hidden patterns that govern communications and behavior has never been more crucial. This book is part of my research journey that began with a compelling idea: that nature and society, people and their interactions, hold profound mining the can inspire the design of the future of communication systems.

Since the beginning of my research studies, I have sought to explore the deep interconnections between different disciplines, driven by an instinct that there is something inherently meaningful, perhaps even essential, in the way biological and social systems operate. Nature, after all, has optimized its processes over millions of years, and by observing its strategies, we can gain critical insights into how to build more resilient, adaptive, and intelligent technological and communication infrastructures. This perspective has guided my work everyday at the intersection of complex systems theory, bio-inspired computation, and the dynamics of social interactions to better study and design the communication networks.

The digital world we are building is not separate from the real one, but it is entangled with it. Communication networks, particularly the mobile and wireless ecosystems that shape our daily lives, are deeply embedded in social fabrics. People use them not only to exchange data but to construct meaning, sustain relationships, receive treatments, disappear loneliness, share opinion, inform other people and globally influence collective behavior. Consequently, the design of these systems must be informed not only by measurable metrics but also by those dynamics that remain unseen, residing in the space between engineered structures and emergent social phenomena.

This book is based on the concept of complexity, seen as a lens through which we can better understand the dynamics of modern communication. It brings together theoretical tools such as multilayer and higher-order networks, graph theory, evolutionary game models, and epidemic dynamics, integrating them with modern computational paradigms such as machine learning and artificial intelligence. Together, these methods allow us to map, simulate, and even predict how behaviors propagate and evolve in the interconnected systems we are building.

My aim is to give a framework that is as much about systems design as it is about systemic understanding. In doing so, I hope to contribute to a new kind of communication engineering—one that acknowledges and leverages the mutual feedback between social behavior and digital infrastructure.

At the heart of this vision is a call for interdisciplinarity: to build bridges between engineering and the social sciences, between data and meaning, between what we can measure and what we must still learn to perceive. The future of telecommunications lies not merely in faster speeds or smarter devices, but in our capacity to understand and model the forces that shape interaction, both human and artificial.

By blending scientific rigor with systems thinking, and by drawing inspiration from the natural world, we can design technologies that are not only efficient, but also adaptive, inclusive, and resilient in the face of global challenges such as climate change, pandemics, and social disruption. This book, I hope, is a small but meaningful part of this complex and interacting world.

Catania, Italy
June 2025

Marialisa Scatà

Competing Interests The author have no conflicts of interest to declare that are relevant to the content of this chapter.

Ethics Approval No ethics approval was required.

Contents

Chapter 1
Complex Systems in Communication Networks

Abstract The first chapter introduces the mission of the book which is the understanding of why complex systems and an interdisciplinary approach can have a crucial role in designing future networks. By discussing the evolution from 1G to 6G, the following chapter shows how integrating artificial intelligence, social sciences, and biology addresses emerging challenges such as energy efficiency, security, and complexity management. This innovative perspective enables the development of a resilient and adaptive network that can meet the future demands of the upcoming and ever-evolving digital society.

1.1 Introduction

To fully understand the disruptive impact of each new generation of mobile communications, it is essential to trace its historical evolution from 1G to 6G [1, 2]. Examining this progression enables us to identify the critical challenges that have emerged, analyze their evolution, and recognize why interdisciplinary integration with complex science can enhance the future of communications and society itself [3]. Mobile communication has transformed daily lives, shaping the global economy and human society. This transformation underscores the symbiotic relationship between technology and society, and now, society feeds what influences it most, the network. Over the past five decades, mobile telephony has revolutionized the way people live, work, and interact, leaving an important mark on global socioeconomic development [4]. The ability to communicate and access information on the move has introduced unprecedented modes of interaction and community evolution. With each new generation of wireless technology, its applications have become more deeply integrated into daily life, becoming a pivotal part of modern existence. The advent of 6G is expected to further redefine society by ensuring seamless, ubiquitous connectivity, enabling constant access to information, anytime and anywhere [5, 6]. In a world facing multiple unprecedented challenges, the evolution of human society and humanity's long-term survival are at risk, making universal access to information and communication so important [7, 8]. Addressing crises such as climate change [9], global pandemics

M. Scatà, *The Power of Complex Systems*, Studies in Systems, Decision and Control 630,
https://doi.org/10.1007/978-3-032-04243-9_1

[10], misinformation [11], and social inequalities [12] requires sustainable digital transformation on a global scale. Emerging digital technologies, with wireless networks at their core, will serve as the foundation for this transformation. In the coming decade, network evolution will facilitate the widespread adoption of solutions that tackle these challenges while delivering economic and social benefits at significantly lower operational costs. Just as the internet and mobile broadband once reshaped our interactions, working ways, and daily habits, the true social impact of the recent 6G will only fully materialize over time [13]. However, its potential is already becoming apparent through the lens of social and economic trends projected for 2030 and beyond. These trends, along with key regulatory and technological advancements, will be examined in the upcoming sections to provide a comprehensive framework for designing and deploying future networks. This book aims to give instruments for strategically thinking on a novel paradigm shift by embedding the principles of complex systems science into the analysis and design of communication networks, with a particular focus on wireless mobile infrastructures. Unlike traditional static infrastructures, wireless mobile networks are inherently dynamic and continuously evolving. They are not only deeply influenced by technological advancements, economic drivers, and social behavior, but they also have a reciprocal influence on how individuals and communities interact, access information, and shape digital ecosystems. Their pervasiveness and sensitivity to external factors make them suitable for analysis through the lens of complex systems.

By taking into consideration this perspective, we can uncover emergent behaviors, optimize resource distribution, and enhance the adaptability of mobile networks to meet changing demands. This systemic approach enables a more holistic understanding of the intricate, multilayered interdependence that defines modern communication ecosystems and their broader social implications. Beyond complexity science, this book gives a starting point for a transdisciplinary vision that brings together complex systems theory, bio-inspired computational models, and collective dynamics to inform the design and evolution of telecommunications and network infrastructures. This integrated framework offers an innovative lens for shaping future communication technologies by leveraging insights from network science and behavioral dynamics. The aim is to build bridges between the digital and physical worlds, enabling novel methodologies applicable across domains such as smart cities, digital health, industrial automation, sustainable infrastructure development, and cybersecurity. Wireless mobile networks play a pivotal role in this transformation. Their ability to both shape and be shaped by social behavior makes them particularly suitable for analysis through the lens of complex systems. In the coming years, their capacity to address real-world challenges will be essential for enabling widespread adoption and achieving long-term objectives such as sustainability, resilience, and digital trust next-generation 6G networks will act as catalysts for innovation across agriculture, logistics, manufacturing, and urban planning. They promise not only to reduce environmental impact and promote sustainable practices, but also to ensure equitable access to critical services, from education and healthcare to disaster response. Their inherent flexibility will allow them to maintain high service quality even under extreme conditions–an essential requirement for safety-critical applications and social continuity. The convergence of complexity science, network theory, and socio-technological dynamics allows for a truly comprehensive approach to future technological ecosystems. This systemic perspective is vital for tackling pressing global issues, digitalization, climate change, pandemics, and structural social inequality. Embedding these principles in the design of future networks ensures systems that are not only technologically advanced but also ethically aligned, socially inclusive, and adaptive to a rapidly evolving world. This holistic vision will define next-generation mobile communications as a transformative enabler of global well-being and sustainable progress.

This book is intended for a broad readership, including researchers, engineers, policymakers, and industry professionals. The opening chapter, titled "Complex Systems in Communication Networks," first traces the historical evolution of wireless mobile networks, setting the stage with an analysis of technological transitions from 5G to 6G. Then, the emerging challenges and future scenarios are discussed before delving into how engineering, biology, and social sciences converge in the design of tomorrow's communication networks. Finally, we introduce and shed light on the concept of complex systems, in terms of tools and definition, in the context of mobile networks. The chapter concludes with the outline, the structure, and the goals of this book.

1.2 From 1G to 6G

Mobile communications made their debut over forty years ago, in the 1980s, with what was known as first-generation networks, or 1G networks [6]. Table 1.1 summarizes the generational evolution of wireless technologies, from the analog voice services of 1G to the anticipated AI-native architectures and ultra-high-speed connectivity envisioned for 6G networks.

The first generation of wireless technology relied on analog systems and was limited to voice communication services. Various systems fell under the 1G, such as the Advanced Mobile Phone System (AMPS) in North America, the Nordic Mobile Telephone (NMT) in the Nordic countries, and the Total Access Communication System (TACS) in the United Kingdom. The launch of 1G represented a step toward greater connectivity, marking the beginning of the mobile telephony era. During that time, society was already undergoing significant technological changes, with the spread of personal computers and the growth of global networks, undoubtedly laying the foundation for what would later be known as the time of complex and interconnected systems. The disruptive aspect of 1G was offering users the freedom to make phone calls from any location within the coverage area, a benefit previously tied exclusively to stationary devices. However, the transformative potential of 1G was highly limited, and several factors contributed to this, such as the high cost of mobile phones, their large size and heavy weight, which made them unsuitable for carrying, and as a consequence, there wasn't a shift in behavior and habits. This created an initial social gap, although the 1980s were characterized by a consumer-driven cul-

Table 1.1 Evolution of Wireless Generations from 1G to 6G: Overview of the technological evolution from 1G to 6G, highlighting major advancements in data rate, frequency, bandwidth, and key applications. The 6G projections are based on current research directions in AI-native networking and THz communications. Acronyms: AMPS = Advanced Mobile Phone System, GSM = Global System for Mobile Communications, CDMA = Code Division Multiple Access, WCDMA = Wideband CDMA, OFDMA = Orthogonal Frequency Division Multiple Access, SC-FDMA = Single Carrier Frequency Division Multiple Access, LTE = Long Term Evolution, NR = New Radio, LDPC = Low Density Parity Check, FBMC = Filter Bank Multi-Carrier. The 6G projections are based on ongoing research in THz communication, AI-native networking, and semantic-aware applications

Generation	Access technology	Data rate	Frequency band	Bandwidth	Switching/Coding	Applications
1G	AMPS, NMT	2.4 kbps	800 MHz	30 kHz	Circuit/NA	Analog voice
2G	GSM, CDMA	10–200 kbps	850–1900 MHz	200 kHz–1.25 MHz	Circuit/Packet	Voice, SMS, basic data
3G	UMTS, WCDMA, CDMA2000	384 kbps–30 Mbps	800–2100 MHz	5 MHz	Circuit/Turbo codes	Web browsing, video calls
4G	LTE, WiMAX	DL up to 1 Gbps, UL 500 Mbps	1.8–3.5 GHz	1.4 MHz–20 MHz	Packet/Turbo codes	HD streaming, gaming, VoIP
5G	NR, FBMC	10–50 Gbps (expected)	1.8, 2.6 GHz, and mm wave (30–300 GHz)	Up to 60 GHz	Packet/LDPC	UHD video, VR/AR, IoT
6G	THz, OAM, RIS, LIS, quantum-assisted	≥100 Gbps–1 Tbps (expected)	100 GHz–3 THz	100 MHz–1 THz (selective)	AI-native, quantum-secured	Holographic telepresence, AI-native networking, digital twins, tactile internet, XR

ture, where technological devices served as status symbols. Despite this, collective behaviors were shifting as society gradually adapted to mobile connectivity, paving the way for the large-scale social transformations that later generations of wireless technology would inevitably bring. Meanwhile, complexity studies were beginning to gain ground, particularly in fields such as economics and ecology [14, 15]. The interconnectedness of technological, social, and economic processes, although not fully understood at the time, was becoming increasingly evident. Thus, while the impact of 1G remained modest, the seeds of a broader systemic change, in particular, linked to communication networks, social interactions, and technological progress, had already been planted.

The revolution in mobile communications occurred in the 1990s with the introduction of 2G networks. This generation introduces a crucial change from the analog systems to digital communication, by changing interactions among people. Differently from the previous generation, 2G enabled the transmission of text messages (SMS) and limited data services, leading to a pivotal transformation in everyday life. It provided better voice quality, greater energy efficiency, a reduced dimensions of mobile devices, making mobile phones portable, lightweight, and pocket-friendly. During the 2G era, the Global System for Mobile Communications (GSM) was introduced, initially in Europe and later adopted worldwide, establishing a standard that revolutionized global communication by allowing users to travel internationally and stay connected. The key innovation of 2G was in its transition from analog to digital technology, which made mobile phones more accessible to a larger number of people. This significantly reduced communication costs and increased mobile device penetration rates. In the mid-1990s, mobile phones became a common personal accessory with the introduction of the capability of making calls and sending text messages from anywhere. This marked a turning point when new technologies began introducing a new collective and social norm centered on availability and instant mobile communication. The growth of free-market economies and the mass consumer culture of the 1990s favored the accessibility and convenience of mobile phones. If we analyze the phenomenon from a complex systems perspective, the 2G period demonstrated how the scalability of technological adoption could trigger network effects that amplified social impact. As the value of the network grew exponentially, the system became more useful. This phenomenon amplified collective behaviors, such as the dissemination of information and the timing of events and at the same time we also had the development of converging digital ecosystems. Services such as mobile banking and early data services allowed users to perform simple online transactions, laying the ground for the future mobile-based digital economies. The concept of "being on the move" became synonymous with being connected, giving rise to the idea of permanent availability. The 2G showed how technological and social systems could co-evolve, creating feedback mechanisms that gave a boost to continuous innovation and large-scale collective transformations, preparing at the same time the basis for the development of the 3G and later.

Mobile communications experienced a major breakthrough with the introduction of 3G that shifted the focus of this generation from simple voice and text-based communication to data-driven services. Unlike previous generations, 3G offered the first integration between mobile telephony and the internet, giving a boost to a mobile broadband revolution that would transform how people accessed information and interacted with the world. The radio access technology behind 3G relied on code-division multiple access (CDMA) and connections achieved transmission rates close to the theoretical maximum described by Shannon's capacity limit, enabling high-speed data transfers. These networks were widely implemented, enabling significant technological advances and supporting features such as video calls, multimedia messaging, and web browsing on mobile devices. In the early years, the mobile internet ecosystem was still at an early stage, and most mobile phones of the time lacked touch screens, with operating systems that were not suitable for browsing and for online content. Furthermore, much of the internet content was designed for desktops and is not suitable for small screens and the limited processing power of mobile devices. The result was that early mobile internet usage was slow and not easy to use, with long loading times and poorly designed user interfaces. In 2007, with the introduction of smartphones with a touch screen interface, many things changed as it was an innovation that completely redefined how users interacted with their devices [16]. This innovation has been specifically designed for mobile internet use, integrating browsing, multimedia applications, and online services. 3G networks and smartphones profoundly influenced social and collective behaviors. In the early 2000s, mobile phones were still primarily viewed as communication tools, but with the rise of apps like social networking, instant messaging apps that replaced traditional SMS, they transformed into multifunctional personal hubs. The concept of "always-on connectivity" was becoming more real and was allowing users to consume information in real time, participate in virtual communities, and remain socially engaged from anywhere. A new era of digital socialization emerged and, at the same time, to be constantly connected has created the first trace of a digital overload that would later become important in both eras of 4G and 5G, where, what hgas been defined as hyper-connectivity, would become a crucial part of our daily life. As more users adopted smartphones, the demand for better apps, faster networks, and more content grew, creating a self-reinforcing cycle of innovation. The innovations introduced during the 3G were fully realized and scaled globally with the enhanced capabilities of 4G LTE, often indicated as the golden age of mobile internet, with higher speeds and reduced latency.

The 4G, launched in 2009, was specifically designed for mobile broadband, offering higher data rates, lower latency, and greater reliability. Thanks to the transition from 3G's CDMA technology to orthogonal frequency-division multiplexing (OFDM) was crucial in achieving this [17]. OFDM was chosen for its robustness in managing interference, flexibility across different channel conditions, and compatibility with multiple-input multiple-output (MIMO) systems, enabling more efficient data transmission through multiple antennas. One of the key innovations of 4G was its ability to leverage full-dimensional MIMO channels, making it possible to adapt to varying signal conditions and optimize real-time transmission rates. 4G provided high-speed internet capable of supporting a wide range of services, such as video streaming, real-time gaming, and high-definition video calls. The dominant 4G system globally was Long-Term Evolution (LTE), developed by the 3rd

Generation Partnership Project (3GPP) and first released as LTE (Release 8). LTE evolved through multiple stages, including LTE-Advanced (LTE-A) (Release 10) and LTE-A Pro (Release 13) [18]. Each evolution brought significant technological improvements that boosted performance and enabled a wider range of applications. These advancements made it possible to access applications like real-time gaming, 4K video streaming, and cloud-based services, transforming smartphones into powerful computing tools capable of competing with desktops and laptops. The 4G era influenced social behavior, business models, and daily habits [19]. With the proliferation of 4G networks, video streaming has grown with a lot of services. Social media platforms became central to people's lives, with users sharing videos, live streaming, and consuming vast amounts of multimedia content directly from their smartphones. From a complex systems perspective, the rapid spread of 4G networks demonstrated how technological advancements could trigger cascading effects across interconnected systems [20]. As bandwidth increased, there was an appearance of new apps and services, driving a self-reinforcing innovation feedback loop: faster networks encouraged the development of bandwidth-intensive applications, which in turn drove further investments in infrastructure.

One of the most significant developments during the 4G was support for Vehicle-to-Everything (V2X) communications, a technology designed to enable real-time communication between vehicles, infrastructure, pedestrians, and networks [21]. V2X communication was an important step toward the realization of autonomous vehicles and smart transportation systems. Machine-Type Communications (MTC) also emerged as a key component of the 4G ecosystem, laying the foundation for the Internet of Things (IoT) [22]. This enabled millions of connected devices, from smart home gadgets to wearable health devices, to communicate over mobile networks. While the 4G era brought unprecedented convenience and connectivity, it also introduced new social challenges. The "always-on" culture, fueled by the constant availability of high-speed internet, led to increased screen time and digital stress [23, 24]. The same era also witnessed the rise of cyberbullying and the spread of disinformation [25], as the speed at which information spreads is more than the society's ability to verify its accuracy, and at the same time, the rapid expansion of mobile connectivity helped bridge the digital divide. The 4G era can be considered a crucial moment in the evolution of mobile communications, bridging the gap between early mobile internet and the hyperconnected world of the future, paving the way for 5G networks that would bring even higher connectivity levels. By transforming smartphones into all-in-one devices for communication, entertainment, and productivity, 4G reshaped modern and mobile life and acted as a catalyst for social changes, redefining how individuals and businesses interact with the digital world.

In 2019, the world witnessed the launch of the first 5G networks, based on 3GPP's 5G System technology (Release 15) [26, 27]. Unlike previous generations, 5G was conceived as a flexible, software-native network, designed to support advanced mobile internet, mission-critical communications, and massive Internet of Things (IoT) connectivity. Its goal was not just to increase data transmission rates, but to transform network functionality by introducing service-oriented architecture, network slicing, and edge computing. With the ability to connect billions of devices and support diverse applications across industries, 5G represents a milestone in the evolution of mobile networks, paving the way for an era of hyperconnectivity and integrated artificial intelligence. During this period, society was already undergoing a rapid technological and social transformation, characterized by the rise of smart cities, autonomous vehicles, virtual reality (VR), and industrial automation, which relied on low-latency, high-reliability communication capabilities offered by 5G. At the same time, large-scale data management was becoming crucial for urban planning, public health monitoring, and environmental sustainability initiatives. The 5G system introduced two key innovations: a new core network (5G Core) and a new radio interface (5G New Radio (NR)). The 5G Core was designed to support service-based architecture (SBA), software-defined networking (SDN), network function virtualization (NFV), and network slicing. These technologies allow to dynamic allocation of network resources based on user requirements, enabling virtual networks optimized for specific applications. The radio access network (RAN) also brought significant innovations, integrating technologies such as millimeter wave (mmWave) communications, massive MIMO (multiple-input, multiple-output), small cells, edge computing, and private networks. With the development of 5G it has been introduced the support for non-terrestrial networks (NTN), including satellite-based communications and unmanned aerial systems (UAS). By providing connectivity between terrestrial, aerial, and satellite networks, 5G began to reduce the line between physical and digital spaces, creating opportunities for applications such as real-time remote surgery, precision agriculture using drones, and emergency response coordination. Release 18 (5G-Advanced) introduced the integration with artificial intelligence (AI) and machine learning (ML) into network design [28]. This marked a turning point where networks could learn, adapt, and optimize themselves based on real-time data. Both algorithms were embedded into the core and access networks in order to make predictions on a network and dynamically allocate bandwidth, optimize energy consumption, improve security, and the service quality. If we consider the case of the smart cities application, AI-powered 5G networks could prioritize traffic signals for autonomous vehicles, optimize power grids, and ensure connectivity for public safety systems [29]. The advent of 5G networks occurred with a shift in social behavior and collective habits [30]. With the increasing of popularity of streaming services, remote work, and cloud gaming, 5G provides the low latency and high bandwidth necessary to support real-time experiences, from augmented reality (AR) applications to multi-user virtual meetings. The proliferation of wearable devices and health monitoring systems allowed individuals to track fitness, monitor chronic conditions. 5G supported a dynamic digital ecosystem where devices, applications, and networks continuously interacted and evolved. Feedback loops became crucial in driving innovation: as users demanded more data-intensive services, networks adapted, and new applications emerged to meet those demands. The system continue to be self-reinforcing and for example, the growth of autonomous vehicles led to improvements in real-time data processing and edge computing, which in turn enabled better performance for smart city applications. By combining high-speed connectivity, large-scale device integration, and AI-driven intelligence, 5G has not only transformed current applications but also laid the foundation for future advancements as 6G, which is expected to change the boundaries of ubiquitous intelligence, immersive experiences, and sustainable communications [31].

The first commercial 6G networks are expected to be operational by 2030, marking the next step in the continuous evolution of mobile communications [1]. Currently in the development phase, with standardization set to begin in 2025, 6G promises to redefine global connectivity through advanced technologies that go beyond the limits of previous generations. Key developments include terahertz (THz) communications, ultra-massive MIMO, reconfigurable intelligent surfaces (RISs), ultra-dense networks (UDNs), and integrated terrestrial and non-terrestrial networks (INTNs). 6G will be the "Intelligent Network of Everything", integrating devices, sensors, vehicles, drones, and robots in real-time, and supporting not only communication but also computation, sensing, and energy management [6]. 6G aims to revolutionize how devices interact, learn, and decide autonomously, paving the way for mobile intelligence, also known as wireless or connected intelligence. This mobile intelligence is expected to reshape society at multiple levels, creating intelligent ecosystems that can adapt, evolve, and make autonomous decisions. 6G is designed to overcome the limitations of 5G by addressing the needs of a fully digital society. One of the most important developments is terahertz (THz) communication, which operates at extremely high frequencies and provides ultra-high bandwidth and data rates in the terabit-per-second range. This will enable instantaneous transmission of high-definition multimedia content, holographic communications, and real-time data exchanges between autonomous systems, supporting future use cases in areas such as telemedicine and immersive entertainment. Among the main features of 6G, the ultra-massive MIMO that leverages arrays of thousands of antennas to increase spectral efficiency and expand network capacity. Other key innovations are the reconfigurable intelligent surfaces (RISs) and Ultra-dense networks (UDNs). Furthermore, the integrated sensing and communication (ISAC) that reduce the lines between communication and environmental sensing, enabling networks to detect objects, movements, and changes in surroundings and the Integrated terrestrial and non-terrestrial networks (INTNs).

The defining feature of 6G will be mobile intelligence, representing the convergence of pervasive artificial intelligence (AI), advanced machine learning (ML), and enhanced connectivity. Unlike 5G, where AI is mainly used for network optimization, 6G will embed AI at the core of its architecture, extending it to the edge and even the air interface. This integration will enable networks to self-optimize, learn from real-time data, and make autonomous decisions, transforming how devices and systems interact, through the AI-driven network, the edge intelligence [32]. The rise of 6G will bring profound changes in social behavior and collective dynamics, as pervasive connectivity redefines how people live, work, and interact [33]. The 6G era will require new regulatory frameworks to ensure equitable access and the ethical use of technology, preventing social divides and digital exclusion. The ubiquitous Internet of Everything (IoE), encompassing sensors, machines, and intelligent devices, will create an ecosystem of interdependencies that accelerates technological progress. As 6G networks emerge, they will become the backbone of an AI-driven society, where mobile intelligence powers virtually every aspect of life. From autonomous vehicles seamlessly communicating with smart traffic systems to AI-assisted healthcare, 6G will deliver unprecedented levels of automation and personalization. The fusion of physical and digital realities will redefine what it means to be "connected" making 6G much more than a technological upgrade. It will be a shift in human-computer interaction and collective intelligence paradigms. By 2030, 6G is set to reshape economies, industries, and social norms, paving the way for an era where intelligent networks continuously adapt to changing needs and offer sustainable, inclusive, and transformative solutions for future challenges.

The illustration in Fig. 1.1 offers a synthetic yet detailed representation of the historical evolution of mobile generations from 1G to 6G. Such a visualization supports the holistic perspective adopted throughout this chapter, where 6G is understood as a socio-technical transition toward connected intelligence, immersive communication, and human-centric network design.

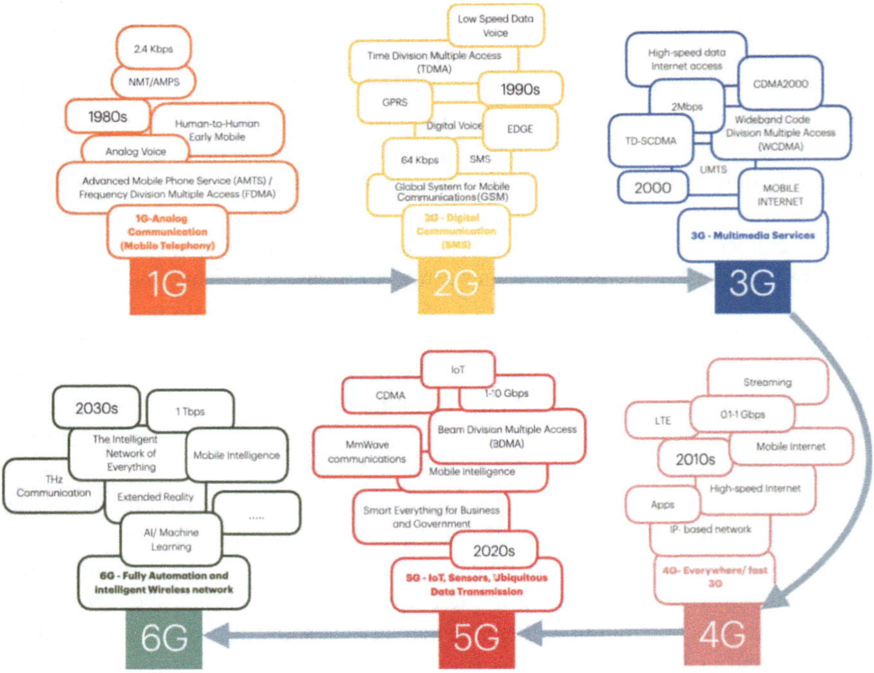

Fig. 1.1 Flow chart illustrating the evolution of mobile communication technologies from 1G to 6G. Each generation is represented by a colored box with key features and technologies. Arrows indicate the progression from one generation to the next

1.3 Evolving Paradigm, Standards and Trends: From 5G to 6G

Since the launch of 5G standardization efforts by 3GPP in early 2016, important advancements have been achieved in both technical specifications and global deployments. As illustrated in Table 1.2, the taxonomy of emerging 6G paradigms highlights a convergence of disruptive communication technologies (e.g., Terahertz, Optical Wireless), innovative architectures (e.g., cell-less and virtualized networks), and AI-driven enablers such as Big Data analytics and intelligent communication. Furthermore, the taxonomy expands into key application domains like Industry 5.0, smart healthcare, and virtual reality, emphasizing the challenges of security, privacy, and decentralized intelligence frameworks such as distributed machine learning.

The initial steps toward 5G standardization by 3GPP began in 2016. By the end of 2018, 3GPP had finalized the first 5G System standard, Release 15 [26], marking the initial phase of 5G development. The second phase, defined by Release 16 [34], was completed in mid-2020. The evolution continued with Release 17 [35],

Table 1.2 Enabling technologies, AI, use cases, and applications for 6G

Enabling technologies	• **Disruptive Technologies**: Terahertz (THz) communication, Optical Wireless Communication (OWC), full-duplex radios, channel estimation, network localization, advanced massive MIMO • **Innovative architectures**: cell-less architectures, 3D/Non-Terrestrial Networks (NTNs), Virtualization (NFV/SDN), Smart backhaul, energy gathering, blockchain, Wireless Information and Energy Transfer (WIET) • **AI-driven intelligence**: edge/fog intelligence, federated learning, self-organizing networks, reinforcement learning
AI technologies	• Big data analytics • Predictive optimization and control • Intelligent resource management • Semantic and goal-oriented communication • Explainable AI (XAI) for network transparency
Potential use cases	• **Extended Reality (XR)**: holographic telepresence, tactile internet, mixed reality collaboration • **Smart Healthcare**: Brain-Computer Interfaces (BCIs), digital twins of human body, real-time remote surgery • **Industrial IoT/Industry 5.0**: digital twins of factories, autonomous vehicles, UAV swarms, smart manufacturing • **Smart environments**: smart cities, environmental sensing, sustainable urban planning
Applications and research challenges	• Privacy, SECURITY, AND TRUST in AI-native networks • Distributed and decentralized learning at the edge • Social impacts and ethical frameworks • Quantum-safe networking and ultra-low latency • Resilience and robustness in socio-technical infrastructures

which was finalized in early 2022. Release 18, also referred to as 5G-Advanced [28, 31, 36], reached completion at the end of 2023. Currently, work is on Release 19, which is scheduled for finalization by mid-2025, indicated as 5G-Advanced Evolution [37]. These advancements have addressed the growing demand for enhanced mobile broadband and supported the expansion of 5G into diverse scenarios such as automotive, satellite communications, and the Internet of Things (IoT). Over three major releases, the 5G New Radio (NR) standard has established a comprehensive framework designed to support a wide array of use cases across an extensive range of frequency bands, by enabling flexible deployments in various environments, including indoor and outdoor settings, macro and small cell networks [38]. The 5G-Advanced is a term officially adopted by 3GPP in April 2021. Built on the solid foundation of existing 5G technologies. 5G-Advanced introduced a range of new capabilities designed to enhance overall performance, broaden the aim of supported use cases favoring the integration of emerging assets into the 5G ecosystem. It is crucial to underline that 5G-Advanced served as a critical bridge, laying the foundation for the development of 6G networks and the next for mobile communication. The 5G-Advanced evolution [37] takes a significant step forward with Release 19 (January 2024–June 2025). A key area of focus for Release 19 is expanding and refining various assets, including advancements in Non-Terrestrial Networks (NTN) for New Radio (NR) and IoT, femto cells, wireless access backhaul, low-power wake-up signals, ambient IoT, and multi-hop relay networks. In this context more significant efforts will be directed at further improving the enhanced Mobile Broadband (eMBB) experience, targeting areas such as extended reality (XR), mobility management, MIMO (Multiple-Input Multiple-Output) optimization, and duplexing techniques. In the 6G development roadmap, enhancements in key performance metrics will be done, including improved network energy efficiency, integration of AI/ML for air interface optimization and mobility management, next-generation Radio Access Networks (NG-RAN), self-organizing networks (SON), and minimization of drive tests (MDT). These developments collectively bridge the gap between current 5G capabilities and future 6G innovations [31]. The development of 6G is following a similar procedure to that of 5G. The ITU-R has taken on the responsibility of shaping the future of IMT systems for 2030 and beyond, referred to as IMT-2030, by defining key performance benchmarks, service requirements, and evaluation criteria. It also assesses potential Radio Interface Technologies (RITs) and selects systems that meet the IMT-2030 standards, which are expected to encompass 6G by 2030. In late 2022, ITU-R gived its initial technological report on IMT-2030, emphasizing future technology directions, innovative services, enabling technologies, and advancements in network and radio interface systems. In 2023, ITU-R completed its strategic vision with the release of a comprehensive framework detailing the objectives, scenarios, capabilities, and technological developments projected for IMT-2030 networks. The upcoming phase of IMT-2030 development, scheduled between 2024 and 2026, will focus on defining the performance and service requirements, along with the evaluation criteria and methodologies. The final phase, set for 2027–2030, will involve assessing and selecting Radio Interface Technologies (RITs) and finalizing the complete IMT-2030 specifications. At the same time, ITU-R will continue to define the frequency bands for IMT-2030 through its World Radio Conferences (WRCs). WRC 2023 has already identified potential spectrum bands, including 4.4–4.8, 7.1–8.4, and 14.8–15.3 GHz, with final allocations expected at upcoming WRCs, particularly WRC 2027. Additionally, 3GPP plays a critical role in the standardization of 6G, with efforts gradually shifting from 5G as its evolution reaches its limits. The 6G standardization process is expected to start in 2025, with Release 20 (scheduled from mid-2025 to the end of 2026) marking the initial studies on 6G and Release 21 (from early 2027 to mid-2028) delivering the first official 6G standards. The global rollout and commercialization of 6G networks are anticipated to begin around 2030.

In 2015, the United Nations established 17 interlinked Sustainable Development Goals (SDGs) to be achieved by 2030 and beyond [39, 40]. Since then, the Information and Communication Technology (ICT) and wireless network sectors have played a vital role in addressing key challenges, such as reducing poverty and CO2 emissions. As we move toward 2030, there is a broad consensus among industry, academia that future network technologies must accelerate sustainable development, enabling a more efficient usage of resources and fostering new, sustainable life. Wireless networks have long driven global economic growth, and 5G and future networks are designed to continue this process. By 2035, they could generate 13.2 trillion in global sales, accounting for 5% of global GDP, and create 22.3 million jobs globally. Beyond ICT, this kind of networks will also impact various assets such as physical, digital, and human. With advanced technological capabilities and human-centric design, future networks will not only support current business models but inspire new ones, fostering innovation and growth across sectors. Effective spectrum management will be crucial as governments explore diverse spectrum bands and access models. Addressing public concerns, particularly regarding electromagnetic field (EMF) exposure, will can be key, with guidelines provided by the International Commission on Non-Ionizing Radiation Protection (ICNIRP). Future networks will involve dense small-cell deployments, requiring refined regulatory frameworks to streamline approval processes and frequency allocations, ensuring seamless deployment and innovation. In a discussion on trends, security plays an important role, in fact, as connectivity underpins critical services and industries, network performance, reliability, and security become crucial [41]. End-to-end security considerations will be embedded in the design, ensuring resilient systems. There is the same interest on sustainability and energy consumption: future networks will address environmental sustainability by optimizing energy consumption, reducing CO2 emissions, and utilizing scarce resources efficiently [42]. Beyond direct impacts, networks will enable sustainable practice and efficient industrial processes. Mitigating rebound effects, where technological improvements lead to higher resource consumption, will be critical. Disruptive technologies are driving innovative trends in future communication networks, facing various challenges involving a multitude of sectors. Mobile network technology has continuously advanced, boosting performance and capacity while reshaping the global economy and enabling seamless communication worldwide. Future advancements will go beyond connectivity, embedding technology deeper into every aspect of modern society. As demands grow more diverse and resource-intensive, network architecture is being transformed into a re-imagined architecture to become more scalable, dynamic, and efficient. Building on 5G's service-based architecture (SBA), future networks will extend virtualization into the radio access network (RAN) for greater flexibility. The 6G architecture will decompose into platforms, functions, orchestration, and specialization, supported by open, scalable, and data-centric cloud environments with hardware acceleration [5]. AI-driven automation will enable autonomous network operations, supporting extreme slicing and specialized sub-networks. Another trend is the enhanced network capacity, given that future bandwidth requires utilizing high-frequency bands, such as sub-THz (100–300 GHz). However, challenges such as signal attenuation, power inefficiency, and susceptibility to failures must be overcome to ensure reliable coverage and performance. In this context, it is crucial to consider the new devices and interfaces, to design future networks to support a wide array of devices, beyond traditional phones and computers, enabling new human–machine interactions. IoT will expand its characteristics, driving a ubiquitous and sustainable deployment of connected systems [43]. The future will also see the interconnected physical and virtual networks across various scales, supporting local needs through private networks and edge processing, across the designing and analysis of the Networks of Networks (NoN) [44]. Digital twins [45] and dense sensor networks [46] will enable real-time data aggregation, automation, and enhanced quality of service (QoS), often in balancing with security [47]. Digital twins will create virtual replicas of physical objects and processes, providing real-time monitoring, predictive maintenance, and optimization. Their accuracy will rely on reliable wireless sensing and control, seamlessly integrating cyber-physical systems to enhance decision-making and operational efficiency. Future networks will also integrate high-precision localization (centimeter-level), sensing, and imaging (millimeter-level) using high-band signals ($>$100 GHz) and simultaneous localization and mapping (SLAM). Advanced algorithms will optimize communications and sensing, supporting applications like autonomous navigation and environmental monitoring [48]. Convergence, intelligence, and massive twinning are three major pivotal aspects in driving innovative trends across the 6G revolution [49]. AI and machine learning will revolutionize network design and operations. Moving beyond task-specific applications, general-purpose AI will enable distributed intelligence across networks, optimizing air interfaces, data processing, and resource management. AI-driven network reprogramming will allow real-time optimization, transforming how networks operate and respond to changing demands [32, 50]. These technological innovations will not only drive unprecedented performance improvements but will also fundamentally reshape network architecture and enable sustainable, secure, and intelligent systems. Together, they will form the backbone of the next-generation digital world, powering global progress and innovation.

1.4 Key Use Cases in 6G

6G is expected to significantly expand the capabilities of mobile networks, enhancing traditional communication features while introducing new functionalities that extend beyond conventional communication. To achieve this, 6G will address various performance dimensions such as capacity, latency, reliability, density, coverage, and mobility, enabling a wide range of applications and innovative. Previous generations of mobile networks had straightforward use cases–initially focusing on voice communication, and later, on data transmission with higher bit rates. With 5G, the scope expanded significantly beyond enhanced mobile broadband (eMBB) to include massive machine-type communications (mMTC) for low-power, low-data-rate devices and ultra-reliable low-latency communications (URLLC) for applications requiring high reliability and minimal latency [51, 52]. The introduction of 5G-Advanced brought new transformative use cases like extended reality (XR), blending virtual reality (VR) and augmented reality (AR) to create immersive digital experiences. In the 6G era, these and other use cases will evolve and expand–mobile broadband will reach remote regions via non-terrestrial networks (NTN) and satellites, while XR applications will transition from stationary, hotspot-based use to mobile, outdoor scenarios. This evolution will rely heavily on computational offloading, with processing tasks shifted from devices to the network, improving performance. Beyond XR, existing use cases will extend their reach, covering more people and devices under more demanding conditions [53]. However, 6G will also introduce revolutionary use cases, driven by new functionalities such as joint communication and sensing [54], integrated AI [55], and energy-neutral devices powered by ambient energy gathering [56]. Collaborations across sectors will give rise to "6G for sustainability" applications, helping reduce environmental impact, drive efficiency, and create entirely new roles for networks beyond traditional markets [13]. The evolution of 6G use cases reflects social, economic, and technological shifts. These use cases, categorized into broad and generic families, will span both evolutionary improvements and revolutionary transformations [53]. The EU-funded H2020 project Hexa-X, recognized as the European Commission's flagship 6G initiative, aims to define the role of 6G in shaping the future evolution of society [5]. Hexa-X bridges the human, physical, and digital worlds through an intelligent and interconnected technological ecosystem. The project outlines six core research challenges that serve as the foundation for forecasting key 6G use cases, while also accounting for significant social and economic trends. In this [5], an initial, yet non-exhaustive, set of use cases is identified to provide a baseline for guiding future 6G research. These use cases are informed by ongoing European research efforts under the 5G PPP and 6G-IA initiatives. The identified use cases span a broad spectrum–from evolutionary scenarios that extend and enhance 5G capabilities, to disruptive applications that introduce transformative possibilities and new social benefits, redefining connectivity. By integrating sensing, AI, and advanced compute capabilities, 6G will also introduce unprecedented applications, transforming industries and improving lives globally.

Novel use cases such as sensing and positioning, AI-driven Network, and Energy-Efficient and Sustainable Devices. High-precision localization, environmental sensing, and real-time tracking will open doors to applications in autonomous systems, smart infrastructure, and immersive virtual environments. AI-powered networks will enable real-time optimization, adaptive service delivery, and autonomous operation, transforming industries from logistics to entertainment. Zero-energy, ambient-powered devices will support large-scale IoT deployments, revolutionizing monitoring, asset management, and environmental protection. The use cases are categorized into distinct clusters based on their nature, technological focus, and the research challenges and values they address. These clusters offer a roadmap to align future 6G advancements with key innovation areas and social needs [55]. The use cases can be classified into communication-oriented and beyond communication [6].

1.4.1 6G Communication-Oriented Use Cases

In the communication-oriented group of use cases for 6G, we find out the following:

- Ultra-Broadband Multimedia Communications: 6G will provide enhanced support for immersive multimedia experiences across diverse environments, from dense urban to rural areas. Applications such as multi-sensory XR, holographic communications, information showers, and ultra-broadband mobile internet will require high capacity, low latency, and robust reliability. To achieve this, 6G will leverage expanded radio resources, fast network mechanisms, and resilient communication protocols. These advancements will further enhance telepresence experiences, enabling seamless XR-based gaming, virtual travel, remote education, and professional collaborations. In healthcare, telepresence will support virtual consultations, remote surgeries, and industrial training simulations, where advanced localization, sensing, and computational offloading will ensure high-fidelity interactions.
- Extreme Time-Sensitive and Mission-Critical Communications: For mission-critical environments such as smart factories, autonomous vehicle systems, emergency response operations, and healthcare systems, 6G will deliver extremely low latency and ultra-reliable connectivity. Advanced coding techniques, link diversity methods, and streamlined transmission processes will ensure rapid data delivery with minimal errors, supporting real-time decision-making in critical scenarios. This will enable seamless collaboration between humans and robots, transforming industrial environments into smart, adaptive ecosystems. As robots evolve into cobots (collaborative robots), they will interact with humans in real-time, enhancing productivity in healthcare, manufacturing, logistics, and smart city environments. These interactions will require precise sensing, minimal latency, and high bit rates to ensure secure and efficient collaboration.
- Ultra-Massive Communications 6G will enable the connectivity of an unprecedented number of devices, spanning IoT sensors, wearable technologies, smart appliances, and industrial robots. This connectivity will support smart environ-

ments, including smart cities, and automated warehouses, as well as large-scale IoT deployments in agriculture, environmental monitoring, and infrastructure management. Flexible spectrum management, efficient resource allocation, and massive grant-free network access will ensure seamless connectivity in dense device environments.

- Global-Scale Communications: To address the digital divide and expand connectivity to underserved regions, 6G will integrate terrestrial and non-terrestrial networks, combining satellite, aerial, and mobile communications. This will enable global coverage in remote land areas, maritime zones, and airspace. Applications such as remote mobile internet, environmental monitoring, and industrial tracking will become more accessible worldwide, aligning with the United Nations Sustainable Development Goals.
- Hyper-Mobility Communications: 6G will enhance connectivity in high-speed environments, such as high-speed trains and aerial transport systems, enabling uninterrupted communication at velocities reaching 500–1000 km/h. Flexible spectrum usage, fast handover techniques, and improved mobility management strategies will ensure stable connections even in rapidly changing network conditions.

1.4.2 6G Beyond Communication Use Cases

Within the group of beyond communication, we find the following:

- Network Intelligence: Artificial intelligence (AI) and machine learning (ML) will play a crucial role in enhancing 6G network capabilities. AI will enable autonomous network management, resource optimization, and intelligent decision-making. Applications such as intelligence-as-a-service, network-assisted smart systems, and computation offloading will rely on powerful computing resources and efficient AI model training and inference. By combining AI with edge caching and efficient data acquisition, 6G will provide smarter, faster, and more adaptive networks.
- Network Sensing: 6G will expand cellular network capabilities to enable advanced sensing functionalities. This will include object detection, activity recognition, movement tracking, and environmental sensing, supporting applications like autonomous vehicles, smart surveillance, and healthcare monitoring. The integration of communication and sensing will enhance situational awareness and improve safety in both public and industrial settings.
- Network Energy: Wireless energy transfer (WET) will be a key innovation in 6G, enabling energy-neutral devices that operate without traditional battery systems. By gathering ambient energy and employing energy beamforming techniques, 6G will power IoT sensors, smart devices, and low-energy systems in remote

and hard-to-reach environments. This will enable sustainable networks capable of supporting long-lasting and autonomous operations.

- Massive Twinning: 6G will expand digital twin technology, creating real-time, dynamic digital replicas of physical systems. These digital twins will enhance manufacturing, urban planning, logistics, and autonomous systems by enabling predictive maintenance, remote monitoring, and process optimization. Enhanced low-latency communication, AI-driven analytics, and edge computing will ensure that digital twins operate seamlessly in interconnected environments.
- Trusted Embedded Networks: 6G will introduce secure, autonomous, and infrastructure-free embedded networks to support decentralized communication. These networks will enable flexible, adaptable connectivity models ideal for micro-networks, localized coverage, and mission-critical applications. This decentralized design will enhance security, privacy, and resilience in production, healthcare, and emergency response systems.
- Hyperconnected Resilient Network Infrastructures: 6G will redefine network resilience with intelligent, self-healing systems that ensure continuous operation under dynamic conditions. AI-assisted V2X (Vehicle-to-Everything), interconnected IoT micro-networks, and automated disaster response systems will support real-time monitoring, secure mission-critical operations, and enhanced public safety. These adaptable networks will offer improved fault tolerance and uninterrupted communication in complex environments.
- Enabling sustainability: sustainability will be a fundamental principle in 6G network design. By leveraging AI, automation, and energy-efficient architectures, 6G will minimize environmental impact while supporting vital applications such as e-health services, autonomous supply chains, and sustainable agriculture. Energy-efficient network models will reduce carbon footprints, improve infrastructure resilience, and enable green connectivity initiatives that promote global sustainability.

1.5 Challenges in 6G

The transition from 5G to 6G represents a significant shift in mobile communications, introducing the concept of "Connected Intelligence," where artificial intelligence (AI), machine learning (ML), and advanced sensing technologies are deeply integrated into network infrastructures. This integration will enable networks to perform self-optimization, predictive maintenance, and real-time adaptation, enhancing both efficiency and performance. Devices and infrastructure will communicate, learn, and autonomously adjust operations based on environmental and user-specific contexts, blurring the lines between computation and communication. Processing data closer to the source will reduce latency and improve energy efficiency, supporting applications such as autonomous systems, real-time digital twins, immersive extended reality (XR), and large-scale human-machine collaboration. The seamless integration of terrestrial, aerial, and satellite networks will create a hyper-connected world where devices, sensors, and intelligent systems operate as a unified ecosystem. However, achieving this vision presents several challenges, including managing distributed intelligence, ensuring security, maintaining sustainability, and achieving interoperability across diverse networks. One major challenge is supporting ultra-high data rates through Terahertz (THz) spectrum utilization. 6G aims to deliver peak data rates of up to 1 terabit per second (Tbps) by leveraging THz frequencies. However, THz signals suffer from high propagation losses, atmospheric absorption, and limited penetration capabilities, making long-distance signal transmission challenging. Solutions such as beamforming, intelligent reflecting surfaces (IRS), ultra-massive multiple-input multiple-output (MIMO), and adaptive waveform design will be essential to optimize spectral efficiency and ensure reliable high-speed communication [57]. Another challenge lies in achieving extremely low latency for real-time applications such as tactile internet, remote robotic control, and holographic communication. These applications will require latency reductions below 1 millisecond, which traditional cloud-based architectures may struggle to achieve. To meet these demands, 6G will rely on distributed computing models like mobile edge computing (MEC), enabling processing closer to end devices to reduce delays and improve response times [55].

6G will also introduce a three-dimensional (3D) networking paradigm, expanding beyond terrestrial networks to integrate non-terrestrial networks (NTNs) such as satellites, high-altitude platforms (HAPs), and unmanned aerial vehicles (UAVs). This multi-layered connectivity will ensure global coverage but will require sophisticated coordination strategies to manage multi-hop relays, inter-layer handovers, and dynamic resource allocation while minimizing interference and optimizing spectrum use [58]. AI-driven network management is another defining feature of 6G. By leveraging AI and ML for network automation, self-organizing networks (SONs) will predict traffic congestion, dynamically allocate resources, and improve fault detection and recovery. However, integrating distributed AI models into network operations introduces risks such as model transparency issues, algorithmic bias, security vulnerabilities, and computational overhead. Federated learning and decentralized AI architectures will be essential to maintaining network efficiency while protecting user privacy [59].

Security and privacy will be a great challenge since they become increasingly complex in 6G due to its reliance on distributed intelligence and autonomous decision-making. Threats such as AI-powered cyberattacks, data poisoning, and adversarial machine learning could compromise network integrity. To mitigate these risks, advanced security techniques such as quantum-resistant encryption, AI-driven anomaly detection, and zero-trust security models will be required to safeguard sensitive data and infrastructure [60]. Energy efficiency and sustainability will also be critical in 6G networks, especially given the increased complexity introduced by dense cell networks, THz communication, and AI-driven processing. Green AI models, energy-gathering IoT devices, and low-power hardware architectures will be essential to balance high-performance demands with sustainable practices. Strategies such as dynamic energy allocation, infrastructure sharing, and intelligent energy gathering will minimize the environmental impact of 6G deployment [13]. Spectrum scarcity and interference management will become increasingly challenging as higher frequency bands are used. Efficient spectrum allocation, dynamic spectrum sharing, and interference mitigation techniques will be critical for maintaining seamless connectivity across different network layers. Techniques such as cognitive radio and intelligent spectrum sensing will help optimize spectrum use without compromising signal quality in congested environments [61].

In conclusion, 6G is will revolutionize mobile communications by integrating AI, sensing, and ultra-reliable connectivity into an intelligent, self-optimizing ecosystem. However, the transition from 5G to introduces significant challenges and to address these, researchers must develop innovative solutions in design, network management, and sustainable deployment strategies, applications to enhance connectivity, revolutionize industries, and improve social well-being worldwide.

1.6 Engineering, Social Sciences and Biology

Today's systems are growing at an unprecedented rate, incorporating an ever-increasing number of components while experiencing rapid surges in information flow. Mobile communication networks, especially 5G and emerging 6G systems, are at the forefront of supporting this expansion, yet managing their complexity, characterized by highly interdependent and heterogeneous components, has become a major challenge that could disrupt the ongoing information revolution. Addressing this challenge demands innovative solutions for designing, controlling, modeling, and monitoring these dynamic and distributed systems. Today's 6G networks will operate as socio-technical environments, a hyper-networked ecosystem where humans and machines interact within a cyber-physical framework. This complexity calls for the integration of diverse mathematical tools and interdisciplinary methods to address shared properties across various systems. This multi-domain approach can discover similarities with multiple disciplines, including biology [62, 63], communication networks [64–66], social networks [67, 68], spreading dynamics [10, 69], air transportation [70], and brain computing systems [71]. These systems share common underlying principles, such as emergent behaviors [72], connectivity patterns, and dynamic interactions, which are central to understanding and optimizing communication networks. The ability to transfer insights across these domains enables the development of generalizable models and methods for 6G network design [64, 73]. Modern communication networks have evolved from closed, semi-hierarchical structures to open, distributed, and highly interconnected ecosystems, requiring new paradigms for self-organization, self-adaptation, and continuous optimization to meet dynamic user demands. With dense deployments, an increasing number of base stations, and billions of connected devices, the integration of Artificial Intelligence (AI), Machine Learning (ML), and Terahertz (THz) communications in 6G further intensifies network complexity. The softwarization of network functions and the dynamic orchestration of services introduce additional layers of abstraction, making traditional information theories insufficient for capturing the behavior and evolution of these networks.

Complex systems theory provides an important mathematical approach for addressing these challenges, offering models, mechanisms, and methodologies to better understand and manage highly dynamic and adaptive communication networks. As networks transition towards 6G and beyond, new approaches that go beyond traditional performance metrics like latency, reliability, and capacity are necessary [74]. The evolution of 6G is not merely a technical upgrade but rather the creation of a fully interconnected ecosystem, seamlessly linking the human, physical, and digital worlds. This vision supports applications such as autonomous robotics, multisensory extended reality (XR), distributed sensing, and highly personalized user experiences, all requiring an intelligent, multi-layered architecture for efficient resource management and dynamic service orchestration. To achieve this vision, interdisciplinary solutions integrating engineering, social sciences, and biology are essential. Engineering disciplines provide the foundation for radio system design, cloud-edge computational architectures, and network management, while machine learning and AI drive automation, adaptive decision-making, and real-time network optimization [64, 67]. These technologies improve the dynamic allocation of network resources, making complex decisions regarding network slicing, load balancing, and quality of service (QoS) based on changing conditions [75]. However, purely technical solutions are insufficient. Incorporating biological and social insights can lead to more resilient and adaptive network designs [62, 76, 77]. Biological systems exhibit efficient mechanisms for resource allocation, adaptation, and cooperative behavior, which can inspire bio-inspired algorithms for network function placement and resource management [64]. Similarly, evolutionary game theory (EGT) can model strategic interactions in dynamic environments, offering insights into competition, cooperation, and self-organizing behavior among network nodes [78]. Beyond engineering and biology, social sciences play a crucial role in shaping user-centered network designs by analyzing human behavior, social interactions, and trust dynamics within digital environments [63]. As 6G networks aim to provide highly personalized services, insights from sociology and behavioral sciences will help optimize service delivery, ensuring they meet social needs while enhancing user experience, security, and inclusivity [79]. From an interdisciplinary perspective, cooperative decision-making models emerge as a vital tool, integrating biological adaptation mechanisms, engineering principles of optimization, and social feedback dynamics [80]. This hybrid approach, often called bio-inspired or socially inspired networking, allows networks to function as intelligent, self-organizing systems capable of adaptation and growth. As future networks expand in scale and complexity, integrating diverse fields will be key to designing sustainable, scalable, and efficient communication infrastructures that not only meet technological demands but also align with humanity's broader social and environmental goals.

Complex systems are formed by interacting constituents whose collective behavior leads to emergent phenomena, outcomes that cannot be predicted by analyzing individual components in isolation [81, 82]. Graph theory, multiplex networks, higher-order networks, or motifs can provide robust analytical frameworks and tools to deeply understand communication networks. These methods and representations enable the study of interactions between nodes in a single graph [83] or across multiple layers [84], each representing different network dimensions, that could represent radio access, the core network, or various service layers. The complex networks and their methodologies can reveal richer structures and interactions, patterns, and dynamics, offering deeper insights into how localized changes propagate across the entire system. This representation helps quantify complex properties such as information flow, resilience, and emergent dynamics, building a complex tool for optimizing end-to-end performance in real-time [64]. The layered interactions within multiplex networks allow us to model and anticipate network-wide phenomena, such as traffic congestion, fault recovery, and collaborative resource allocation. By analyzing community structures within the network, it is possible to detect clusters of nodes that play critical roles in maintaining connectivity and balancing resource demands accordingly. This capability is particularly important for maintaining reliable communication under varying conditions and supporting diverse use cases, from real-time autonomous vehicles to ultra-low-latency industrial applications.

Another key aspect of complex systems applied to the communication networks is their ability to self-organize and adapt, which will be central to 6G's evolutionary dynamics of competition and cooperation [72, 80]. Evolutionary Game Theory (EGT) [85] is a tool that offers a powerful framework for modeling these interactions, treating network nodes as strategic players that compete or collaborate to achieve optimal or sub-optimal solutions. Nodes dynamically adjust their strategies based on real-time feedback, allowing the system to converge toward a common, globally optimal or sub-optimal state. For instance, EGT can be applied for resource allocation problems, where nodes negotiate bandwidth, processing power, or energy based on mutual benefits and constraints. By simulating these strategic interactions, 6G networks can dynamically allocate resources in a way that maximizes efficiency and reliability.

Moreover, competition and cooperation modeled through EGT help in balancing the conflicting objectives, such as maintaining high throughput while minimizing energy consumption or ensuring low latency during peak traffic periods. As the network becomes more complex, hybrid approaches combining EGT with machine learning algorithms will further enhance adaptability by predicting network behavior and preemptively addressing potential bottlenecks or failures. The spreading dynamics represent tools and methods crucial in capturing the impact that the dynamics of collective behaviors [86]. In networks, a multitude of phenomena spread interpersonally, and following classical epidemiological models, or studies about the coevolution of different spreading processes, this aspect involves several research fields in network science. Thanks to this analysis it is possible to detect the real nature of social ties that have a key role in a phenomenon on social networks. Several studies investigate the role of awareness in the spreading process and interactions, disclosing its effect on networked individuals. Many studies have explored the spreading and competition to unveil the dynamic nature of networks [10, 65, 69]. The integration of complex network modeling, spreading processes analysis, and evolutionary game theory will create a self-sustaining, data-driven ecosystem within 6G. By continuously gathering and analyzing data and dynamics under certain kinds of events (such as pandemic or war) and across multiple domains–including radio access, core network, and application layers–the system will make intelligent, decentralized and also strategic decisions in real time. This cooperative, layered approach ensures that 6G networks remain resilient to disruptions, scalable to accommodate growing demands, and adaptive to changing environmental and user needs. For this reason, an interdisciplinary integration of network science, complexity theory, and AI will enable 6G networks to go beyond traditional limitations [87]. By leveraging self-organization, cooperation, and the emergent properties of complex systems, future networks will dynamically evolve to meet the demands of future applications, from autonomous systems and smart cities to personalized digital experiences and sustainability-driven solutions. Through this approach, it will not only deliver unparalleled performance but also act as a catalyst for social and technological advancement. This calls for a formal modeling framework grounded in the science of complex systems. In the following section, we explore how complexity theory provides the mathematical and conceptual tools to address the dynamic, multi-scale nature of future network ecosystems.

1.7 More on Complexity in Networks

Biological, social, and technological systems share fundamental features arising from the richness of interactions among their entities, which evolve and serve as a measure of their complexity [62]. In recent years, extensive research has enabled the study of various complex systems, modeled as networks where pairs of nodes interact and are connected through links. These networks span human communication systems, chemical reactions, ecological networks, and urban infrastructures [88]. Interactions in these systems may involve pairs of nodes, described in dyadic terms, or groups of three or more nodes, where interactions cannot be reduced to simple pairwise relationships [89]. Networks are pervasive in our surroundings, and we function as network units, interconnected through social relationships and part of biochemical reaction systems at the biological level. Networks can be tangible, with a physical structure, such as electrical grids, telecommunication networks, highways, railway systems, metro infrastructures, and neural networks. They can also be abstract, emerging from conceptual or relational structures, such as co-authorship networks in research publications or collaborative networks among professionals [90].

The study of networks has historically been associated with discrete mathematics, particularly graph theory. Swiss mathematician Leonhard Euler is credited with founding graph theory in 1736. He solved the Seven Bridges of Königsberg problem, proving that no closed path could traverse each bridge exactly once [91]. Over the decades, the mathematical theory of graphs has played a pivotal role in studying networks, enabling substantial progress in multiple fields. Social network analysis emerged in the 1920s, focusing on the relationships among social entities and revealing common structural patterns across various systems, including communication within groups, trade exchanges between nations, and economic transactions between companies [92]. Recent research has extensively analyzed complex networks, whose structure is highly irregular, intricate, and dynamically evolving. Initially, studies focused on small networks, but research has progressively expanded to include large-scale systems with millions of nodes [93]. Two groundbreaking contributions that shaped the modern study of complex networks are Watts and Strogatz's small-world network model [94], which demonstrated that many real-world networks exhibit short path lengths and high clustering, facilitating efficient information flow, and Barabási and Albert's scale-free network model [95], which revealed that numerous networks follow a power-law degree distribution, where a few nodes (hubs) have an exceptionally high number of connections, significantly affecting network dynamics.

Advancements in computational power and the availability of large-scale real-world datasets have enabled in-depth analysis of these structures and their underlying dynamical behaviors [96, 97]. Using complex network theory, several real-world systems can be modeled, including transportation networks such as air traffic [70], metro systems, and highway networks [98], telecommunication infrastructures [64], collaboration and citation networks in scientific research [90], biological and clinical networks such as protein interactions [99], epidemiological models [10], and genetic pathways, and neural networks [100], both biological (brain connectivity) [101] and artificial (deep learning architectures) [102]. A central finding in complex network studies is that many diverse systems share common structural and dynamic properties. One of the most important features is degree distribution, which describes how connections are distributed across nodes. In real-world networks, the degree distribution $P(k)$, defined as the probability that a randomly selected node has degree k, follows a power-law (scale-free) distribution: $P(k) \sim k^{-\gamma}$, where the exponent γ typically ranges between 2 and 3 [95].

Other significant properties include degree correlations, influencing how nodes with similar or different degrees connect, the small-world phenomenon, ensuring the existence of shorter paths between most node pairs, enhancing connectivity efficiency, and high clustering and motif structures [103, 104], indicating that real-world networks are non-random and often exhibit hierarchical or modular organization [105]. This theory has driven the development of new network models, as traditional random graph theory has often not been suitable in capturing the full complexity of real-world systems. Studying complex networks remains a fundamental research area, bridging mathematics, physics, computer science, and engineering, with wide-ranging applications across scientific and technological domains [88].

Understanding and modeling the structure of a complex network leads to a deeper knowledge of the mechanisms that have shaped it over time and, most importantly, to a better comprehension of its dynamic and functional behavior. The structure of a network represents the architecture of interactions. It allows for the evaluation of network robustness, its ability to respond to external perturbations, including failures, attacks, and unexpected disruptive events [106]. Furthermore, by observing systems and their behavior, interest has grown in studying the dynamic behavior of large interacting systems, emphasizing the role of topology in influencing the emergence of collective dynamic behaviors such as epidemic spreading, information propagation, and misinformation diffusion [107]. The theory of complex systems emerged in the 1970s as a framework for describing systems that cannot be understood through a reductionist approach [108]. The technological revolution, which has undeniably shaped the evolution of our society over the years, has further established the holistic approach as a fundamental way to describe and analyze these systems. The contribution of physics has been significant in providing a precise definition of complex systems, characterizing them as systems composed of many interacting components that collectively give rise to emergent behaviors.

In *More is Different* [109], Anderson argues that the behavior of large and complex aggregates of elementary particles cannot be understood merely by extrapolating the properties of a small number of particles. At each level of complexity, there is an emerging of new properties, and understanding these emergent behaviors requires novel research approaches. Anderson suggests that scientific disciplines can be approximately arranged in a hierarchical structure, where the fundamental entities of science X obey the laws of science Y. However, this hierarchy does not imply that science X is simply an application of science Y.

Although the definition of complex systems has historically been widely debated among scientists and extensively discussed in the literature, recent scientific consolidation has led to a broadly accepted definition: a complex system consists of many interacting elements that give rise to emergent phenomena [88]. The most intriguing aspects lie in observing such complex systems, understanding them, and determining how to translate them into applied scenarios to design new systems capable of performing similarly to those initially studied. Several fundamental questions arise: how many elements are necessary for emergent phenomena to manifest? What kind of behavior can emerge? What dynamics are expected to be observed? How can emerging dynamics be leveraged to improve certain aspects of the connection between network elements? How can performance be improved by embedding interaction rules within the network itself, enabling it to self-organize, learn, and enhance its condition by continuously responding to contextual changes? Complexity emerges from the interplay between randomness and order. The underlying topology of a complex system is intrinsically linked to a degree of randomness in the interaction network among its elements, following a network approach [110].

The study of complexity science has significant challenges in understanding its impact on society across various domains. Many of these aspects, with profound social implications, increasingly require integration of an interdisciplinary approach, incorporating bio-inspired and natural system models. Climate change, for example, is an area that urgently demands a highly interdisciplinary approach and a perspective from the social sciences, as understanding how society responds to climate change can inform strategies for improving global cooperation, which is essential for the planet's future [9]. Another crucial example is pandemic management, as demonstrated by the recent COVID-19 crisis [10]. Complexity science helps to understand and predict the evolution of such phenomena, supporting policymakers by assessing risks and modeling disease spread. Network science had already identified a high probability that the world would face a similar challenge, particularly when considering transportation systems as primary vectors for global transmission. Although this knowledge emerged from an interdisciplinary effort to understand complex phenomena, we were not ready for the scale and impact of COVID-19. To check its evolution, vast datasets on human mobility were integrated with predictive models to inform policymakers and support strategic decision-making, despite uncertainties regarding the biological evolution of the virus [111, 112].

Biology and complexity science have also enabled significant advancements in precision medicine [113, 114]. One of the greatest challenges is designing an interdisciplinary framework that integrates network science, machine learning, and artificial intelligence with neuroscience and molecular biology [63, 64]. It is now widely recognized that most diseases are complex, and understanding them requires studying the heterogeneity and complexity of the interactive networks within cells. In neuroscience, brain research continues to advance. Yet, the intrinsic multiscale complexity of the system, enriched at the molecular level by an extraordinary diversity of interactions, remains an important challenge for achieving a comprehensive understanding of brain function [115, 116]. In the next two decades, complexity will play a fundamental role in laying the foundations for the quantum Internet, which will require combining advances in quantum information theory with our understanding of classical complex communication systems, such as the current Internet [117]. One of the greatest challenges will be communication, which needs enhanced integration across different disciplines and research sectors. Interdisciplinary research has the potential not only to drive significant scientific advancements but also to disseminate and promote these innovations across multiple scientific domains. This facilitates a deeper understanding and the ability to predict, design, and control the interconnected socio-technical systems that underpin the digital and social ecosystems of today. A major contribution of complexity science is its ability to study and analyze these systems, helping to address global challenges through an interdisciplinary framework. We often refer to both networks and complex systems throughout this work. However, it is important to emphasize that networks serve as the mathematical skeleton of a complex system, where many parts (vertices) interact (through edges) and generate emergent behaviors. Networks represent real-world instances of phenomena that can be modeled using graph theory. This is particularly relevant in the context of next-generation networks such as 5G and 6G, given that features developed in recent generations require particular attention, as they strongly motivate a paradigm shift and the integration of complexity science approaches.

1.8 Overview of the Book

Mobile communications, wireless networks, and the technological advancements achieved during these decades have revolutionized the daily lives of billions of people. Over time, communications have not only supported but actively shaped the evolution of society, from social and behavioral to economic and infrastructural dimensions. Every sector has been touched, and their effects are evident across all critical layers of contemporary life. Today's world, recovering from a pandemic, must also confront climate change, social inequalities, misinformation, discrimination, and growing cyber threats. In this scenario, it becomes imperative to ask: how have the evolving structures of communication networks transformed society? And how are these very networks, now powered by artificial intelligence and machine learning algorithms that adapt based on our behaviors, guiding the way our collective future unfolds? The answer lies in recognizing that many aspects of our interconnected world can only be understood through the study of interactions and interdependencies. What we observe is not a single system, but a web of systems, interconnected worlds where interactions and relationships define the nature of the whole. Communication networks (Internet, wireless systems, mobile infrastructures), ecosystems, transport networks, financial markets, biological circuits, and social systems are all governed by similar underlying principles. These are, by their very nature, complex systems. The decision to focus this book primarily on mobile communications stems from the recognition that they inhabit a critical intersection between technological progress, societal evolution, and human behavior. Not only have mobile systems changed how we live and interact–they also offer a unique lens through which to observe the complex, emergent properties of modern human societies. They are both technological enablers and tools of behavioral insight. Through mobile devices, researchers can track spatial and temporal dynamics, analyze social interactions, monitor opinion trends, and simulate real-world phenomena like epidemic spreading, information cascades, and awareness diffusion. In doing so, mobile networks become key instruments in applying the principles of complex systems to real-world problems. They enable the study of how collective behavior emerges, how information spreads, and how social and artificial systems influence one another. This dual role, observational and transformative, places mobile communications at the heart of this book's vision: as both subjects and enablers of complexity science.

This book is structured to progressively build a transdisciplinary framework that integrates complexity science with mobile communications:

- **In this chapter** lays the groundwork by retracing the historical and technological evolution of mobile networks, and by introducing the key motivations for embedding complexity science into their analysis. It concludes with an important discussion on the relevance of complex systems in this domain.
- **Chapter** 2 delves into the structure and dynamics of complex networks. It introduces theoretical models, such as small-world, scale-free, and multilayer networks, and applies them to both technological and social systems.
- **Chapter** 3 focuses on communication systems as complex adaptive infrastructures. It discusses emergent paradigms like network-of-networks, edge intelligence, federated learning, and AI-native architectures, illustrating how these can be understood through complexity-aware approaches.
- **Chapter** 4 introduces the concept of Digital Anthropology, examining how social behaviors and human interactions shape and are shaped by communication tech-

nologies. It explores challenges such as misinformation, trust, and emergent dynamics in digital ecosystems.

- **Chapter** 5 offers a brief comparative evaluation of computational tools, namely R and Python, for modeling complex systems in communication contexts.

Throughout the book, examples, case studies, and graphical representations support the reader in exploring this new integrated perspective and approach. The ultimate goal is to offer a toolkit for rethinking and designing communication networks that are not only technically advanced but also socially responsive, ethically informed, and aligned with the evolving needs of a complex and dynamic world.

References

1. Jiang, W., Han, B., Habibi, M.A., Schotten, H.D.: The road towards 6G: a comprehensive survey. IEEE Open J. Commun. Soc. **2**, 334–366 (2021)
2. Gupta, A., Jha, R.K.: A survey of 5G network: architecture and emerging technologies. IEEE Access **3**, 1206–1232 (2015)
3. Jiang, W., Han, B.: Cellular Communication Networks and Standards: The Evolution from 1G to 6G. Springer Nature (2024)
4. Castells, M., Fernández-Ardèvol, M., Qiu, J.L., Sey, A.: Mobile Communication and Society: A Global Perspective. MIT Press (2009)
5. 5GPPP.: 6G Strategic Research and Innovation Agenda (2022). Retrieved from https://5g-ppp.eu/6g-sria/
6. Pennanen, H., Hänninen, T., Tervo, O., Tölli, A., Latva-Aho, M.: 6G: the intelligent network of everything. IEEE Access (2024)
7. Beniiche, A., Rostami, S., Maier, M.: Society 5.0: internet as if people mattered. IEEE Wirel. Commun. **29**(6), 160–168 (2022)
8. Nair, M.M., Tyagi, A.K., Sreenath, N.: The future with industry 4.0 at the core of society 5.0: open issues, future opportunities and challenges. In: International Conference on Computer Communication and Informatics (ICCCI) (2021)
9. George, G., Merrill, R.K., Schillebeeckx, S.J.: Digital sustainability and entrepreneurship: how digital innovations are helping tackle climate change and sustainable development. Entrep. Theory Pract. (2021)
10. Scatà, M., Attanasio, B., Aiosa, G.V., La Corte, A.: The dynamical interplay of collective attention, awareness and epidemics spreading in the multiplex social networks during COVID-19. IEEE Access **8**, 189203–189223 (2020)
11. Vicario, M.D., Quattrociocchi, W., Scala, A., Zollo, F.: Polarization and fake news: early warning of potential misinformation targets. ACM Trans. Web (TWEB) **13**(2), 10 (2019)
12. Tóth, G., Wachs, J., Di Clemente, R., Jakobi, Á., Ságvári, B., Kertész, J., Lengyel, B.: Inequality is rising where social network segregation interacts with urban topology. Nat. Commun. (2021)
13. Imoize, A.L., Adedeji, O., Tandiya, N., Shetty, S.: 6G enabled smart infrastructure for sustainable society: opportunities, challenges, and research roadmap. Sensors (MDPI) (2021)
14. Pippenger, N.: Complexity theory. Sci. Am. **238**(6), 114-125B (1978)
15. Bernstein, B.B.: Ecology and economics: complex systems in changing environments. Annu. Rev. Ecol. Syst. **12**, 309–330 (1981)
16. Park, Y., Chen, J.V.: Acceptance and adoption of the innovative use of smartphone. Ind. Manag. Data Syst. **107**(9), 1349–1365 (2007)
17. Goldsmith, A.: Wireless Communications. Cambridge University Press, Cambridge (2005)

18. Dahlman, E., Parkvall, S., Sköld, J.: 4G, LTE-Advanced Pro and the Road to 5G. Academic Press (2016)
19. Frattasi, S., Fathi, H., Gimmler, A., Fitzek, F.H.P., Prasad, R.: Designing socially robust 4G wireless services. IEEE Technol. Soc. Mag. **25**(2), 51–64 (2006)
20. Gomez, K., Goratti, L., Rasheed, T., Reynaud, L.: Enabling disaster-resilient 4G mobile communication networks. IEEE Commun. Mag. **52**(12), 66–73 (2014)
21. Garcia-Roger, D., González, E.E., Martín-Sacristán, D., Monserrat, J.F.: V2X support in 3GPP specifications: trom 4G to 5G and beyond. IEEE Access **8**, 190946–190963 (2020)
22. Ratasuk, R., Prasad, A., Li, Z., Ghosh, A., Uusitalo, M.A.: Recent advancements in M2M communications in 4G networks and evolution towards 5G. In: Proceedings of the 18th International Conference Intelligence in Next Generation Networks, pp. 52–57 (2015)
23. McDowall, A., Kinman, G.: The new nowhere land? a research and practice agenda for the "always on" culture. J. Organ. Eff. People Perform. **4**(3), 256–266 (2017)
24. Steele, R.G., Hall, J.A., Christofferson, J.L.: Conceptualizing digital stress in adolescents and young adults: toward the development of an empirically based model. Clin. Child. Fam. Psychol. Rev. **23**(1), 15–26 (2020)
25. Rezayi, S., Balakrishnan, V., Arabnia, S., Arabnia, H.R.: Fake news and cyberbullying in the modern era. In: Proceedings of the International Conference Computer Science Computer Intelligence (CSCI), pp. 7–12 (2018)
26. 3GPP.: Release 15 description: summary of Rel-15 work items. TR 21.915 V15.0.0 (2019)
27. Ghosh, A., Maeder, A., Baker, M., Chandramouli, D.: 5G evolution: a view on 5G cellular technology beyond 3GPP release 15. IEEE Access **7**, 127639–127651 (2019)
28. Lin, X.: An overview of 5G advanced evolution in 3GPP release 18. IEEE Commun. Stand. Mag. **6**(3), 77–83 (2022)
29. Gohar, A., Nencioni, G.: The role of 5G technologies in a smart city: the case for intelligent transportation system. Sustainability **13**(9), 5188 (2021)
30. Narayanan, A., Korium, M.S., Melgarejo, D.C., Hussain, H.M., Sena, A.S., Silva, P.E.G., Gutierrez-Rojas, D., Ullah, M., Nezhad, A.E., Rasti, M., et al.: Collective intelligence using 5G: concepts, applications, and challenges in sociotechnical environments. IEEE Access **10**, 70394–70417 (2022)
31. Chen, W., Lin, X., Lee, J., Toskala, A., Sun, S., Chiasserini, C.F., Liu, L.: 5G-advanced toward 6G: past, present, and future. IEEE J. Sel. Areas Commun. **41**(6), 1592–1619 (2023)
32. Letaief, K.B., Chen, W., Shi, Y., Zhang, J., Zhang, Y.J.A.: The roadmap to 6G: AI empowered wireless networks. IEEE Commun. Mag. **57**(8), 84–90 (2019)
33. Chukhno, O., Galinina, O., Andreev, S., Molinaro, A., Iera, A.: Interplay of user behavior, communication, and computing in immersive reality 6G applications. IEEE Commun. Mag. **60**(12), 28–34 (2022)
34. 3GPP.: Release 16 description: summary of Rel-16 work items. TR 21.916 V16.2.0 (2020)
35. 3GPP.: Release 17 description: summary of Rel-17 work items. TR 21.917 V17.0.1 (2021)
36. Liberg, O., Hoymann, C., Tidestav, C., Larsson, D.C., Rahman, I., Blasco, R., Falahati, S., Blankenship, Y.: Introducing 5G advanced. IEEE Commun. Stand. Mag. **8**(1), 52–57 (2024)
37. Lin, X.: The bridge toward 6G: 5G-advanced evolution in 3GPP release 19. IEEE Commun. Mag. **9**(1), 28–35 (2025)
38. Parkvall, S., Dahlman, E., Furuskar, A., Frenne, M.: NR: the new 5G radio access technology. IEEE Commun. Stand. Mag. **1**(4), 24–30 (2017)
39. United Nations (UN).: Transforming our world: the 2030 agenda for sustainable development. Accessed 7 Mar 2015. https://sdgs.un.org/2030agenda
40. Bulakcı, O., Li, X., Gramaglia, M., Gavras, A., Uusitalo, M., Rugeland, P., Boldi, M.: Towards sustainable and trustworthy 6G: challenges, enablers, and architectural design. Now Publishers (2023)
41. Suomalainen, J., Ahmad, I., Shajan, A., Savunen, T.: Cybersecurity for tactical 6G networks: threats, architecture, and intelligence. Futur. Gener. Comput. Syst. (2025)
42. Kamran, R., Kiran, S., Jha, P., Karandikar, A., Chaporkar, P.: Green 6G: energy awareness in design. In: 2024 16th International Conference on COMmunication Systems and NETworkS (COMSNETS), pp. 1122–1125. IEEE (2024)

43. Alwakeel, A.M., Alnaim, A.K.: Network slicing in 6G: a strategic framework for IoT in smart cities. Sensors **24**(13), 4254 (2024)
44. Rothkötter, M., Weise, J., Mostaghim, S.: Towards a new model for a 6G network-of-networks. In: 2023 IEEE Conference on Artificial Intelligence (CAI), pp. 255–256. IEEE (2023)
45. Masaracchia, A., Sharma, V., Canberk, B., Dobre, O.A., Duong, T.Q.: Digital twin for 6G: taxonomy, research challenges, and the road ahead. IEEE Open J. Commun. Soc. (2022)
46. Liu, F., Cui, Y., Masouros, C., Xu, J., Han, T.X., Eldar, Y.C., Buzzi, S.: Integrated sensing and communications: toward dual-functional wireless networks for 6G and beyond. IEEE J. Sel. Areas Commun. **40**(6), 1728–1767 (2022)
47. Fadlullah, Z.M., Mao, B., Kato, N.: Balancing QoS and security in the edge: existing practices, challenges, and 6G opportunities with machine learning. IEEE Commun. Surv. Tutor. **24**(4), 2419–2448 (2022)
48. Wymeersch, H., Shrestha, D., De Lima, C.M., Yajnanarayana, V., Richerzhagen, B., Keskin, M.F., Schindhelm, K., Ramirez, A., Wolfgang, A., De Guzman, M.F., et al.: Integration of communication and sensing in 6G: a joint industrial and academic perspective. In: 2021 IEEE 32nd Annual International Symposium on Personal, Indoor and Mobile Radio Communications (PIMRC), pp. 1–7. IEEE (2021)
49. Chowdhury, M.Z., Shahjalal, M., Ahmed, S., Jang, Y.M.: 6G wireless communication systems: applications, requirements, technologies, challenges, and research directions. IEEE Open J. Commun. Soc. **1**, 957–975 (2020). https://doi.org/10.1109/OJCOMS.2020.3010270
50. Matera, F., Settembre, M., Vizzarri, A., Mazzenga, F., Llorca, J., Tulino, A., Detti, A., Ronzani, D., Barbarossa, S.: Opportunities and challenges for the integration of AI in network evolution toward 6G. In: 2024 AEIT International Annual Conference (AEIT), pp. 1–6. IEEE (2024)
51. Sohaib, R.M., Onireti, O., Sambo, Y., Swash, R., Ansari, S., Imran, M.A.: Intelligent resource management for eMBB and URLLC in 5G and beyond wireless networks. IEEE Access **11**, 65205–65221 (2023)
52. Goshi, E., Mehmeti, F., La Porta, T.F., Kellerer, W.: Modeling and analysis of mMTC traffic in 5G core networks. IEEE Trans. Netw. Serv. Manag. (2024)
53. Hong, E.K., Lee, I., Shim, B., Ko, Y.C., Kim, S.H., Pack, S., Lee, K., Kim, S., Kim, J.H., Shin, Y., et al.: 6G R&D vision: requirements and candidate technologies. J. Commun. Netw. **24**(2), 232–245 (2022)
54. Jiang, W., Zhou, Q., He, J., Habibi, M.A., Melnyk, S., El-Absi, M., Han, B., Di Renzo, M., Schotten, H.D., Luo, F.L., et al.: Terahertz communications and sensing for 6G and beyond: a comprehensive review. IEEE Commun. Surv. Tutor. (2024)
55. Wang, S., Qureshi, M.A., Miralles-Pechuán, L., Huynh-The, T., Gadekallu, T.R., Liyanage, M.: Explainable AI for 6G use cases: technical aspects and research challenges. IEEE Open J. Commun. Soc. (2024)
56. López, O., Singh, R.K., Phan-Huy, D.T., Katranaras, E., Mazloum, N., Ruttik, K., Jäntti, R., Khan, H., Rosabal, O., Alexias, P., et al.: Zero-energy devices for 6G: technical enablers at a glance. IEEE Internet Things Mag. **8**(3), 14–22 (2025)
57. Rappaport, T.S., Xing, Y., Kanhere, O., Ju, S., Madanayake, A., Mandal, S., Alkhateeb, A., Trichopoulos, G.C.: Wireless communications and applications above 100 GHz: opportunities and challenges for 6G and beyond. IEEE Access **9**, 3680–3704 (2021)
58. Azari, M.M., Solanki, S., Chatzinotas, S., Kodheli, O., Sallouha, H., Colpaert, A., Montoya, J.F.M., Pollin, S., Haqiqatnejad, A., Mostaani, A., et al.: Evolution of non-terrestrial networks from 5G to 6G: a survey. IEEE Commun. Surv. Tutor. **24**(4), 2633–2672 (2022). https://doi.org/10.1109/COMST.2022.3200436
59. Letaief, K.B., Chen, W., Shi, Y., Zhang, J., Zhang, Y.J.A.: The roadmap to 6G: AI-empowered wireless networks. IEEE Commun. Mag. **57**(8), 84–90 (2024)
60. Rifa-Pous, H., Garcia-Font, V., Nunez-Gomez, C., Salas, J.: Security, trust and privacy challenges in AI-driven 6G networks (2024). arXiv:2409.10337
61. Muzaffar, M.U., Sharqi, R.: A review of spectrum sensing in modern cognitive radio networks. Telecommun. Syst. **85**(2), 347–363 (2024). https://doi.org/10.1007/s11235-023-01066-2

62. Scatà, M., Di Stefano, A., Giacchi, E., La Corte, A., Liò, P.: The bio-inspired and social evolution of node and data in a multilayer network. In: 2014 5th International Conference on Data Communication Networking (DCNET), pp. 1–6. IEEE (2014)
63. Di Stefano, A., Scatà, M., Vijayakumar, S., Angione, C., La Corte, A., Liò, P.: Social dynamics modeling of chrono-nutrition. PLoS Comput. Biol. **15**(1), e1006714 (2019). https://doi.org/10.1371/journal.pcbi.1006714
64. Scatà, M., La Corte, A., Marotta, A., Graziosi, F., Cassioli, D.: A complex network and evolutionary game theory framework for 6G function placement. IEEE Open J. Commun. Soc. (2024). https://doi.org/10.1109/OJCOMS.2024.XXXXXXX
65. Scatà, M., Di Stefano, A., La Corte, A., Liò, P.: A multiplex social contagion dynamics model to shape and discriminate D2D content dissemination. IEEE Trans. Cogn. Commun. Netw. (2020)
66. Sergiou, C., Lestas, M., Antoniou, P., Liaskos, C., Pitsillides, A.: Complex systems: a communication networks perspective towards 6G. IEEE Access **8**, 89007–89030 (2020)
67. Giacchi, E., Di Stefano, A., La Corte, A., Scatà, M.: A dynamic context-aware multiple criteria decision making model in social networks. In: International Conference on Information Society (i-Society 2014), pp. 157–162. IEEE (2014)
68. Xian, J., Yang, D., Pan, L., Wang, W., Wang, Z.: Misinformation spreading on correlated multiplex networks. Chaos: Interdiscip. J. Nonlinear Sci. **29**(11), 113123 (2019). https://doi.org/10.1063/1.5116750
69. Scatà, M., Di Stefano, A., La Corte, A., Liò, P.: Quantifying the propagation of distress and mental disorders in social networks. Sci. Rep. **8**(1), 5005 (2018). https://doi.org/10.1038/s41598-018-23397-7
70. Varga, I.: Weighted multiplex network of air transportation. Eur. Phys. J. B **89**, 1–5 (2016). https://doi.org/10.1140/epjb/e2016-70365-9
71. Battiston, F., Nicosia, V., Chavez, M., Latora, V.: Multilayer motif analysis of brain networks. Chaos: Interdiscip. J. Nonlinear Sci. **27**(4), 047404 (2017). https://doi.org/10.1063/1.4979282
72. Attanasio, B., La Corte, A., Scatà, M.: Evolutionary dynamics of MEC's organization in a 6G scenario through EGT and temporal multiplex social network. ICT Express **7**(2), 138–142 (2021). https://doi.org/10.1016/j.icte.2020.09.003
73. Scatà, M., Attanasio, B., La Corte, A.: Cognitive profiling of nodes in 6G through multiplex social network and evolutionary collective dynamics. Futur. Internet **13**(5), 135 (2021). https://doi.org/10.3390/fi13050135
74. Gupta, M., Jha, R.K.: Advanced network design for 6G: leveraging graph theory and slicing for edge stability. Simul. Model. Pract. Theory **138**, 103029 (2025). https://doi.org/10.1016/j.simpat.2024.103029
75. Wu, W., Zhou, C., Li, M., Wu, H., Zhou, H., Zhang, N., Shen, X.S., Zhuang, W.: AI-native network slicing for 6G networks. IEEE Wirel. Commun. **29**(1), 96–103 (2022). https://doi.org/10.1109/MWC.001.2100240
76. Forrest, S., Perelson, A.S., Allen, L., Cherukuri, R.: Self-nonself discrimination in a computer. In: Proceedings of 1994 IEEE Computer Society Symposium on Research in Security and Privacy, pp. 202–212. IEEE (1994)
77. Tero, A., Takagi, S., Saigusa, T., Ito, K., Bebber, D.P., Fricker, M.D., Yumiki, K., Kobayashi, R., Nakagaki, T.: Rules for biologically inspired adaptive network design. Science **327**(5964), 439–442 (2010). https://doi.org/10.1126/science.1177894
78. Di Stefano, A., Scatà, M., La Corte, A., Das, S.K., Liò, P.: Improving QoE in multi-layer social sensing: a cognitive architecture and game theoretic model. In: Proceedings of the Fourth International Workshop on Social Sensing, pp. 18–23. ACM (2019)
79. Wymeersch, H., Chen, H., Guo, H., Keskin, M.F., Khorsandi, B.M., Moghaddam, M.H., Ramirez, A., Schindhelm, K., Stavridis, A., Svensson, T., et al.: 6G positioning and sensing through the lens of sustainability, inclusiveness, and trustworthiness. IEEE Wirel. Commun. **32**(1), 68–75 (2025)
80. Di Stefano, A., Scatà, M., La Corte, A., Liò, P., Catania, E., Guardo, E., Pagano, S.: Quantifying the role of homophily in human cooperation using multiplex evolutionary game theory. PLoS ONE **10**(10), e0140646 (2015)

81. Cohen, R., Havlin, S.: Complex Networks: Structure, Robustness and Function. Cambridge University Press, Cambridge (2010)
82. Boccaletti, S., Latora, V., Moreno, Y., Chavez, M., Hwang, D.U.: Complex networks: structure and dynamics. Phys. Rep. **424**(4), 175–308 (2006)
83. Van Steen, M.: Graph theory and complex networks. An Introd. **144**(1) (2010)
84. Battiston, F., Nicosia, V., Latora, V.: Structural measures for multiplex networks. Phys. Rev. E **89**(3), 032804 (2014). https://doi.org/10.1103/PhysRevE.89.032804
85. Newton, J.: Evolutionary game theory: a renaissance. Games **9**(2), 31 (2018) . https://doi.org/10.3390/g9020031
86. Pastor-Satorras, R., Castellano, C., Van Mieghem, P., Vespignani, A.: Epidemic processes in complex networks. Rev. Mod. Phys. **87**(3), 925 (2015)
87. Cerquitelli, T., Meo, M., Curado, M., Skorin-Kapov, L., Tsiropoulou, E.E.: Machine learning empowered computer networks (2023)
88. Bianconi, G., Arenas, A., Biamonte, J., Carr, L.D., Kahng, B., Kertesz, J., Kurths, J., Lü, L., Masoller, C., Motter, A.E., et al.: Complex systems in the spotlight: next steps after the 2021 Nobel Prize in physics. J. Phys. Complex **4**(1), 010201 (2023)
89. Battiston, F., Cencetti, G., Iacopini, I., Latora, V., Lucas, M., Patania, A., Young, J.G., Petri, G.: Networks beyond pairwise interactions: structure and dynamics. Phys. Rep. **874**, 1–92 (2020). https://doi.org/10.1016/j.physrep.2020.05.004
90. Menichetti, G., Remondini, D., Panzarasa, P., Mondragón, R.J., Bianconi, G.: Weighted multiplex networks. PLoS One **9**(6), e97857 (2014). https://doi.org/10.1371/journal.pone.0097857
91. Shields, R.: Cultural topology: The seven bridges of Königsburg, 1736. Theory, Cult. Soc. **29**(4–5), 43–57 (2012). https://doi.org/10.1177/0263276412451163
92. Wasserman, S., Faust, K.: Social Network Analysis: Methods and Applications. Cambridge University Press, Cambridge (1994)
93. Newman, M.E.J.: The structure and function of complex networks. SIAM Rev. **45**(2), 167–256 (2003). https://doi.org/10.1137/S003614450342480
94. Watts, D.J., Strogatz, S.H.: Collective dynamics of 'small-world' networks. Nature **393**(6684), 440–442 (1998). https://doi.org/10.1038/30918
95. Barabási, A.L., Albert, R.: Emergence of scaling in random networks. Science **286**(5439), 509–512 (1999). https://doi.org/10.1126/science.286.5439.509
96. Gonzalez, M.C., Hidalgo, C.A., Barabási, A.L.: Understanding individual human mobility patterns. Nature **453**(7196), 779–782 (2008). https://doi.org/10.1038/nature06958
97. Mao, Y., You, C., Zhang, J., Huang, K., Letaief, K.B.: A survey on mobile edge computing: the communication perspective. IEEE Commun. Surv. Tutor. **19**(4), 2322–2358 (2017). https://doi.org/10.1109/COMST.2017.2745201
98. Ding, R., Ujang, N., Hamid, H.B., Manan, M.S.A., Li, R., Albadareen, S.S.M., Nochian, A., Wu, J.: Application of complex networks theory in urban traffic network researches. Netw. Spat. Econ. **19**, 1281–1317 (2019). https://doi.org/10.1007/s11067-019-09447-5
99. Jeong, H., Tombor, B., Albert, R., Oltvai, Z.N., Barabási, A.L.: The large-scale organization of metabolic networks. Nature **407**(6804), 651–654 (2000). https://doi.org/10.1038/35036627
100. Edwards, R., Glass, L.: Combinatorial explosion in model gene networks. Chaos: Interdiscip. J. Nonlinear Sci. **10**(3), 691–704 (2000). https://doi.org/10.1063/1.1287067
101. Presigny, C., Fallani, F.D.V.: Colloquium: multiscale modeling of brain network organization. Rev. Mod. Phys. **94**(3), 031002 (2022). https://doi.org/10.1103/RevModPhys.94.031002
102. Trabelsi, C., Bilaniuk, O., Zhang, Y., Serdyuk, D., Subramanian, S., Santos, J.F., Mehri, S., Rostamzadeh, N., Bengio, Y., Pal, C.J.: Deep complex networks (2017). arXiv:1705.09792
103. Lee, C., Wilkinson, D.J.: A review of stochastic block models and extensions for graph clustering. Appl. Netw. Sci. **4**(1), 1–50 (2019). https://doi.org/10.1007/s41109-019-0232-2
104. Lee, G., Ko, J., Shin, K.: Hypergraph motifs: Concepts, algorithms, and discoveries. Proc. VLDB Endow. **13**(12), 2256–2269 (2020). https://doi.org/10.14778/3407790.3407823
105. Murtagh, F., Contreras, P.: Algorithms for hierarchical clustering: an overview. Wiley Interdiscip. Rev.: Data Min. Knowl. Discov. **2**(1), 86–97 (2012). https://doi.org/10.1002/widm.53

106. Artime, O., Grassia, M., De Domenico, M., Gleeson, J.P., Makse, H.A., Mangioni, G., Perc, M., Radicchi, F.: Robustness and resilience of complex networks. Nat. Rev. Phys. **6**(2), 114–131 (2024). https://doi.org/10.1038/s42254-023-00655-3
107. Aïmeur, E., Amri, S., Brassard, G.: Fake news, disinformation and misinformation in social media: a review. Soc. Netw. Anal. Min. **13**(1), 30 (2023). https://doi.org/10.1007/s13278-023-01023-5
108. Simon, H.A.: The organization of complex systems. In: Models of Discovery: And Other Topics in the Methods of Science, pp. 245–261. Springer, Berlin (1977)
109. Anderson, P.W.: More is different: broken symmetry and the nature of the hierarchical structure of science. Science **177**(4047), 393–396 (1972). https://doi.org/10.1126/science.177.4047.393
110. Albert, R., Barabási, A.L.: Statistical mechanics of complex networks. Rev. Mod. Phys. **74**(1), 47 (2002). https://doi.org/10.1103/RevModPhys.74.47
111. Kraemer, M.U.G., Yang, C.H., Gutierrez, B., Wu, C.H., Klein, B., Pigott, D.M., Open COVID-19 Data Working Group, et al.: The effect of human mobility and control measures on the COVID-19 epidemic in China. Science **368**(6490), 493–497 (2020). https://doi.org/10.1126/science.abb4218
112. Chinazzi, M., Davis, J.T., Ajelli, M., Gioannini, C., Litvinova, M., Merler, S., Viboud, C., Vespignani, A.: The effect of travel restrictions on the spread of the 2019 novel coronavirus (COVID-19) outbreak. Science **368**(6489), 395–400 (2020). https://doi.org/10.1126/science.aba9757
113. Barabási, A.L., Gulbahce, N., Loscalzo, J.: Network medicine: a network-based approach to human disease. Nat. Rev. Genet. **12**(1), 56–68 (2011). https://doi.org/10.1038/nrg2918
114. Topol, E.: High-performance medicine: the convergence of human and artificial intelligence. Nat. Med. **25**(1), 44–56 (2019). https://doi.org/10.1038/s41591-018-0300-7
115. Bassett, D.S., Sporns, O.: Network neuroscience. Nat. Neurosci. **20**(3), 353–364 (2017). https://doi.org/10.1038/nn.4502
116. Kringelbach, M.L., Deco, G.: Brain states and transitions: insights from computational neuroscience. Cell Rep. **32**(10), 108128 (2020). https://doi.org/10.1016/j.celrep.2020.108128
117. Rozenman, G.G., Kundu, N.K., Liu, R., Zhang, L., Maslennikov, A., Reches, Y., Youm, H.Y.: The quantum internet: A synergy of quantum information technologies and 6G networks. IET Quantum Commun. **4**(4), 147–166 (2023). https://doi.org/10.1049/qtc2.12051

Chapter 2
Fundamentals of Complex Systems

Abstract This chapter introduces the basic concepts of complex systems as an important step in mapping the interdisciplinary space, finding out the link between complexity and communication networks. As the book aims to explore how emerging technologies, such as artificial intelligence, machine learning, and enabling paradigms for 6G, intersect with complex systems, a solid understanding of their main features is essential. Complex systems theory offers powerful tools to model emergent behaviors, nonlinearity, and adaptive dynamics, which increasingly shape the design and behavior of modern communication infrastructures. This chapter enables readers to recognize how complexity science can reshape key aspects of networks. The discussion covers perspectives, key mathematical formalisms, and the evolution of complexity as a transdisciplinary paradigm. This chapter sets the stage for understanding how complexity can lead to novel approaches in designing and managing next-generation communication systems.

2.1 Introduction

The awarding of the 2021 Nobel Prize in Physics marked a transformative milestone for complexity science, formally recognizing stochastic dynamics and disordered systems as fundamental components of modern physics. This acknowledgment reinforced the scientific importance of complex systems research across diverse domains, including climate change, spin-glass theory, network science, biology, and engineering, where emergent behaviors arise from interactions ruled by uncertainty and randomness [1]. This recognition has significantly improved public awareness, reshaped educational interests, and stimulated dedicated research funding. For the first time, complexity science was explicitly centered at the core of physics, strengthening its relevance in addressing real-world phenomena such as climate change, misinformation, and socio-technical dynamics. The prominence of spin-glass theory in the Nobel discussion reflects a broader evolution of complexity science. Spin-glass models have since catalyzed methodological advances with wide applicability in network theory, biological systems, and information science [2]. This progress highlights how

M. Scatà, *The Power of Complex Systems*, Studies in Systems, Decision and Control 630,
https://doi.org/10.1007/978-3-032-04243-9_2

complexity science emerges at the intersection of physics, applied mathematics, and computing, going beyond the boundaries established in each field. The emphasis on nonlinear modeling further underscored the methodological paradigm shift embraced by complexity science. Systems characterized by high-dimensional feedback loops and sensitivity to initial conditions are often opposed to deterministic predictions, requiring probabilistic and multiscale approaches. Within this framework, uncertainty is no longer a limitation but a defining property to be modeled and understood. The Nobel recognition further states the growing need for systems-level approaches. For example, in the climate change discussion, addressing global ecological and environmental challenges demands the integration of control theory, ecology, and complexity-based systems thinking [3]. This interdisciplinary convergence is increasingly recognized not as optional, but as essential. Complexity science is gaining institutional interest, as evidenced by its inclusion in academic curricula and national research agendas. This key time reflects a broader acknowledgment of the field's strategic relevance. Moreover, complexity science is extending its reach into novel domains such as the dynamics of online platforms, social networks, and large-scale collective behavior. The increasing availability of massive, interconnected datasets leads to both unprecedented opportunities and theoretical challenges, particularly in modeling emergent phenomena in socio-technical systems [4]. These changes represent a major step forward for complexity science, showing that it is now a mature, useful, and essential field for interdisciplinary research.

Interdisciplinarity represents both a fundamental strength and an ongoing challenge for the progress of complexity science. The field aspires to integrate perspectives from physics, biology, computer science, engineering, and the social sciences, with the goal of developing generalizable models for analyzing diverse real-world systems. This inclusive strategy fosters bottom-up innovation and avoids the rigidity often associated with top-down methodological frameworks. Beyond the academic context, communicating complexity science to broader audiences remains an interesting challenge. There is a pressing need for improved science communication strategies capable of translating abstract models into socially relevant narratives, particularly in domains such as public health, climate science, and technology policy. Applications of complexity science currently cover quantum networks [7], epidemic modeling [8], urban systems [9], and communication infrastructures [10], underscoring its versatility and real-world impact. The future of the field is increasingly envisioned as open, integrative, and globally collaborative. Supporting a research environment that bridges different fields in science and technology is essential for tackling global challenges such as climate change, pandemics, systemic instability, and information disorders. These problems demand systemic thinking, adaptive modeling, and cross-sector cooperation.

Recent discussions in the scientific community have highlighted a few key directions for the future of complexity science [1]. One of the most important is the need to keep developing models that include randomness and non-linear effects, especially in situations where simple cause-and-effect logic doesn't work. The recognition of ideas like replica symmetry breaking and stochastic processes shows that uncertainty and unexpected behaviors are not just challenges to overcome, but essential parts of how complex systems work. This science is gaining recognition as a transformative paradigm, one that offers critical tools for decision-making, resilience planning, and systems-level policy design. In the context of interest of this book, complexity offers a concrete, interdisciplinary framework to decode, model, and redesign communication networks as dynamic, adaptive systems. This book aims to map the convergence between communication technologies and complexity theory, offering a cross-disciplinary perspective that bridges network science, information theory, and systems thinking to understand better and engineer the interconnected world we live in. In this chapter, the theoretical foundations of complex systems is provided, summarizing and discussing the essential tools and conceptual frameworks crucial for understanding and working across disciplines that connect communication networks and complexity science. This study aims to provide greater insight into how these disciplines intersect and contribute to one another.

2.2 The Evolution of Complex Science

The study of complex systems has evolved significantly over recent decades, emerging as a crucial interdisciplinary field that addresses the behavior of systems composed of numerous interacting entities, often referred to as "agents" [11]. Complex systems science has grown in parallel with the rise of data-driven research, advances in computational power, and an increasing interest from both academia and industry. These are generally described as systems in which the collective behavior arising from interactions between their components cannot be easily deduced from the behavior of individual elements alone [12]. This emergent behavior is a distinctive feature resulting in outcomes that are often unpredictable, non-linear, and highly dynamic [13].

Today, complex systems provide a powerful framework for understanding a wide range of real-world phenomena, from ecosystems, economic markets, and neural networks to social interactions, biological systems, and modern communication infrastructures such as the Internet and 6G networks [10, 14]. These systems have become increasingly relevant for addressing global challenges, including pandemic modeling, climate change, socio-technical interactions, and sustainability transitions [15–17]. A contemporary perspective highlights their critical role in shaping emerging technologies. In artificial intelligence, for instance, complex network theory supports the understanding of deep neural architectures, explainability, and robustness [18, 19]. Network medicine and precision healthcare use multilayer modeling to capture interdependencies in biological and clinical systems [20, 21]. Smart cities and critical infrastructures increasingly rely on adaptive complex systems to ensure resilience and scalability in real-time environments [22, 23].

Due to their inherent complexity, these systems present significant analytical challenges. While experimental studies can reveal key insights, the theory primarily emphasizes theoretical and computational modeling to explore emergent dynamics [24, 25]. Modern research generally adopts two complementary approaches [11, 12, 26]. The first approach develops simplified mathematical models to isolate and capture key qualitative features and properties. These models may not fully replicate real-world systems, but they offer fundamental insights into emergent phenomena. Techniques such as dynamical systems theory, game theory, information theory, cellular automata, and network theory are commonly employed in this context. The second approach emphasizes detailed computational simulations that approximate real-world complexity. Researchers simulate behavior at the micro-level and observe system-wide phenomena by using methods such as Monte Carlo simulations and agent-based modeling (ABM). This makes such models particularly effective for dynamic systems like epidemic spreading, human mobility, and human-machine ecosystems [27–29]. As digital transformation continues, complex systems science is increasingly applied to artificial intelligence, edge computing, and self-organizing cyber-physical systems. By combining theoretical insights with advanced simulations, researchers develop adaptive strategies to design resilient systems capable of reacting to real-world uncertainty and heterogeneity [30]. To gain advanced insights, recent works explore both modeling techniques and real-world applications [26, 31, 32]. Many works shed light on complex adaptive systems, where adaptation and learning mechanisms generate emergent properties, especially in the socio-economic and evolutionary applications fields [33, 34].

Complex networks have advanced through the integration of theoretical models and empirical studies. While complex systems encompass a broad range of interacting components governed by nonlinear dynamics, complex networks specifically focus on the topological structure of interactions among system elements, offering a powerful lens through which to analyze and model connectivity-driven phenomena [35]. Key contributions on both topics include scale-free and small-world networks, modular structures, and multilayer topologies that reflect realistic patterns of interaction [36, 37]. Network topology is now considered a fundamental determinant of dynamic processes such as information spreading, diffusion dynamics, and collective behavior [28, 38, 39]. A growing body of literature explores the role of dynamical systems theory in modeling the temporal evolution of complex systems. Mathematical tools such as differential equations capture nonlinear, chaotic behaviors typical of natural and engineered systems. From synchronization to bifurcations and critical transitions, both visual and formal representations have enriched the scientific understanding of complex dynamics [40–42]. Discrete-time models, particularly cellular automata and iterated maps, enable researchers to explore the transition from order to chaos, revealing phenomena such as complex behavior, pattern formation, and emergent computation [43–45]. Recent progress in scaling, criticality, and phase transitions has revealed common principles shared by many fields. Mechanisms such as preferential attachment and self-organized criticality explain robust patterns across ecosystems, social networks, and urban systems [46–48]. Adaptation and evolution remain central to complex systems. From natural selection to artificial changes, models like genetic algorithms and evolutionary game theory provide essential tools for exploring learning and optimization in decentralized systems [49–53]. Information theory, originally developed for communication systems, has become essential in analyzing uncertainty, structure, and entropy in complex systems. Measures such as mutual information, entropy rate, and algorithmic complexity play a critical role in evaluating emergent behavior [54, 55]. Computational complexity theory complements this by classifying the solvability of problems, helping to identify computational limits that exist even in natural processes [56, 57]. Finally, agent-based modeling also represents one of the most versatile approaches in the field. By simulating individual behaviors and local rules, ABMs generate system-level insights across domains such as ecology, economics, urban studies, and social science [58, 59]. Recent developments have extended ABMs to digital twins, human-in-the-loop systems, and hybrid simulation platforms [60, 61].

In summary, complex systems science has developed into a multifaceted discipline driven by a synthesis of theoretical insights, computational innovation, and real-world applications. It offers a set of powerful tools for understanding emergent patterns and adaptive behaviors that characterize both natural and artificial worlds.

2.3 Understanding Complexity Through Interdisciplinary Approaches

2.3.1 Defining Complex Systems

A shared understanding has emerged across scientific disciplines, recognizing complex systems as those systems composed of multiple interacting components whose collective behavior results in emergent, often unpredictable phenomena. These systems display rich dynamics across various scales, and their behavior cannot be explained by analyzing their single parts. Emergent phenomena are not exclusive to large-scale systems. Even systems, such as the neural network of the worm *Caenorhabditis elegans*, with only a few hundred neurons, or minimal cells with a few hundred genes, can sustain life and exhibit complex behaviors. This indicates that the number of components alone does not determine complexity. Instead, the pattern and nature of interactions among components play a decisive role [62, 63]. Topological organization has been identified as central to the emergence of complexity. Networks of interactions, whether random, regular, or scale-free, influence the pathways through which local interactions propagate and generate global structures. The balance between randomness and order within these networks contributes to the system's capacity for self-organization and adaptation. Complex systems are typically described as decentralized structures where local interactions lead to emergent global behavior. This describes a wide range of systems, from cellular regulatory networks and social structures to infrastructure and communication systems. The emergence of macroscopic patterns from localized interactions is one of the key ideas in complexity science. Approaches based on reductionism—focused on analyzing systems through their simplest elements—struggle to account for the emergent and nonlinear behaviors typical of complex systems. In many cases, the number of interaction states grows in a combinatorial way, rendering an analytical approach hard to complete. This is exemplified by modern quantum systems, such as quantum processors, whose configuration space exceeds even the number of particles in the observable universe. Such challenges demand theoretical frameworks that go beyond traditional bottom-up modeling [7]. In this context, computational complexity theory provides essential tools by classifying problems based on their solvability and resource requirements. It bridges the gap between fundamental descriptions and practical modeling, particularly in systems characterized by a combinatorial explosion or non-trivial behavior. Mathematical models of complex systems increasingly adopt multilayer and time-varying network structures to represent their evolution [64, 65]. These models often rely on higher-order interactions captured through tensors or hypergraphs, which allow the encoding of temporal, spatial, or functional layers. Recurrent structural patterns identified in these models across diverse disciplines suggest the existence of universal organizing principles [12, 66].

Prediction in complex systems is inherently limited, not only due to incomplete information but also because of the intrinsic uncertainty emerging from nonlinear, feedback-driven interactions. A probabilistic perspective becomes essential for modeling such systems, where outcomes cannot be deterministically derived from initial conditions alone. A set of recurring features for these models are the presence of many interacting components, nonlinear dynamics, feedback loops, self-organization, emergence, and sensitivity to initial conditions. Additional properties are hierarchy, adaptability, and modularity that often characterize complex systems. An important distinction must be drawn between complicated and complex systems. While the former may involve many parts, they can still be decomposed and understood through deterministic analysis. In contrast, complex systems resist reductionist decomposition due to the non-trivial interplay of their components and the nonlinearity of their dynamics. The conceptual evolution of complexity has expanded the scope of study from low-dimensional systems to high-dimensional networks with vast numbers of interacting entities. This shift has enabled the modeling and analysis of real-world phenomena such as financial crises [67], epidemic outbreaks [68], brain activity [69], climate dynamics [70], and power grid stability [71]. Complex systems exhibit multiscale behavior, where phenomena at different spatial or temporal scales influence one another. This interplay can give rise to long transients, extreme events, or memory effects. These systems can also display multistable behavior, where several stable configurations coexist and the system's trajectory depends on its starting point or perturbations [72]. The distinction between structural and functional complexity is equally important. Structurally identical systems, such as graphite and diamond, composed of the same atoms, may exhibit drastically different behaviors due to variations in their topological configuration [73]. Systems are often modeled as co-evolving networks, where both the state of the elements and their interactions change over time. In others, complexity emerges from evolutionary dynamics, as in mobile crowdsensing and decentralized decision-making [52, 53]. Social systems further exemplify emergent behavior and unpredictability, with small interventions sometimes triggering large-scale transformations, a phenomenon evident in urban dynamics, policy-making, and digital networks.

Finally, the strength of complexity science lies in its integrative capability. It combines methodological tools from nonlinear dynamics, network theory, probabilistic modeling, and computational complexity to address fundamental questions of emergence, adaptation, and collective intelligence in both natural and artificial systems.

2.3.2 *Challenges and Emerging Directions in Complexity Science*

Complexity science is playing an increasingly central role in addressing the most critical challenges of the 21st century. The 2021 Nobel Prize in Physics, awarded to Manabe, Hasselmann, and Parisi, marked a turning point in the recognition of the field, underlining its relevance in explaining complex phenomena such as climate dynamics and disordered systems. This recognition also emphasizes the urgency of adopting complexity-based approaches to address global issues like climate change and system-level risk propagation, as has been well explained in [1]. In the domain of climate research, complexity science contributes to integrative frameworks that merge environmental models with socio-technical dynamics. Understanding human behavioral responses to environmental threats is crucial for fostering cooperation and developing effective mitigation strategies [74]. These considerations naturally extend to other global systems exposed to cascading failures and long-range dependencies, such as transportation and mobility networks. In the context of the recent pandemic as the COVID-19, this event has exposed the vulnerability of global transportation and mobility networks, which exhibit scale-free properties long identified as relevant to epidemic spreading [8, 68]. Integrating real-world mobility data with predictive models enhances epidemic forecasting and supports data-driven public policy. Beyond public health, similar complexity-informed approaches are proving vital in other domains where dynamic, interconnected structures govern system behavior. In the biomedical domain, the development of precision medicine increasingly relies on the interplay between network science, machine learning, and artificial intelligence to decode multiscale biological processes. This shift challenges traditional reductionist views, emphasizing the need to model the nonlinear and hierarchical nature of cellular and neural systems [75–77].

Complexity science is also envisioned as a fundamental enabler in the future development of the quantum internet, requiring hybrid models that unify quantum information theories with classical network infrastructures. Theoretical unification remains a central objective. Advancing toward a comprehensive framework for emergent behavior entails integrating tensorial algebra, spectral graph theory, statistical physics, and nonlinear dynamics to capture multiscale synchronization and structural evolution [78]. Within the field of quantum complex networks, studies have demonstrated quantum-specific phenomena, such as subgraph connectivity driven by entanglement, within random network structures [62, 79]. Despite this progress, a complete theory of quantum networks and their topological implications remains an open challenge. Efforts to formalize complexity through physical classifications have highlighted the importance of non-Markovian environments, non-Gaussian distributions, multiscale hierarchies, and structured probabilistic geometries [80].

As the availability of data grows, the ability to extract causality from high-dimensional, noisy datasets emerges as both a methodological and epistemological challenge. Interpretable models that reveal underlying mechanisms are increasingly necessary for both theoretical insights and real-world applications [81]. Further complexity arises in modeling far-from-equilibrium systems such as biological evolution, economic instability, and historical transformations. These systems often follow non-reproducible trajectories, which limits the application of classical ensemble-based approaches. Accounting for rare events and instability remains an ongoing theoretical effort [82, 83]. Multilayer and higher-order network models have been developed to describe systems with diverse types of interactions and evolving structures [8, 38, 84]. However, spatial constraints and volume exclusion effects, particularly in neural and molecular systems, add significant modeling complexity and require specialized mathematical representations [85].

These modeling efforts benefit significantly from interdisciplinary convergence, leveraging techniques across physics, data science, and systems engineering. Methodological convergence is crucial for analyzing large-scale, heterogeneous systems affected by dynamic coupling and non-Gaussian noise. Integrating tools from statistical physics, stochastic modeling, graph theory, machine learning, and nonlinear systems theory enables researchers to analyze infrastructure networks, cyberspace, Earth systems, and disease propagation more effectively [86]. Complexity science must also compare the technical and conceptual challenges associated with managing large-scale, noisy datasets. Extracting meaningful patterns and causal inferences demands moving beyond correlation-based methods, particularly in neuroscience, where pairwise interactions fail to capture the full richness of brain activity [87]. Beyond neuroscience, the integration of AI and network science has begun to influence a broad spectrum of applications, from smart cities to public health and economics. Instead, applications such as urban analytics, public health, and economic forecasting benefit from the fusion of network science with artificial intelligence, offering new pathways for data-driven decision-making [88]. Disinformation dynamics represent another pivotal area of research. Scale-free communication networks drive rapid propagation of misinformation and polarization. Modeling these dynamics is crucial for developing interventions that promote resilience and social cohesion [89]. As deeply discussed in [1], the fragmentation of complexity science into distinct disciplinary parts, such as systems biology, computational sociology, and network neuroscience, presents structural challenges. Consolidating the field through shared frameworks, journals, and academic programs is critical for advancing a solid vision. While its methods are widely adopted in diverse fields, from physics and biology to economics and sociology, the discipline often lacks a unified theoretical core and academic structure. Addressing real-world problems, such as climate change, epidemic forecasting, and social instability, requires the integration of complexity theory with established scientific methodologies and policy frameworks. It depends on building tools that are robust, interpretable, and able to influence decision-making processes at scale. Analyzing the convergence of complexity science and data science, it is transforming the landscape of research. The explosive growth of heterogeneous and time-evolving datasets calls for novel analytical strategies capable of managing high-dimensional and context-rich information. Deriving actionable insights from such data requires combining machine learning techniques with domain-informed modeling, especially in socio-technical, biological, and environmental systems [90–92]. Yet the effective use of these tools also depends on public trust and comprehension, especially in times of crisis. Recent global events, including the COVID-19 pandemic, have demonstrated how misunderstandings about model limitations and probabilistic forecasts can degrade trust in science. Enhancing public understanding of complexity and uncertainty has become a strategic objective for researchers working at the interface of science and society [93]. Differently, regarding AI, advances in artificial intelligence offer valuable tools for complexity science, helping to automate model selection, reveal hidden patterns, and simulate emergent behavior. However, the widespread use of black-box models risks reducing understanding. To ensure clarity and scientific reliability, AI systems must be transparent and explainable, and suited to the dynamics of complex systems. In parallel, various fields are actively exploring how complexity theory can be operationalized to improve design, diagnosis, and collective coordination. Several application domains stand to benefit from complexity methods. These include health sciences, where emergent phenomena can inform diagnosis and treatment [94]; collective human–technology interactions, as in swarm robotics and automated infrastructures [95]; and complexity engineering, which seeks to design systems with specified emergent properties [38, 96]. Promoting systems thinking, data skills, and interdisciplinary learning in education can prepare future scientists to tackle complex challenges [97].

Education, therefore, plays a pivotal role in shaping the next generation of researchers who must navigate this evolving landscape. A central scientific challenge involves understanding macroscopic causality in systems where outcomes emerge from distributed, non-linear interactions. This requires not only the identification of critical parameters but also the mapping of the limits of predictability. A paradigmatic shift may be needed to extend classical physics and embrace frameworks that explicitly account for co-evolution, feedback, and structural adaptation [12]. This complexity becomes even more pronounced in socio-technical ecosystems, where algorithms and infrastructures co-evolve with human behavior. Socio-technical systems, such as those formed by interactions among humans, algorithms, and infrastructures, have introduced new levels of unpredictability and complexity. While these systems may foster collective intelligence and innovation, they also pose risks of systemic failure, disinformation cascades, and algorithmic bias. Managing their dynamics requires experimental designs and methodological tools capable of distinguishing causation from correlation in large-scale, distributed contexts [98]. To meet these demands, methodological innovation must keep pace with the growing intricacy of the systems under observation. Furthermore, methodological advancements are needed to move beyond traditional statistical paradigms. Linear regression models and significance testing are often insufficient for capturing the rich dependencies found in complex systems. Future progress will depend on scalable, ethical, and causal inference frameworks that reflect the intrinsic nature of the systems under study. In sum, navigating the challenges of complexity science in the 21st century calls for a combination of theoretical integration, methodological innovation, and a strong dialogue with societal needs.

2.3.3 Robustness and Vulnerability in Complex Systems

The relationship between robustness and vulnerability represents a central issue in the study of complex systems. While scale-free networks exhibit remarkable resilience to random failures, they are simultaneously vulnerable to targeted attacks, a duality with profound implications for infrastructure resilience and cybersecurity [46]. This paradox has been explored extensively in various domains, including neuroscience, where understanding the propagation of dysfunction in neural networks requires accounting for stochastic dynamics and structural vulnerabilities. Integrating topological and geometrical approaches with network theory offers powerful tools for investigating how structural properties influence system behavior [66]. Robustness and vulnerability also depend on the dynamics of perturbation propagation. Systems exhibit robustness when disruptions fail to spread beyond localized regions, while vulnerability emerges when small disturbances cascade into global failures. This difference is important for infrastructures such as power grids, communication systems, and financial markets. Analytical tools derived from perturbation theory provide insights into the thresholds and structural conditions that govern systemic stability [100]. The concept of "highly optimized tolerance" (HOT) offers a predictive framework for examining these trade-offs. According to this model, systems such as biological organisms or engineered processors maintain robustness through carefully structured configurations that, while efficient, remain sensitive to rare and extreme events [3]. In quantum systems, for instance, noisy intermediate-scale quantum (NISQ) devices are inherently fragile due to rapid decoherence. Nevertheless, innovations such as Goldilocks quantum cellular automata illustrate how optimal circuit design can enhance robustness by minimizing noise exposure [101]. Probabilistic modeling emerges as a necessary approach for capturing the responses of complex systems to perturbations. Due to their nonlinear nature, small variations in initial conditions may lead to drastically different outcomes, and the deterministic frameworks are insufficient for capturing system behavior. Understanding such variability demands models that incorporate distributions of possible outcomes rather than singular predictions.

Evolutionary processes also shape the robustness of complex systems. Systems that have adapted over time often exhibit stable configurations, yet this evolutionary robustness is not uniform. Interdependent networks, for instance, can amplify vulnerabilities: a failure or perturbation in one layer can cascade through others, generating widespread disruption. In social systems, some specific network structures may also preserve negative patterns by reinforcing influential but dysfunctional interactions [102].

Classical tools like Lyapunov stability remain useful for analyzing responses to small perturbations, but there is a growing need for methods capable of assessing the impact of large perturbations. This is particularly relevant in systems such as ecosystems and the climate, which may undergo abrupt transitions once critical thresholds are exceeded [103]. Maintaining a dynamic balance between order and disorder is often essential for stabilizing complex systems. While local irregularities may appear chaotic, global structure can confer resilience. But, exposure to external conditions may trigger critical points or systemic collapse. Identifying the rules that govern this balance is crucial for both prediction and control [99]. Moreover, regarding nonlinearities, while diversity within a system may promote stability, the failure of specific nodes or pathways can lead to cascading effects. Identifying such critical components is essential for strengthening resilience in technological and ecological systems. It is also crucial to note that perturbations may remain localized or propagate dramatically, depending on the network topology and the nature of interdependencies. Identifying structural vulnerabilities, such as hub nodes or weak bridges, allows for targeted interventions [104]. It is also pivotal to have a deep understanding of real-world examples that illustrate the vulnerability of systems that otherwise appear resilient, as the climate [105]. Despite withstanding sustained pressure, complex systems often exhibit abrupt transitions once critical thresholds are exceeded, reflecting their inherent nonlinearity and sensitivity to initial conditions. Following this, what that can influence the equilibrium is the structure of interaction networks. Systems may appear robust under random perturbations or exhibit extreme vulnerability under intentional disruptions. This paradox highlights the importance of examining topological features and their implications for systemic stability [106]. Resilience is often counterbalanced by latent vulnerabilities. Systems such as the brain or ecosystems demonstrate a skill to absorb minor shocks, or remain exposed to catastrophic consequences from seemingly small disturbances [107]. Understanding both the mechanisms that promote resilience and those that compromise it remains a critical goal.

Moreover, the statistical properties of complex systems often follow heavy-tailed distributions, where extreme events are not rare but expected. Structural reconfigurations, such as rewiring of connections, can collapse, yet recovery may be possible if enough systemic redundancy or adaptability exists. Finally, the notion of self-organized criticality explains how systems may hover around critical thresholds, maintaining order while remaining susceptible to dramatic phase transitions. Social systems, in particular, often operate at such thresholds, balancing adaptability and systemic vulnerability, as seen in organizational collapses or large-scale behavioral shifts.

2.3.4 Prediction in Complexity Science

The ability to predict the behavior of complex systems remains a significant scientific challenge. While recent advances in data science and artificial intelligence have enhanced forecasting capabilities, they have raised concerns about interpretability and robustness [108]. Integrating these powerful tools with frameworks based on complexity offers a promising approach to managing uncertainty and achieving depth of understanding [66]. Simple models that incorporate heterogeneous interactions, such as scale-free networks, have demonstrated the ability to capture complex dynamics while preserving interpretability. The nonlinear and often asymmetric nature of complex systems further complicates predictive efforts. Approximate methods offer practical solutions. Notably, epidemiological forecasting has demonstrated that complexity science can produce meaningful predictions, such as those that informed public health responses during the COVID-19 pandemic [109]. As systems approach critical thresholds or undergo phase transitions, predictive accuracy tends to decline. Focusing on tractable observables, rather than modeling the system in its entirety, may offer more robust and interpretable insights [7]. This approach prioritizes identifying key variables that maintain predictive relevance despite underlying complexity.

A probabilistic perspective has become increasingly accepted as a means of accounting for uncertainty. Confidence intervals and distribution-based forecasts, similar to those employed in weather prediction, are now widely used to convey the likelihood of various outcomes. Such methods reflect the intrinsic unpredictability of complex systems and enable more detailed decision-making [110]. In systems with unique historical trajectories, such as biological evolution or macroeconomic development, rare events can dramatically influence dynamics. In these scenarios, ensuring predictive reliability requires interdisciplinary models coupled with real-time data integration, as exemplified by COVID-19 epidemic forecasting [111]. To improve prediction in complexity science, three main approaches have been proposed: detecting critical points early, forecasting stable dynamics over the medium and long term, and creating scenario-based simulations. These methods aim to balance accuracy and computational efficiency [112, 113]. Sensitivity to initial conditions limits long-term prediction, as small changes can cause large differences over time, typical of chaotic systems. In these cases, probabilistic forecasting is not optional but essential. Real-world systems like recommendation systems and behavioral models show how useful probability-based predictions can be [114]. Complex systems are further constrained by high dimensionality and heterogeneity. The unpredictability resulting from non-stationary dynamics and interdependent structures makes it difficult for even advanced machine learning techniques to anticipate rare or abrupt transitions. Nevertheless, methods based on network science and nonlinear data analysis have successfully improved forecasting models in domains such as climate science [115]. Combining data-driven and theory-based models is now considered key to understanding complex systems. These combine the computational power of machine learning with the interpretability and structure of network theory, enhancing the prediction of events such as epidemics or financial crises [116].

In complex systems, it is often impossible to predict exact outcomes, but estimating the chances of different scenarios is still useful and achievable. Predictions are more reliable when the system is far from a critical point. As it gets closer, even small changes can cause big effects, making probabilistic scenario modeling so important. Multiscale structures and chaotic dynamics present additional challenges. Trageted predictions have emerged in areas such as contagion modeling and percolation theory, where structural properties of networks offer insights into systemic resilience [117]. Data noise, limited observation, and insufficient modeling continue to pose major obstacles to accurate long-term forecasting. Robust predictions depend on improving model architectures, expanding datasets, and enhancing methods for quantifying uncertainty [118]. The difference between prediction, control, and understanding is central to complexity science. Rather than aiming for exact forecasts, complexity-based approaches emphasize how to formulate meaningful questions and establish realistic boundaries for what is predictable [119].

Finally, even in social and environmental systems, digital data and computational strength offer new factors for prediction. Behavioral patterns on digital platforms, for instance, serve as valuable proxies for understanding social change and informing policy decisions [120].

2.3.5 Interdisciplinary Research in Complexity Science

Complexity science has evolved as an inherently interdisciplinary field, drawing from physics, biology, computer science, social sciences, and engineering. Its capacity to synthesize knowledge across domains fosters innovation and enables the development of models capable of addressing multifaceted real-world phenomena. This integrative approach supports a bottom-up methodology that avoids the constraints of rigid, top-down paradigms. Despite its strengths, interdisciplinary collaboration presents significant challenges. Establishing common conceptual frameworks and fostering a culture of mutual adaptation remain essential to overcoming these obstacles, as well explained in [1].

Efforts to enhance interdisciplinary integration should become central in as many institutions as possible. Successful collaboration often requires researchers to move beyond disciplinary compartments and cultivate new communicative practices for derived and inspiring actions [101]. In the context of quantum networks, for instance, standard modeling techniques often fail due to the non-classical behavior of quantum phenomena. Addressing these issues requires hybrid strategies that incorporate insights from quantum theory, network science, and information theory. Bridging disciplinary divides also necessitates reforms in education. Curricula that incorporate complexity science, particularly through project-based and research-driven learning, can better prepare students for collaborative, cross-disciplinary work. Communication beyond academic circles presents another critical asset. The abstraction inherent in probabilistic and nonlinear models can obscure their real-world relevance. Effective science communication strategies are needed to articulate the value of complexity science in domains such as public health, climate policy, and technological governance. The development of dedicated venues, such as interdisciplinary conferences, journals, and funding programs, plays a vital role in cultivating a unified research community. These initiatives can support the sharing of ideas and help mitigate the fragmentation that can arise when researcher operate within isolated subfields. Interdisciplinary complexity research also prompts reflection on the formation of new scientific domains. For example, the application of complexity science to urban systems has contributed to the emergence of'city science' as a distinct research area [9]. While such developments attest to the field's vitality, they also underscore the importance of maintaining cohesion across expanding subdomains. Interdisciplinary complexity research also prompts novel approaches to modeling mental health dynamics, as demonstrated by the integration of social contagion, awareness, and multiplex networks to study the spread of distress and suicidal ideation [121].

The interest in interdisciplinary research in complexity science, jointly with social sciences, engineering, and other fields, can increase if we could build environments that support collaboration, mutual understanding, and the translation of theoretical insights into actionable knowledge for societal benefit.

2.4 Exploring Structure and Dynamics

Biological and chemical systems, neural networks, social networks, and more represent examples of systems composed of numerous interconnected dynamic units. To understand the properties of such systems, researchers often model them using graph theory [122–125]. In this approach, the individual dynamic elements are represented as nodes, while the interactions or connections between these units are depicted as links.

From a structural perspective, researchers face several challenges. These include analyzing and characterizing the intricate topology that defines the system's connectivity, identifying the fundamental principles that rule the organization of real-world networks, and developing theoretical models that can accurately simulate the network's growth and represent its structural features [126]. Beyond structure, another significant research focus is understanding the behavior of dynamic processes that occur on these complex networks. This involves investigating how a large number of interconnected systems, influenced by the underlying network architecture, can collectively evolve and exhibit emerging behavior. Studying these dynamics is crucial for uncovering emergent phenomena that arise from the complex interplay between network structure and system behavior.

Networks are ubiquitous in both natural and artificial systems, characterizing various aspects of life and technology. Humans are embedded in a vast network of social relationships, while their biological functions rely on intricate networks of biochemical interactions. Networks manifest physically in Euclidean spaces, such as power grids, transportation systems like highways and subway networks, as well as neural pathways. Alternatively, they can exist in abstract spaces, such as social networks. Historically, the study of networks has been rooted in graph theory, a branch of discrete mathematics established by Leonhard Euler in 1736 [127]. From this, the graph theory has since expanded to address numerous practical problems. Beyond mathematics, network studies have developed in the social sciences since the early 20th century. Social network analysis [128] emerged as a field that was focused on the patterns of communication between individuals, playing a key role in understanding relationships within complex social systems. In recent decades, the field of network science has experienced rapid growth, in particular with the emergence of complex networks. These systems are characterized by irregular and evolving structures, often composed of thousands or even millions of nodes. Unlike earlier efforts that concentrated on small-scale networks, contemporary research has shifted toward analyzing large-scale systems with dynamic interactions [129]. This growing interest was driven by two defining studies: the 1998 paper by Watts and Strogatz on small-world networks [130], and the 1999 work by Barabási and Albert introducing the concept of scale-free networks [46]. These important contributions generated strong interest from the physics community, partly thanks to advances in computing and the increasing availability of real-world network data.

The new capacity to examine large-scale networks across various domains has revealed unexpected insights. The primary challenge in this research lies in defining new metrics and concepts to describe the intricate topology of real networks. One of the key outcomes has been the identification of unifying statistical properties that appear across diverse network types. A notable discovery is the behavior of node connectivity, or degree distribution, denoted as $P(k)$. This function represents the probability that a randomly chosen node possesses exactly k connections. Unlike the Poisson distribution typical of random graphs, real-world networks often display a power-law distribution with an exponent between 2 and 3. Such distributions, characteristic of scale-free networks, imply that most nodes have relatively few connections, while a few hubs possess high connectivity [131]. Real networks further exhibit distinctive features such as degree correlations, the small-world property (where most nodes are separated by only a few steps), and the presence of numerous short cycles and motifs—localized patterns of connectivity that appear repeatedly. Understanding a network's structure is crucial since its architecture directly influences its functionality and resilience. For example, the structure of connections plays a fundamental role in determining the system's robustness to random failures or targeted attacks [132]. Furthermore, network structure strongly affects collective behaviors such as synchronization and significantly shapes dynamic processes, including epidemic spreading, information diffusion, and rumor propagation [68]. This deeper exploration into network dynamics has provided powerful insights into how complex systems behave, highlighting the intricate interplay between structure and function in various domains. As network science continues to evolve, it remains a vital framework for investigating the collective behavior of interconnected systems, both natural and artificial.

In the following sections, I provide a concise overview of the mathematical foundations underlying several key properties and formulations, primarily based on [126] and complemented by many other acclaimed and top-tier references in the field.

These concepts are central to this book, offering a rigorous framework for inspiring innovative approaches in engineering and communication systems.

2.4.1 Structure

In this section, we summarize the main parameters and metrics of the graph-theoretical principles and provide insights into various aspects of network topology, such as connectivity, reachability, clustering, and spectral characteristics. The following Table 2.1 summarizes the main structural metrics and parameters used in the analysis of complex networks and discussed in the following subsections.

2.4.1.1 Definition and Notations

Graph theory [133–135] serves as the fundamental mathematical framework for analyzing complex networks. Formally, a complex network can be represented as a graph. An undirected (or directed) graph $G = (\mathcal{N}, \mathcal{L})$ is defined through two sets: $\mathcal{N}$, the set of nodes (also called vertices), and $\mathcal{L}$, the set of links (also known as edges). In this context, $\mathcal{N} \neq \emptyset$ and $\mathcal{L}$ consist of unordered (or ordered) pairs of elements from $\mathcal{N}$. The cardinality of $\mathcal{N}$ and $\mathcal{L}$ is denoted as N and K, representing the number of nodes and links, respectively. For clarity, the notation $G(N, K)$ or $G_{N,K}$ is often used to express the network size and link count. Each node in the network is represented by i in the set N. In an undirected graph, a link between nodes i and j is represented as (i, j) or l_{ij}, indicating a bidirectional connection. Such nodes are referred to as “adjacent” or “neighboring”. Conversely, in a directed graph, the order of nodes is important, and a directed link from node i to node j is denoted l_{ij}, with $l_{ij} \neq l_{ji}$.

Graph visualization typically involves representing nodes as dots and links as connecting lines. While the specific layout of this representation may change, the structure is defined by the presence or not of links between node pairs [126]. In graph theory, the number of edges K in a graph G of size N has a minimum value of zero and a maximum of $N(N-1)/2$, corresponding to a fully connected network where every pair of nodes is linked. Graphs are classified based on their edge density: a graph is considered *sparse* if $K > N$, and *dense* if $K = O(N^2)$. A complete graph with N nodes, denoted K_N, is a fully connected graph in which $K = \binom{N}{2} = N(N-1)/2$. Notably, the complete graph with three nodes (K_3) is commonly called a *triangle*. A *subgraph* $G' = (N', L')$ of a graph $G = (N, L)$ is defined such that $N' \subseteq N$ and $L' \subseteq L$. If G' includes all links in G that connect pairs of nodes in N', it is called an *induced subgraph* and is denoted $G' = G[N']$. A subgraph is considered *maximal* concerning a given property if extending the subgraph causes it to lose that property. A significant type of subgraph is the neighborhood subgraph of a given node i, denoted G_i. This subgraph includes all nodes adjacent to i and the links connecting them: $G_i = G[N_i]$.

Table 2.1 Structural metrics and parameters in complex network analysis

Metric/Parameter	Definition	Description
Number of Nodes (N)	Total number of nodes in the network	Indicates the network size; each node represents an entity in the system
Number of Links (K)	Total number of edges or connections	Determines how densely nodes are interconnected
Degree (k_i)	Number of links incident to node i	Measures local connectivity of a node; higher degree implies more interactions
Degree Distribution ($P(k)$)	Probability distribution of node degrees	Describes how degrees are spread over the network; useful for identifying scale-free or homogeneous structures
Average Degree ($\langle k \rangle$)	Mean degree over all nodes	Provides a global sense of connectivity
Second Moment ($\langle k^2 \rangle$)	Expectation of squared degrees	Captures variability in node degrees; relevant for epidemic thresholds
Shortest Path Length (L)	Average geodesic distance between all node pairs	Indicates efficiency of information or resource flow across the network
Diameter (Diam(G))	Longest shortest path in the network	Represents the maximum extent of the network
Efficiency (E)	Harmonic mean of geodesic distances	Reflects the network's overall communication efficiency
Betweenness Centrality (b_i)	Fraction of shortest paths passing through node i	Quantifies a node's control over flow and communication in the network
Clustering Coefficient (C, c_i)	Density of triangles around a node or network-wide	Indicates the level of transitivity or local cohesion
Transitivity (T)	Global ratio of triangles to connected triples	Measures the prevalence of closed triplets in the network
Local Efficiency (E_{loc})	Average efficiency of each node's neighborhood	Assesses robustness and fault tolerance in local regions
Motif Frequency (n_M)	Count of specific small subgraph patterns	Used to identify recurrent local structures with functional relevance
Community Structure	Groups of nodes with dense internal connectivity	Reveals mesoscopic organization and functional modules
Adjacency Matrix (A)	Binary matrix encoding node connections	Basic mathematical representation of the network
Spectral Radius ($\lambda_{\max}$)	Largest eigenvalue of the adjacency matrix	Related to dynamics such as spreading processes and stability
Laplacian Matrix (L)	Difference between degree matrix and adjacency matrix	Useful for analyzing synchronization, flow, and connectivity
Algebraic Connectivity (λ_2)	Second smallest eigenvalue of Laplacian	Indicates robustness and cohesion; higher values imply greater resilience

A fundamental concept in graph theory is *reachability*, which describes whether a path exists between two nodes. Even if nodes are not directly connected, they may still be reachable through intermediate nodes. Several path-related terms are defined as follows:

- A *walk* is a sequence of nodes and connecting edges, starting at node i and ending at node j, with length defined by the number of edges in the sequence.
- A *trail* is a walk where no edge is repeated.
- A *path* is a walk in which no node is visited more than once.
- The *shortest path* (or *geodesic*) is the minimal-length path between two nodes.

A *cycle* is a closed walk containing at least three nodes where no edge is repeated. Cycles are classified by length: a 3-cycle (C_3) corresponds to a triangle (K_3), a 4-cycle (C_4) is a quadrilateral, a 5-cycle (C_5) is a pentagon, and so on. A graph is defined as *connected* if there is a path between every pair of distinct nodes; otherwise, it is *unconnected* or *disconnected.* A *component* is a maximally connected subgraph, while a *giant component* is a connected subgraph whose size scales comparably to the total number of nodes N. Graphs can be efficiently represented using matrix notation. The *adjacency matrix* A is an $N \times N$ matrix where the entry $a_{ij} = 1$ if there is a link between nodes i and j, and zero otherwise. The diagonal elements of A are zero, and the matrix is symmetric for undirected graphs. An alternative representation is the *incidence matrix*, which provides a different mathematical description of the network's structure.

2.4.1.2 Degree Distribution and Correlations

A key concept in characterizing complex networks is the *degree* (or connectivity) of a node, which quantifies the number of links incident with that node. Formally, the degree of a node i is defined using the adjacency matrix A as follows:

$$k_i = \sum_{j \in N} a_{ij} \tag{2.1}$$

In directed networks, the degree is divided into two components: the *out-degree*, $k_i^{out} = \sum_j a_{ij}$, which counts the outgoing links from node i, and the *in-degree*, $k_i^{in} = \sum_j a_{ji}$, which counts the incoming links. The total degree in this case is the sum of these two components:

$$k_i = k_i^{out} + k_i^{in} \tag{2.2}$$

The complete set of degrees for all nodes in a graph is known as the *degree sequence*. One of the most fundamental topological properties of a network is its *degree distribution*, $P(k)$, which defines the probability that a randomly chosen node has degree k. This can be interpreted as the fraction of nodes in the network with degree k. For directed networks, the degree distribution requires two functions: $P(k^{in})$ and $P(k^{out})$. The degree distribution can also be characterized through its statistical moments. The n-th moment of the distribution is defined as:

$$\langle k^n \rangle = \sum_k k^n P(k) \tag{2.3}$$

The first moment, $\langle k \rangle$, represents the mean degree of the network, while the second moment, $\langle k^2 \rangle$, quantifies the fluctuations in the degree distribution. Notably, when $\langle k^2 \rangle$ diverges in large networks, it significantly modify the behavior of dynamical processes, such as epidemic spreading or synchronization. In real networks, the connectivity structure is often not random; nodes with certain degrees are more likely to connect with others following specific patterns. This effect is captured by the *conditional probability* $P(k'|k)$, which defines the probability that a node with degree k is connected to a node with degree k'.

$$\sum_{k'} P(k'|k) = 1 \tag{2.4}$$

For *uncorrelated networks*, $P(k'|k)$ is independent of k and can be expressed as:

$$P(k'|k) = \frac{k' P(k')}{\langle k \rangle} \tag{2.5}$$

Since calculating $P(k'|k)$ directly from real data often produces noisy results due to finite network size, an alternative measure is the *average nearest-neighbors degree* for a node i, denoted $k_{nn,i}$:

$$k_{nn,i} = \frac{1}{k_i} \sum_{j \in N_i} k_j \tag{2.6}$$

This measure allows one to estimate the average degree of the nearest neighbors for nodes of degree k, written as:

$$k_{nn}(k) = \sum_{k'} k' P(k'|k) \tag{2.7}$$

Based on the behavior of $k_{nn}(k)$, networks can be categorized as:

- **Assortative networks**: $k_{nn}(k)$ is an increasing function of k, indicating that nodes with high degrees have the tendency to connect to other highly connected nodes.
- **Disassortative networks**: $k_{nn}(k)$ is a decreasing function of k, where low-degree nodes are more likely to connect to highly connected nodes.

The presence and strength of degree correlations are often quantified using the Pearson correlation coefficient of node degrees at either end of a link, providing a concise measure of the network's correlation structure [126, 134].

2.4.1.3 Shortest Paths and Betweenness Centrality in Complex Networks

Shortest paths are crucial for understanding transport and communication processes in complex networks. In the context of networking, the shortest (or geodesic) path between two nodes ensures efficient data packet transmission, improving efficiency and reducing computational load [136, 137]. Consequently, shortest paths are also valuable in characterizing the internal structure of a network [134]. To quantify shortest paths across a network, one can define a distance matrix D, where the entry d_{ij} represents the geodesic distance between nodes i and j. The maximum element in this matrix is known as the *diameter* of the graph, denoted as $\mathrm{Diam}(G)$, which measures the greatest shortest path in the network. An alternative metric for quantifying node distance is the *average shortest path length*, or *characteristic path length*, defined as:

$$L = \frac{1}{N(N-1)} \sum_{i \neq j} d_{ij} \tag{2.8}$$

This metric can become challenging in networks with disconnected components, as L diverges when nodes cannot be reached. To mitigate this, the summation can be restricted to nodes within the largest connected component [130]. An alternative measure that avoids this divergence is the *efficiency* of the network, defined as the harmonic mean of the geodesic distances:

$$E = \frac{1}{N(N-1)} \sum_{i \neq j} \frac{1}{d_{ij}} \tag{2.9}$$

This measure captures the network's capacity to efficiently transmit information, while ensuring that disconnected nodes contribute zero to the total summation, preventing divergence [138]. Another key concept for evaluating network structure is *betweenness centrality*, which measures the significance of a node in connecting other parts of the network. The betweenness b_i of node i is calculated as:

$$b_i = \sum_{j \neq k} \frac{n_{jk}(i)}{n_{jk}} \tag{2.10}$$

Here, n_{jk} denotes the total number of shortest paths between nodes j and k, while $n_{jk}(i)$ is the number of those paths that also pass through node i. Nodes with high betweenness play a critical role in maintaining communication and controlling network flow [139].

The *Edge betweenness* is an extension of this concept, calculated as the number of shortest paths that pass through a particular edge. This measure is useful for detecting network vulnerabilities and analyzing the impact of edge removal [140]. Efficient algorithms are essential for computing shortest paths and betweenness in large networks. Dijkstra's algorithm and breadth-first search are commonly employed for shortest path identification [141]. Recently developed fast algorithms have significantly improved the efficiency of betweenness calculations, enhancing scalability in large-scale networks [142].

2.4.1.4 Clustering in Complex Networks

Clustering, often referred to as *transitivity*, is a fundamental property commonly observed in social networks. It reflects the tendency that if two individuals share a mutual neighbor, they are more likely to be connected, they are likely to know each other [36]. In graph theory, this feature is a measure of the significant presence of triangles within the network. The transitivity T of a network quantifies this property by measuring the proportion of connected triples (triads) that also form triangles:

$$T = \frac{3 \times \# \text{ of triangles in G}}{\# \text{ of connected triples of vertices in G}}. \tag{2.11}$$

The factor 3 accounts for the fact that each triangle contributes three connected triples, one centered at each node in the triangle. Consequently, the transitivity is between 0 and 1, with $T = 1$ representing a complete graph K_N [143].

An alternative and widely used metric is the *clustering coefficient* C, introduced by Watts and Strogatz [130]. This measure reflects the local density of triangles within the network. The local clustering coefficient c_i for node i is defined as:

$$c_i = \frac{2e_i}{k_i(k_i - 1)} = \frac{\sum_{j,m} a_{ij}a_{jm}a_{mi}}{k_i(k_i - 1)} \tag{2.12}$$

where e_i is the number of actual links between the neighbors of node i, and k_i is the degree of node i. The local clustering coefficient measures how interconnected a node's neighbors are.

The global clustering coefficient C is then obtained as the average of the local clustering coefficients across all nodes:

$$C = \frac{1}{N} \sum_{i \in N} c_i \tag{2.13}$$

Both c_i and C are bounded between 0 and 1, with $C = 1$ in fully connected networks.

Several extensions of clustering coefficients have been proposed to capture higher-order structures. Some variants include the k-clustering coefficient, which accounts for k-neighbors [144], and measures based on four-node cycles or generalized cycle counts [145].

Another important metric related to clustering is the *local efficiency*, which evaluates the efficiency of communication within the local neighborhood of each node [138]. The local efficiency E_{loc} is defined as:

$$E_{loc} = \frac{1}{N} \sum_{i \in N} E(G_i) \tag{2.14}$$

Here, $E(G_i)$ represents the efficiency of the subgraph G_i, calculated using the harmonic mean of the geodesic distances as described earlier.

2.4.1.5 Motifs and Community Structures in Complex Networks

Motifs are distinctive patterns of interconnections that appear in complex networks significantly more often than in their randomized counterparts. Such randomized networks retain the same number of nodes, links, and degree distribution as the original network, but with edges randomly distributed. Formally, a motif M is defined as a connected subgraph composed of n nodes that appears frequently within a given graph G. In [146], the study of motifs has been introduced in biological and other complex networks. Their research introduced a method to detect motifs by counting the total occurrences of each n-node subgraph both in the original network and in randomized ones.

The statistical significance of a motif M is quantified by the Z-score:

$$Z_M = \frac{n_M - \langle n_M^{rand} \rangle}{\sigma_{n_M}^{rand}}, \tag{2.15}$$

where n_M is the number of times the motif appears in the original network, $\langle n_M^{rand} \rangle$ is the average number of appearances in the randomized network ensemble, and $\sigma_{n_M}^{rand}$ is the corresponding standard deviation [147]. Motifs are particularly relevant in biological systems, where they often reveal fundamental functional modules, such as feedback loops, feedforward systems, or signal transduction pathways. However, even structurally identical motifs, such as bi-fan configurations, can display different dynamics depending on internal parameters, indicating that structure alone does not determine function [148]. The concept of *community* in networks emerges from social sciences and refers to groups of nodes that are more densely connected internally than with the rest of the network [149]. A community (or cluster) within a graph $G(N, L)$ is a subgraph $G'(N', L')$ where nodes exhibit strong connectivity patterns. Several definitions of community structures exist based on different connectivity criteria. The most restrictive definition of a community is a *clique*, a maximal complete subgraph in which every node is directly connected to every other node in the subgraph. Extending this concept, an n-clique is defined as a maximal subgraph in which the maximum geodesic distance between any two nodes is no greater than n. An alternative definition of community structures is the *k-plex*, a maximal subgraph of n nodes where each node is connected to no fewer than $n - k$ nodes in the subgraph. This definition relaxes the strict adjacency condition required by cliques. Communities can also be identified by their link density. A common definition states that a subgraph G' is a community if the sum of the internal degrees within G' is greater than the total number of links connecting nodes in G' to nodes outside the community [150].

This density-based approach is particularly suitable for identifying overlapping in networks where connections are inherently sparse. Identifying motifs and communities is crucial in understanding the structural and functional organization of complex systems. In biological networks, motifs can reveal some mechanisms linked to evolution, while community detection in social networks helps identify groups with shared interests or behaviors. Both concepts have been effective in domains, such as sociology, biology, computer science, and engineering.

2.4.1.6 Graph Spectra in Complex Networks

The spectrum of a graph provides essential insights into its structure and connectivity. The spectrum refers to the set of eigenvalues of the graph's adjacency matrix A [151]. For a graph $G_{N,K}$ with N nodes and K links, there are N eigenvalues λ_i (where $i = 1, 2, \ldots, N$) and corresponding eigenvectors v_i. In undirected graphs without loops or multiple edges, the adjacency matrix A is real and symmetric, resulting in real eigenvalues that can be ordered as $\lambda_1 \leq \lambda_2 \leq \cdots \leq \lambda_N$, and their associated eigenvectors are orthogonal.

For directed graphs, the eigenvalues may include imaginary components, making their ordering and properties more complex. The Perron–Frobenius theorem ensures that any graph (even directed) has a largest real eigenvalue λ_N with an associated non-negative eigenvector. In connected graphs, λ_N has multiplicity one, and all other eigenvalues satisfy $|\lambda_i| < \lambda_N$.

For large graphs, the spectral density function is defined as:

$$\rho(\lambda) = \frac{1}{N} \sum_{i=1}^{N} \delta(\lambda - \lambda_i), \tag{2.16}$$

where δ is the Dirac delta function. As $N \to \infty$, the spectral density converges to a continuous function, offering insights into the network's macroscopic behavior [152].

The eigenvalues are closely tied to the graph's structural features, in fact:

- The sum of all eigenvalues is zero since $\mathrm{Tr}(A) = 0$.
- The sum of eigenvalue pairs corresponds to the number of links K.
- The sum of products of triples of eigenvalues equals twice the number of triangles (3-cycles).
- Higher-order products relate to cycles of longer lengths (e.g., quadrilaterals or pentagons).

Furthermore, the elements of the matrix A^k count the number of walks of length k between pairs of nodes, and the trace of A^k provides the total number of such walks [152].

The k-th moment of the spectral density is defined as:

$$M_k = \frac{1}{N}\sum_{i=1}^{N}\lambda_i^k = \frac{1}{N}\mathrm{Tr}(A^k), \tag{2.17}$$

which reflects the total number of closed walks of length k in the network.

The graph's diameter is constrained by its eigenvalues. Specifically, for a connected graph, the diameter $\mathrm{Diam}(G)$ is upper-bounded by the number of distinct eigenvalues. For k-regular graphs (where each node has the same degree k), an improved upper bound is obtained using the second-largest eigenvalue λ_{N-1}:

$$\mathrm{Diam}(G) \leq \frac{\ln(N-1)}{\ln\left(\frac{k}{\lambda_{N-1}}\right)}. \tag{2.18}$$

Additional insights arise from the graph's Laplacian and normal matrices. The Laplacian matrix L, also known as the Kirchhoff matrix, is defined as:

$$L = D - A, \tag{2.19}$$

where D is the diagonal matrix of node degrees. The smallest Laplacian eigenvalue is zero, with multiplicity equal to the number of connected components in the graph. A larger second-smallest eigenvalue λ_2 indicates stronger connectivity and greater robustness to node removal [151].

The normal matrix is defined as:

$$N = D^{-1}A, \tag{2.20}$$

and is often used to investigate network stability, synchronization, and flow dynamics. Spectral analysis is widely used for understanding the structure and dynamics of networks. It is particularly crucial in identifying cohesive subgroups, revealing hierarchical structures, and detecting communities. Spectral techniques are also important in representing networks and analyzing diffusion processes across nodes [126].

2.4.2 Networks Models

2.4.2.1 Real Networks

The graph theory approach simplifies complex interactions by reducing time-dependent, spatial, and multi-variable dependencies to a binary relationship (link or no link); it often offers an insightful representation of the system's global structure. Despite its limitations, this abstraction has proven useful for characterizing and understanding diverse real-world networks.

Remarkable advancements have been made in the exploration of network topologies, driven by three key developments:

- The availability of extensive real-world datasets,
- Improved computational resources with enhanced processing capabilities,
- The development of sophisticated data analysis tools.

These advancements have enabled researchers to investigate the structural properties of various real-world networks, including communication systems [36], social networks [143], and biological networks [153].

Studies have revealed that, despite their inherent differences, many real networks share key structural characteristics:

- **Small path lengths**: Real networks often exhibit short average distances between nodes, enhancing communication efficiency.
- **High clustering coefficients**: Networks tend to form tightly-knit groups where nodes are highly interconnected.
- **Fat-tailed degree distributions**: Many real networks follow a power-law distribution, where a few nodes have exceptionally high connectivity.
- **Degree correlations**: Networks may display assortative or disassortative mixing patterns, depending on whether highly connected nodes prefer linking to similarly connected nodes or less connected ones.
- **Motifs and community structures**: Networks commonly exhibit recurring patterns of connectivity (motifs) and tightly-bound clusters of nodes (communities).

The identification of these shared features has set real-world networks apart from traditional models like random graphs or regular lattices, which do not exhibit such complex structural characteristics. CAs a result, considerable effort has been directed toward developing new network models that reproduce the empirical features of real-world systems. These models seek to reflect the underlying mechanisms driving network evolution, offering deeper insights into their dynamic behavior.

2.4.2.2 Random Graphs

The study of random graphs began with the work of Erdős and Rényi in 1959 who introduced a probabilistic framework to investigate how structural properties emerge as the number of random connections increases [154]. The term *random graph* reflects the disordered nature of the connections between nodes. Erdős and Rényi proposed two fundamental models for generating random graphs:

- **Model 1 ($\mathbf{GER_{N,K}}$)**: Starting with N disconnected nodes, pairs of randomly chosen nodes are connected until the graph reaches exactly K edges. This model is analogous to the *canonical ensemble* in statistical mechanics.

- **Model 2 ($GER_{N,p}$)**: Each possible pair of nodes is connected with a probability p. This method defines an ensemble of graphs with varying link counts and is similar to the *grand canonical ensemble* in statistical mechanics [133].

In both models, as the network size $N \to \infty$, they become statistically equivalent for fixed average degree $\langle k \rangle = p(N-1)$. ER random graphs exhibit distinct structural behaviors depending on the value of the connection probability p. Erdős and Rényi identified a critical threshold $p_c = 1/N$, corresponding to a critical average degree $\langle k \rangle_c = 1$. This phase transition has important implications for the network's connectivity:

- **For** $p < p_c$: The network is fragmented into small components, with no cluster larger than $O(\ln N)$.
- **For** $p = p_c$: The largest component scales as $O(N^{2/3})$.
- **For** $p > p_c$: A *giant component* emerges, containing $O(N)$ nodes and several cycles, while remaining components are small and sparse.

This phase transition resembles a second-order phase transition in statistical mechanics, with the size of the largest component acting as the order parameter.

In ER random graphs, the degree distribution follows a binomial form:

$$P(k) = \binom{N-1}{k} p^k (1-p)^{N-1-k}. \tag{2.21}$$

For large N and fixed $\langle k \rangle$, this binomial distribution converges to a Poisson distribution:

$$P(k) \approx e^{-\langle k \rangle} \frac{\langle k \rangle^k}{k!}. \tag{2.22}$$

Due to the random nature of the connections, ER networks are inherently uncorrelated, meaning that the probability $P(k'|k)$ of a node with degree k connecting to a node of degree k' is independent of k.

For sufficiently large N and $p > p_c$, ER networks tend to be fully connected with a characteristic diameter scaling as:

$$\text{Diam}(G) \sim \frac{\ln N}{\ln \langle k \rangle}. \tag{2.23}$$

Similarly, the average shortest path length L exhibits the same logarithmic dependence:

$$L \sim \frac{\ln N}{\ln \langle k \rangle}. \tag{2.24}$$

In ER networks, the clustering coefficient C is approximately equal to p, which decreases as the network size grows, resulting in a vanishing clustering coefficient in the limit $N \to \infty$. The spectral density of ER random graphs converges to Wigner's semicircle law for symmetric uncorrelated random matrices:

$$\rho(\lambda) = \begin{cases} \frac{\sqrt{4Np(1-p)-\lambda^2}}{2\pi Np(1-p)} & \text{if } |\lambda| < 2\sqrt{Np(1-p)} \\ 0 & \text{otherwise.} \end{cases} \tag{2.25}$$

The largest eigenvalue λ_N is isolated from the spectral bulk and scales as $\lambda_N \sim pN$. For $p < p_c$, the spectral density deviates from the semicircle law, and the odd moments vanish, indicating the disconnected nature of the network.

2.4.2.3 Generalized Random Graphs

The ER model does not always capture the heterogeneity observed in real-world networks. Consequently, generalized random graphs have been proposed to account for diverse structural properties, particularly non-Poisson degree distributions.

The *configuration model*, introduced by Bender and Canfield in [155], generates random graphs with a specified degree sequence $D = \{k_1, k_2, \ldots, k_N\}$, where $K = \frac{1}{2}\sum_i k_i$ is the total number of links. Each node i is assigned k_i stubs (or half-edges), and pairs of stubs are randomly matched to form full edges. This process generates random graphs that preserve the desired degree distribution $P(k)$ in the limit of large N [156].

The condition determines the emergence of a giant component in such graphs:

$$Q = \sum_k k(k-2)P(k) > 0, \tag{2.26}$$

provided that the maximum degree $k_{\max}$ is not too large [157]. Conversely, if $Q < 0$, the graph's largest component scales as $O(k_{\max}^2 \ln N)$.

Newman et al. proposed an alternative model where node degrees are independent random integers drawn from a distribution $P(k)$. They derived an approximate formula for the average shortest path length:

$$L \approx \frac{\ln(N/z_1)}{\ln(z_2/z_1)} + 1, \tag{2.27}$$

where z_m denotes the average number of neighbors at distance m [157]. This expression reduces to the ER result when $z_1 = \langle k \rangle$ and $z_2 = \langle k^2 \rangle$.

The clustering coefficient in the configuration model is given by:

$$C = \frac{\langle k \rangle}{N} \frac{\langle k^2 \rangle - \langle k \rangle}{\langle k \rangle^2}. \tag{2.28}$$

Although $C \to 0$ for large N, finite-size networks with skewed degree distributions may exhibit significant clustering values.

2.4.2.4 Small-World Networks

The concept of the small-world property emerged from the observation that many real-world networks possess *shortcuts*, which are bridging links that connect otherwise distant nodes. These shortcuts significantly improve the efficiency of communication across the network. In a regular hypercubic lattice of dimension D, the average number of nodes one must traverse to reach a randomly chosen target node scales as $N^{1/D}$ with the network size N. However, in most real networks, the average path length between nodes grows logarithmically with the network size, $L \sim \log N$. This feature is known as the *small-world property*, which highlights the remarkable connectivity observed in real networks [130]. The concept was first identified by Stanley Milgram in the 1960s through a groundbreaking social experiment and then replicated with other ones [158, 159]. In this study, Milgram instructed randomly selected individuals in Nebraska to send letters to a target individual in Boston, identified only by their name, occupation, and approximate location. Each participant could forward the letter only to someone they knew personally, presumably bringing the message closer to the target. Contrary to expectations of lengthy chains, Milgram found that the average path length was approximately six steps, coining the popular notion of "six degrees of separation".

The small-world property is not limited to social systems but has also been observed in biological, technological, and information networks [36, 136]. While random graphs also exhibit short average path lengths, real networks often combine this feature with high levels of clustering. To formalize this duality, Watts and Strogatz [130] proposed a model to describe *small-world networks*, which are characterized by a low path length L, similar to random graphs, and a high clustering coefficient C, similar to regular lattices. In the context of network efficiency, this concept translates to networks exhibiting both high global efficiency and high local efficiency [138]. Such networks demonstrate exceptional efficiency in transmitting information on a large scale and within local communities.

The *Watts–Strogatz* (WS) model was introduced to describe networks that simultaneously exhibit the small-world property and a high clustering coefficient [130]. This model starts with a regular ring lattice of N nodes, each connected to $2m$ nearest neighbors. Each link is then rewired with probability p, creating shortcuts that reduce the path length L.

- For $p = 0$, the graph is a regular lattice with large L and high clustering C.
- For $p = 1$, the graph becomes a random network, exhibiting small L and low C.
- For intermediate values of p, the network retains high clustering yet achieves a logarithmic scaling of L, ensuring both local and global efficiency.

The WS model reveals that the transition from large L to logarithmic scaling occurs rapidly for small but nonzero values of p. This reflects the nonlinear effect of introducing shortcuts, which drastically improves the network's connectivity.

The scaling behavior near the transition point can be described as:

$$L(N, p) \sim p^{-\alpha} f(Np^{\beta}), \tag{2.29}$$

where the scaling function $f(x)$ follows:

$$f(x) \sim \begin{cases} x & \text{if}\, x > 1 \\ \ln x & \text{if}\, x < 1 \end{cases}$$

Barrat and Weigt determined that $\alpha = 1$, which implies that the scaling length diverges as p^{-1} [160].

The clustering coefficient $C(p)$ decreases gradually with p following the relation:

$$C(p) \sim \frac{3(m-1)}{2(2m-1)}(1-p)^3. \tag{2.30}$$

In the WS model:

- For $p = 0$, the degree distribution is a delta function centered at $2m$.
- For $p = 1$, the degree distribution resembles that of an ER network.

For intermediate values of p, the degree distribution is given by:

$$P(k) = \sum_{i=0}^{\min(k-m,m)} \binom{m}{i} (1-p)^i p^{m-i} \frac{(pm)^{k-m-i}}{(k-m-i)!} e^{-pm}. \tag{2.31}$$

Spectral properties have also been studied extensively. As p increases, the spectral distribution gradually aligns with that of a random graph [152]. Monasson analyzed the Laplacian operator's spectral properties to better understand connectivity and cohesion in the WS model [161].

2.4.2.5 Scale-Free Networks

For many years, scientific research primarily focused on homogeneous networks, characterized by nodes with similar connectivity patterns. Examples include regular lattices and random graphs, where each of the $N(N-1)/2$ possible links appears with equal probability. Consequently, the degree distribution in such networks follows a binomial or Poisson distribution within the limit of large graph sizes. However, empirical studies of real-world networks revealed a surprising difference from that behavior. Contrary to earlier assumptions, many real networks exhibit a degree distribution following a power-law formula:

$$P(k) \sim Ak^{-\gamma}, \tag{2.32}$$

where the exponent γ generally is within the range $2 < \gamma < 3$ [46]. Networks characterized by such distributions are referred to as *scale-free networks* due to the property of self-similarity inherent to power-law functions. Power-law distributions are unique because they maintain their mathematical form under rescaling, described by the equation:

$$f(\alpha x) = \alpha^{-\gamma} f(x). \tag{2.33}$$

In scale-free networks, the average degree $\langle k \rangle$ is finite and well-defined. However, the variance,

$$\sigma^2 = \langle k^2 \rangle - \langle k \rangle^2, \tag{2.34}$$

is heavily influenced by the upper cutoff k_{max}. This divergence occurs because the second moment scales with the upper limit as:

$$\langle k^2 \rangle \sim k_{max}^{3-\gamma}. \tag{2.35}$$

This behavior implies that the connectivity distribution is dominated by nodes with exceptionally high degrees, known as *hubs*. These hubs play a crucial role in improving connectivity and the overall robustness of the network. In practice, real networks often present finite-size effects that introduce natural cutoffs in the degree distribution [162]. Additionally, real-world data can be affected by noise, particularly in the distribution tails. To address this, researchers often compute the *cumulative degree distribution*:

$$P_{cum}(k) = \sum_{k'=k}^{\infty} P(k'). \tag{2.36}$$

The cumulative distribution smooths statistical fluctuations and provides a more stable measure of the power-law exponent. In this case, the exponent can be derived from $P_{cum}(k)$ in a log-log plot:

$$\gamma = 1 + \alpha_{cum}. \tag{2.37}$$

Another alternative for improving data clarity, which efficiently manages data with heavy-tailed distributions, is discussed in [163].

The need to construct graphs with power-law degree distributions emerged following the identification of scale-free properties in real-world networks. Static scale-free networks are distinguished from evolving models by their fixed structure. Aiello et al. introduced a two-parameter model, denoted $G_{\lambda,\beta}$, which assigns uniform probability to all graphs with a degree sequence $N(k) \sim e^{-\lambda} k^{-\beta}$ [164]. In this model, λ defines the logarithm of nodes with degree $k = 1$, while $e^{\lambda/\beta}$ represents the maximum degree in the graph. This model predicts a giant component if $\beta < \beta_c \approx 3.47875$, with smaller connected components scaling as $O(\log N)$ for $2 < \beta < \beta_c$ and $O(1)$ for $1 < \beta < 2$.

Newman in [36] derived an expression for the clustering coefficient in power-law graphs:

$$C \sim N^{(3-7)/(\beta-1)}. \tag{2.38}$$

For large networks with $\beta > 7/3$, $C \to 0$, while for $\beta < 7/3$, the clustering coefficient increases with system.

The *fitness model* by Goh et al. assigns each node i a fitness value $w_i = i^{-\alpha}$, where $\alpha \in [0, 1)$ is a tunable parameter. Two nodes i and j are connected with probability proportional to their respective weights. This model generates networks with a degree distribution $P(k) \sim k^{-\gamma}$, where $\gamma = 1 + 1/\alpha$ [165].

A similar model by Caldarelli et al. assigns each node a fitness value drawn from a given distribution, and edges are formed if the combined fitness of two nodes exceeds a threshold [166]. Gradient networks are constructed by assigning a scalar field h_i to each node i. Links are directed from nodes to their neighbors with the highest potential. If the underlying substrate is scale-free, the gradient network inherits this feature, often resulting in power-law in-degree distributions [167].

While static models describe fixed structures, evolving models capture growth dynamics. The Barabási-Albert (BA) model is one of the most widely studied in this class [46]. The BA model incorporates two mechanisms:

- **Growth**: Starting from m_0 isolated nodes, a new node with m links is added at each time step.
- **Preferential Attachment**: Each new link connects to an existing node i with probability:

$$\Pi_{j\to i} = \frac{k_i}{\sum_l k_l}. \tag{2.39}$$

This process leads to a scale-free degree distribution $P(k) \sim k^{-3}$. The model exhibits logarithmic growth of the average path length $L \sim \ln N / \ln \ln N$ and a clustering coefficient scaling as $C \sim N^{-0.75}$ [122].

Several modifications to the BA model have been proposed:

- **DMS Model**—Incorporates initial node attractiveness, modifying the attachment probability to $\Pi_{j\to i} = (k_i + k_0)/\sum_l (k_l + k_0)$, where k_0 controls the degree exponent [168].
- **Nonlinear Growth Model**—Introduces a nonlinear attachment rule, $\Pi_{j\to i} \propto k_i^{\alpha}$, where $\alpha < 1$ leads to stretched exponential distributions and $\alpha > 1$ results in networks dominated by a few hubs [169].
- **Fitness-Based Models**—The Bianconi-Barabási (BB) model assigns each node an intrinsic fitness η_i, modifying the attachment probability as:

$$\Pi_{j\to i} = \frac{\eta_i k_i}{\sum_l \eta_l k_l}. \tag{2.40}$$

This model proposes similar elements to Bose-Einstein condensation in network evolution [62].

The Ravasz-Barabási (RB) model introduces hierarchical organization by iteratively constructing clusters with internal and external links. This model effectively reproduces real-world modular structures [170]. The Klemm-Eguíluz (KE) model incorporates node deactivation to enhance clustering properties [171]. In this model, nodes alternate between active and inactive states, ensuring a stable clustering structure. The Sole-Pastor-Satorras-Kepler (SPSK) model mimics biological networks by employing duplication, divergence, and mutation mechanisms [172]. Similarly, the Vázquez-Flammini-Maritan-Vespignani (VFMV) model models duplication and divergence, generating power-law degree distributions and realistic clustering properties [173].

2.4.2.6 Weighted Networks

Most discussions on network properties have focused on unweighted networks, where connections are binary. However, real-world networks often exhibit considerable variability in the strength or intensity of connections. Such weighted networks provide a more accurate representation of systems like social interactions, metabolic pathways, and communication infrastructures [144]. Weighted networks are defined as $G_W = (N, L, W)$, where N represents nodes, L the links, and W the set of positive numerical weights associated with the links. These weights often encode connection strength, capacity, or interaction frequency. In this case, the network is represented by a weight matrix W, where each entry w_{ij} is the weight of the link between nodes i and j. If nodes are disconnected, $w_{ij} = 0$. Several key features and parameters have been defined to characterize weighted networks by extending traditional topological concepts. The node strength s_i generalizes the degree in weighted networks and is defined as:

$$s_i = \sum_{j \in N} w_{ij}. \tag{2.41}$$

The strength combines information on both the number of connections and the intensity of interactions. In networks with uncorrelated weights, the node strength grows proportionally to the degree, following $s(k) \approx \langle w \rangle k$. However, when strong weight-topology correlations are present, this relationship can significantly differ. To assess differences in link weights around a node, the disparity index is defined as:

$$Y_i = \sum_{j \in N_i} \left(\frac{w_{ij}}{s_i} \right)^2, \tag{2.42}$$

where N_i is the set of first neighbors of node i. Values near 1 indicate dominance by a few strong links, while smaller values suggest uniform weights. In weighted networks, defining an optimal path is more complex. The shortest path need not minimize the number of edges, but rather the total path length. Often, the link length is assumed to be inversely proportional to its weight:

$$\ell_{ij} = \frac{1}{w_{ij}}. \tag{2.43}$$

The weighted shortest path is then defined as the path that minimizes the sum of link lengths. To extend the concept of clustering in weighted networks, Barrat et al. defined a weighted clustering coefficient:

$$C_i^W = \frac{1}{s_i(k_i - 1)} \sum_{j,m} (w_{ij} + w_{im}) a_{ij} a_{jm} a_{mi}. \tag{2.44}$$

This measure reflects both the presence of triangles and the strength of the interactions in those triangles. Networks where strong links dominate interconnected triples tend to have $C_i^W > C$, while the opposite behavior is found in systems where weak links cluster more prominently [144].

Motifs in weighted networks can be evaluated using two measures:

$$I(g) = \left(\prod_{(ij)\in L_g} w_{ij} \right)^{1/l}, \tag{2.45}$$

$$Q(g) = \frac{I(g)}{\sum_{(ij)\in L_g} w_{ij}}. \tag{2.46}$$

The intensity $I(g)$ reflects the geometric mean of link weights in the motif, while coherence $Q(g)$ evaluates weight uniformity across the motif. These quantities allow weighted motifs to be distinguished from their topological counterparts [174].

An important extension of weighted network theory involves multiplex structures, where nodes participate in multiple layers of interactions, each representing a different type or context of connection. Menichetti et al. [175] propose a framework to integrate weight heterogeneity across layers by introducing the concept of node activity and inter-layer strength correlations. Their approach distinguishes between node strength vectors across layers and explores how weights are distributed both within and across layers. They show that weights in one layer can be statistically dependent on node properties in another, revealing non-trivial interdependencies that cannot be captured by single-layer analysis. The study further demonstrates that null models preserving intra-layer topology but reshuffling weights across layers fail to reproduce observed weight distributions, highlighting the structural significance of multiplex weight correlations. This perspective is particularly relevant for understanding systems like transportation, communication, and biological networks, where the interaction strength in one domain may influence behavior in another.

2.4.2.7 Spatial Networks

Spatial networks form a distinct class of complex networks in which nodes are embedded in two- or three-dimensional Euclidean space, and links represent physical connections between them. These networks are crucial in various domains such as neuroscience (e.g., neural networks), communication systems, power grids, and transportation networks [176, 177]. Unlike other types of networks, spatial networks are influenced by geographical constraints, which affect their structure and behavior. Below are some features that characterize this kind of network:

- Nodes in spatial networks often follow geographical patterns. For instance, in power grids and transportation systems, node distribution is typically fractal in nature. In [177], Yook et al. demonstrated that Internet routers form a fractal pattern driven by the spatial distribution of population density. Their analysis using box-counting methods revealed that both Internet nodes and population distributions share a fractal dimension of $D_f = 1.5 \pm 0.1$. This correlation arises because regions with higher population density tend to require more Internet infrastructure, leading to a denser router distribution.
- Spatial networks impose limitations on node degrees. Due to physical space limitations, the number of connections a node can sustain is restricted. This constraint is particularly evident in urban street grids and ant colony tunnel systems [178]. In planar graphs—defined as networks in which edges can only cross at vertices—the maximum number of links scales as $3N - 6$ for N nodes [133].
- In spatial networks, the likelihood of two nodes connecting decreases with increasing Euclidean distance, as longer-range connections typically incur higher costs. This property heavily influences the network's small-world behavior, where geographical constraints can limit the presence of shortcuts. Despite these limitations, some spatial networks can still exhibit small-world features if appropriately structured.

Furthermore, spatial networks exhibit distinct clustering behavior. Unlike many complex networks that display hierarchical clustering with $C(k) \sim k^{-1}$ scaling, geographically-constrained networks like power grids and Internet router networks often show no clear dependency between the clustering coefficient $C(k)$ and node degree k [170]. This distinction arises because geographical constraints limit the complexity of local link formation, reducing the potential for hierarchical organization.

2.4.3 Robustness and Resilience

The robustness of a network refers to its capacity to maintain functionality in the face of partial damage, such as the removal or failure of nodes or links. This aspect is critical in ensuring the reliability of systems where processes depend on network integrity, ranging from social structures to technological infrastructures [179, 180].

Two main types of robustness can be distinguished: static and dynamic robustness. Static robustness addresses the network's structural integrity when nodes are removed without the need to adjust or redistribute flows. An example is a social network, where certain connections between individuals are severed, but no reallocation of "load" is needed.

The study of static robustness is often based on analytical methods, notably percolation theory. It allows for assessing how the size of the largest connected component evolves as nodes are progressively removed.

In particular, networks exhibit different responses to two node removal strategies:

- **Random failures**, where nodes are removed randomly, typically have minimal impact on scale-free networks due to the majority of nodes having low degree.
- **Intentional attacks**, where high-degree nodes (hubs) are targeted, can separate the network even after the removal of a small fraction of nodes [181].

Dynamic robustness concerns the network's ability to respond to failures while maintaining function through rerouting. Unlike the static case, analyzing dynamic robustness generally requires computational approaches and simulations, due to the complexity of flow redistribution and network dynamics. Overall, the resilience of networks to failures or attacks depends on both topological features and the specific nature of the processes on top. Understanding robustness is crucial for designing resilient systems, within transportation, communication, and biological regulation contexts. A network's static tolerance to random failures refers to its ability to maintain overall connectivity when a fraction f of its nodes or edges are randomly removed. Conversely, an *attack* targets specific components, such as the most connected nodes (hubs), potentially leading to more severe disruption. Analytically, this problem is often treated using *percolation theory* [182]. In percolation models, one distinguishes between *site* percolation, random removal of nodes, and *bond* percolation, random removal of links. The key is represented by the *giant component*, whose size as a function of f reveals the network's resilience. A *critical threshold* f_c defines the percolation transition: below this value, the network remains connected, while above it becomes fragmented. This framework works well for uncorrelated networks under random removals, but more advanced methods are needed for analyzing attacks, where node selection is not random. Early numerical studies by Albert et al. and Broder et al. investigated the response of real networks to both random failures and targeted attacks [123, 179]. Random failures cause a gradual reduction in the size of the giant component, which persists even at high removal rates. In contrast, targeted removal of high-degree nodes rapidly fragments the network into isolated clusters, that is, an effect more pronounced than in classical random graphs.

For example, if we consider an Erdős Rényi (ER) random graphs with $N = 10^4$ and average degree $\langle k \rangle = 4$, both error and attack scenarios exhibit a threshold-like behavior with $f_c \simeq 0.28$. Conversely, scale-free networks of the same size and average degree show no clear threshold for random failures, but are highly sensitive to targeted attacks, with $f_c \simeq 0.18$.

These findings demonstrate the intrinsic robustness of scale-free networks to random failures, due to their many low-degree nodes, but also their vulnerability to deliberate attacks, which exploit the network's reliance on a few highly connected hubs.

On network robustness, Cohen et al. [132, 183] proposed a theoretical framework based on percolation theory. Taking into consideration a random graph composed of N nodes, a degree distribution $P(k)$, and an average degree $\langle k \rangle$, although a randomly chosen node has degree k with probability $P(k)$, the neighbors are more likely to have higher degrees. The adjusted probability that a vertex reached by following a random edge has degree k is given by:

$$q_k = \frac{kP(k)}{\langle k \rangle}. \tag{2.47}$$

Assuming the graph is sparse, large, and locally tree-like (i.e., negligible clustering), the existence of a giant component depends on whether the average number of second neighbors (z_2) of a randomly selected node exceeds the number of its first neighbors. This condition is expressed as:

$$z_2 = \frac{\langle k^2 \rangle - \langle k \rangle}{\langle k \rangle}. \tag{2.48}$$

When $z_2/\langle k \rangle = 1$, the system reaches the critical value. This leads to the criterion:

$$\sum_k k(k-2)P(k) > 0, \tag{2.49}$$

that is known as the Molloy–Reed criterion [156, 184] indicates that a giant connected component will form if the second moment $\langle k^2 \rangle$ is sufficiently large. Notably, this is always true for scale-free networks with $2 < \gamma < 3$, due to the divergence of $\langle k^2 \rangle$. Now, consider the impact of random node removal (site percolation) where a fraction f of nodes is removed.

The updated degree distribution becomes:

$$P(k') = \sum_{k=k'}^{\infty} P(k) \binom{k}{k'} (1-f)^{k'} f^{k-k'}. \tag{2.50}$$

By evaluating the new first and second moments of the degree distribution and substituting them into the percolation threshold condition, one obtains the critical fraction f_c of nodes that must be removed to collapse the giant component:

$$1 - f_c = \frac{1}{\frac{\langle k^2 \rangle}{\langle k \rangle} - 1}. \tag{2.51}$$

If $\langle k^2 \rangle$ diverges, as in many real-world scale-free networks, then $f_c \to 1$, so these networks are exceptionally robust to random failures.

Further validation of this result has been achieved through more rigorous techniques. One such method, developed by Vázquez et al. [185], derives from statistical mechanics of disordered systems. Another, based on generating functions and introduced by Newman et al. [157], models the probability x that an edge leads to a finite component using the self-consistent equation:

$$x = f + (1 - f) \sum_{k=0}^{\infty} \frac{kP(k)}{\langle k \rangle} x^{k-1}. \tag{2.52}$$

A nontrivial solution $x < 1$ indicates the presence of a giant component.

In real networks, correlations often exist between the degrees of adjacent nodes, meaning that the degree distribution $P(k)$ alone is insufficient to characterize the network structure. To account for this, one introduces the conditional probability $P(k'|k)$, which represents the likelihood that a node of degree k is connected to a node of degree k'. This is derived from the symmetric matrix $E_{k'k}$, which counts the number of edges between nodes of degree k and k' [185, 186].

The percolation problem in such networks can be analyzed using techniques from statistical physics, including mappings to the Ising or Potts models [187]. Alternatively, one can use the generating function formalism, which leads to the following expressions for the size S of the giant component and the probability u_k that a neighbor of a degree-k node does not belong to the giant component:

$$S = 1 - \sum_k P(k) u_k^k, \tag{2.53}$$

$$u_k = f + (1 - f) \sum_{k'} P(k'|k) u_{k'}^{k'-1}. \tag{2.54}$$

In the case of site percolation, where a fraction f of the nodes is randomly removed, these equations become:

$$S = 1 - f - (1 - f) \sum_k P(k) u_k^k, \tag{2.55}$$

$$u_k = f + (1 - f) \sum_{k'} P(k'|k) u_{k'}^{k'-1}. \tag{2.56}$$

The stability of the trivial solution $u_k = 1$ (corresponding to $S = 0$) determines the existence of a giant component. This is governed by the largest eigenvalue $\lambda_{\max}$ of the matrix $C_{kk'} = (k' - 1)P(k'|k)$, leading to the critical threshold condition:

$$(1 - f_c)\lambda_{\max} = 1. \tag{2.57}$$

In bond percolation, where a fraction f of the edges is randomly removed, a similar framework applies, and the same critical condition holds. For uncorrelated networks, the matrix $P(k'|k)$ factorizes, leading to:

$$\lambda_{\max}^{\text{unco}} = \frac{\langle k^2 \rangle}{\langle k \rangle} - 1. \tag{2.58}$$

More generally, one can interpolate between correlated and uncorrelated cases using a mixture model:

$$P(k'|k) = \epsilon q_{k'} + (1 - \epsilon)\tilde{P}(k'|k), \tag{2.59}$$

with $0 < \epsilon < 1$. If the network is robust in the uncorrelated case, it remains robust for any $\epsilon > 0$, provided $\langle k^2 \rangle$ diverges.

However, this divergence is not a sufficient condition for robustness in the presence of correlations. In disassortative networks, where high-degree nodes connect preferentially to low-degree nodes, robustness can be significantly reduced. Consider a degree-dependent attachment probability $g_k = k^{-\alpha}$ with $P(k) \sim k^{-\gamma}$, leading to:

$$\lambda_{\max} \sim \sum_k g_k^2 (k - 1) k P(k), \tag{2.60}$$

which diverges only if $\alpha < (3 - \gamma)/2$. This implies that increasing disassortativity (larger α) can suppress the formation of the giant component, even when $\langle k^2 \rangle$ diverges.

Therefore, disassortative correlations make it more difficult for a giant component to emerge. Even in undamaged networks with power-law degree distributions, the largest component may vanish in the thermodynamic limit, showing that a divergent second moment is not a sufficient condition for network robustness.

Considering the robustness of uncorrelated networks under targeted attacks, a network exhibits a giant component if the ratio $\langle k^2 \rangle / \langle k \rangle \geq 2$ is satisfied. For random failures, the critical threshold is modified accordingly. However, when nodes are removed selectively, specifically the most highly connected ones, the situation changes drastically. Cohen et al. [181] developed a formalism to analyze this case.

Targeted attacks remove not only the nodes but also a substantial number of edges, given that highly connected nodes (hubs) are critical to network connectivity. The degree distribution $P(k)$ is altered, particularly the upper cutoff k_{cut}, which is initially determined by:

$$\int_{k_{\text{cut}}}^{\infty} P(k) dk = \frac{1}{N}. \tag{2.61}$$

After removing a fraction f of the most connected nodes, the new cutoff $\tilde{k}$ satisfies:

$$\int_{\tilde{k}}^{\infty} P(k)dk = f. \tag{2.62}$$

Assuming a continuous degree distribution and a large system size, this yields:

$$\tilde{k} = k_{\min} f^{1/(1-\gamma)}, \tag{2.63}$$

where γ is the power-law exponent and $k_{\min}$ the minimum degree.

Since links from removed nodes are deleted randomly across the network, the impact on the remaining connectivity is equivalent to randomly removing edges, though with a much higher impact. The probability $\tilde{f}$ that an edge leads to a removed node becomes:

$$\tilde{f} = \frac{\int_{\tilde{k}}^{\infty} kP(k)dk}{\langle k \rangle}. \tag{2.64}$$

For $\gamma > 2$, this results in:

$$\tilde{f} = f^{(2-\gamma)/(1-\gamma)}. \tag{2.65}$$

This expression shows that even a small f can lead to the removal of a significant fraction of edges. For example, with $\gamma = 2.2$ and $f = 0.1$, over 70% of the edges may be eliminated. In the extreme case $\gamma = 2$, $\tilde{f}$ scales as $\ln(Nf/k_{\min})$.

Despite these changes, the criterion for the presence of a giant component remains the same, although the network becomes more fragile. A critical fraction f_c can be derived using:

$$\left(\frac{\tilde{k}}{k_{\min}}\right)^{2-\gamma} - \left(\frac{\tilde{k}}{k_{\min}}\right)^{3-\gamma} = 3 - \gamma. \tag{2.66}$$

This equation can be solved numerically to estimate $f_c(k_{\min}, \gamma)$.

Cohen et al. further showed that generating function techniques [157, 180] can be adapted to calculate other properties like the size of the giant component post-attack.

In summary, uncorrelated scale-free networks exhibit extreme vulnerability to targeted attacks [179]. Removing just a few highly connected nodes can fragment the network, a phenomenon not observed in random failures. However, understanding the resilience of correlated networks to such attacks remains an open challenge, due in part to the limited availability of generative models with tunable degree correlations [188].

An intriguing example of static robustness in social networks is provided by Hobbs and Burke's study [189] on "connective recovery" following the death of a friend. Analyzing approximately 15,000 Facebook networks, they observed that while the immediate loss of a friend led to a decrease in interactions, the surviving friends increased their interactions with each other, effectively compensating for the lost connection. This reorganization occurred without any central coordination, illustrating the network's inherent ability to maintain its structural integrity despite node loss. Notably, younger users (ages 18–24) exhibited the most significant increases in interaction, suggesting age-related differences in network resilience. This phenomenon underscores the capacity of social networks to self-heal through localized adjustments, preserving overall connectivity even in the face of individual losses.

2.4.4 Spreading Processes

Epidemic models aim to simulate the transmission of infectious diseases across a population to understand and reproduce real-world dynamics. These models are crucial for designing effective containment strategies and long-term control measures

In contrast, rumor spreading models are often developed to maximize the speed and reach of information dissemination. Typical applications include the optimization of data distribution protocols in computer networks and the formulation of strategies for viral marketing campaigns. Unlike epidemics, in this context, one has control over the design of the dynamics to achieve efficient diffusion.

In both types of models, special attention is given to the influence of network topology on the spreading dynamics. The inclusion of heterogeneous and realistic network structures significantly alters the outcomes predicted by traditional models based on regular lattices or random graphs.

Epidemic models, while originally developed in epidemiology, have become central tools in the study of spreading phenomena over complex networks. Their importance extends well beyond biological contagions: they provide a powerful framework for analyzing the diffusion of information, behaviors, failures, and risks in interconnected systems [8, 38, 96].

In the context of this book, which explores the intersection of complex systems and communication networks, epidemic models offer valuable insights cause they enable the description of how local interactions can lead to global transitions, revealing the nonlinear and emergent behavior of complex systems. Furthermore, these models can simulate the propagation of faults or cyberattacks across networked infrastructures, aiding in the design of robust communication architectures. Moreover, models such as SIS or SIR are widely used to understand how information, memes, or fake news spread in digital social systems. In mobile and distributed networks, epidemic-style protocols are applied for routing, data replication, and fault tolerance in scenarios with limited connectivity. Including epidemic models thus reinforces the systemic perspective of the book and highlights the universality of diffusion mechanisms across domains—from public health to engineering and communication technologies.

2.4.4.1 Epidemic Modeling and Complex Networks

Since Bernoulli's pioneering work in 1760, mathematical models have played a central role in understanding the spread of infectious diseases. The classical compartmental models introduced have been by Kermack and McKendrick (1927), later comprehensively summarized by Anderson and May [190]. These models provide essential tools for analyzing disease dynamics and informing public health strategies. Over time, epidemic modeling has extended beyond biological infections to represent diverse spreading phenomena, such as information diffusion, behavior adoption, and cascading failures [191]. Many of these processes share similar mathematical structures governed by reaction-diffusion dynamics [192]. Recent advances in epidemic modeling, enabled by access to large-scale datasets and high-resolution simulations down to individual levels [193–195]. These data-driven models support real-time decision-making in epidemiology and highlight the influence of individual heterogeneity, spatial structure, and multiple time scales [196, 197]. At the core of these approaches, there is the topology of human interactions and mobility, best captured using network theory [134, 198]. The proliferation of mobile and digital technologies now enables detailed mapping of contact networks and behavioral patterns [199]. These complex networks are typically heterogeneous, modular, and hierarchical, with heavy-tailed degree distributions and community structures [126, 200, 201]. Such complexity significantly impacts contagion dynamics, rendering traditional homogeneous models insufficient. However, the exact resolution of dynamical equations on networks is frequently unfeasible due to their inherent complexity. Thus, the field has developed different mathematical and computational approaches, including mean-field approximations and spectral methods, to study epidemic processes on networks [202, 203].

2.4.4.2 The Mathematical Approach to Epidemic Spreading: Classical Models

The mathematical study of epidemic spreading has developed over more than two centuries into a well-established interdisciplinary context [196, 204–206]. The main approach for the epidemic modeling is the compartmental framework, in which individuals are categorized based on their disease status, such as susceptible (S), infectious (I), and recovered (R). This approach can be extended to incorporate additional compartments, for example, to model immunity or vector-borne transmission (e.g., via mosquitoes in malaria) [190, 196].

Epidemic models describe the time evolution of these compartments by specifying rules for transitions between states. In the simplest setting, assuming a closed population of N individuals, the **SIS model** consists of two transitions:

- $S \to I$: a susceptible becomes infected upon contact with an infectious individual;
- $I \to S$: the infected individual recovers but does not acquire immunity, becoming susceptible again.

This model captures diseases where reinfection is possible and the epidemic can persist indefinitely under certain conditions. A more realistic model for many diseases is the **SIR model**, in which recovery leads to permanent immunity. Here, the transition is:

- $I \rightarrow R$: infected individuals recover and move to the recovered compartment, no longer participating in transmission.

These basic models form the foundation for more complex epidemic dynamics and will serve as the starting point for the theoretical developments. Two foundational models are the SIS (Susceptible-Infected-Susceptible) and SIR (Susceptible-Infected-Recovered) models. In the SIS model, individuals become infected upon contact with infectious individuals and return to the susceptible class upon recovery, with no lasting immunity. In the SIR model, recovered individuals acquire permanent immunity, preventing further disease propagation once no susceptible individuals remain.

The dynamics of these models can be interpreted using reaction processes:

$$S + I \xrightarrow{\beta} 2I, \tag{2.67}$$

$$I \xrightarrow{\mu} S \quad \text{(SIS)}, \quad \text{or} \quad I \xrightarrow{\mu} R \quad \text{(SIR)}. \tag{2.68}$$

The transition from S to I is interaction-driven, depending on contact rates and disease transmissibility. In contrast, recovery (I to S or R) typically follows a Poisson process with rate μ, implying exponential recovery time [207]. The infection rate is characterized by the force of infection α, often modeled as:

$$\alpha = \beta \frac{N_I}{N}, \tag{2.69}$$

where N_I is the number of infected individuals, and β is the transmission rate. This expression is known as the mass-action law [208].

Extensions of these basic frameworks include more compartments and features. For example, the SIRS model includes temporary immunity:

$$R \xrightarrow{\eta} S, \tag{2.70}$$

where η is the rate at which immunity wanes. The SEIR model accounts for an exposed (E) stage where individuals are infected but not yet infectious:

$$S + I \xrightarrow{\beta} E + I, \tag{2.71}$$

$$E \xrightarrow{\epsilon} I, \tag{2.72}$$

$$I \xrightarrow{\mu} R. \tag{2.73}$$

These models can be understood as stochastic reaction-diffusion processes, where individuals act as interacting particles [192]. They may be further generalized to include demographics, age structure, asymptomatic states, and more complex dynamics, offering a closer representation of real-world epidemiology [205].

2.4.4.3 Deterministic Epidemic Modeling: Mean-Field Approximation

Although epidemic spreading is fundamentally a stochastic reaction-diffusion process, classical epidemic modeling often relies on deterministic approaches derived from the continuous-time limit of discrete-time equations. These approaches assume a homogeneous mixing of individuals—each individual in a compartment is indistinguishable from others—closely resembling the mean-field approximation used in statistical physics [209, 210].

Let N_α be the number of individuals in compartment α, and $\rho_\alpha = N_\alpha / N$ its corresponding density. The average time evolution of compartments is governed by differential equations based on the law of mass action, where the change in compartment densities is proportional to the product of interaction rates and densities [208]. For the SIS and SIR models, the dynamics can be described by:

$$\frac{d\rho_I}{dt} = \beta \rho_I \rho_S - \mu \rho_I, \tag{2.74}$$

$$\frac{d\rho_S}{dt} = -\beta \rho_I \rho_S + \alpha \rho_I, \tag{2.75}$$

where $\alpha = \mu$ in the SIS model and $\alpha = 0$ in the SIR model, and β is the infection rate. Normalization constraints apply: $\rho_S = 1 - \rho_I$ (SIS) and $\rho_R = 1 - \rho_S - \rho_I$ (SIR).

Assuming $\rho_I \approx 0$ in the early stage, it is obtained that:

$$\frac{d\rho_I}{dt} \approx (\beta - \mu)\rho_I, \tag{2.76}$$

which integrates to:

$$\rho_I(t) \approx \rho_I(0) e^{(\beta - \mu)t}. \tag{2.77}$$

This yields the basic reproduction number:

$$R_0 = \frac{\beta}{\mu}, \tag{2.78}$$

which determines whether an epidemic outbreak ($R_0 > 1$) occurs or the infection dies out ($R_0 < 1$) [204]. In SIS-like models, $R_0 > 1$ leads to an endemic state; in SIR models, it indicates whether a sizable fraction of the population is affected. While this framework provides valuable insight, it overlooks individual mobility and stochastic effects, which are particularly significant in small populations. In stochastic systems, even if $R_0 > 1$, an outbreak might fail to propagate due to random fluctuations [204]. Moreover, this classic approach assumes Poissonian (memoryless) transitions. However, deviations from exponential waiting times can significantly alter epidemic dynamics. Finally, the homogeneous mixing assumption neglects the individual variability in contact patterns. In structured populations, epidemic thresholds must take into account the network topology. For a network where every individual has $\langle k \rangle$ contacts, it is redefined:

$$R_0 = \frac{\langle k \rangle \beta}{\mu}, \tag{2.79}$$

emphasizing the critical role of connectivity heterogeneity in shaping epidemic outcomes [206].

2.4.4.4 Epidemic Models and Phase Transitions in Statistical Physics

The study of epidemic spreading has attracted growing interest in the statistical physics community, particularly due to its parallels with non-equilibrium phase transitions [210, 211]. A critical concept in epidemiology, the *epidemic threshold*, mirrors that of a phase transition: when a key parameter exceeds a critical point, a qualitative change occurs in the system's behavior. In this context, the density of infected individuals plays the role of an order parameter, while the transmission rate acts as the control parameter [212]. The SIS (Susceptible-Infected-Susceptible) model belongs to the class of non-equilibrium systems with *absorbing states*, where the disease-free state is absorbing. A well-known analogous model is the *contact process*, which exhibits a transition from an active to an absorbing phase as the infection rate crosses a critical value [213]. In SIS dynamics, the active phase sustains a non-zero density of infected individuals, while in the absorbing phase, infections die out. Although SIS and the contact process share core features, they differ in that an infected individual in SIS may infect all its neighbors simultaneously, whereas in the contact process, only one offspring is produced per event. This distinction becomes critical in networks with heterogeneous connectivity.

The SIR (Susceptible-Infected-Recovered) model also displays a transition analogous to percolation, between localized outbreaks and large, scale epidemics. The critical point in this case is linked to the probability of forming a giant connected component of recovered individuals [182, 214]. Mathematically, the SIR process can be mapped onto a bond percolation problem, where the effective percolation probability depends on the disease's transmission and recovery rates.

In both SIS and SIR models, the presence and nature of the epidemic threshold depend strongly on the underlying topology. On scale-free networks, for example, the threshold can vanish due to highly connected hubs, making the disease persist even at low transmission rates [28].

2.4.5 *Collective Dynamics*

2.4.5.1 Information Theory and Complexity

The scientific exploration of complex systems has deepened our understanding of diverse phenomena and has led to the development of valuable theoretical and practical tools [31, 215]. Key notions used to describe complex systems, such as emergence, self-organization, adaptivity, and complexity, have often been applied across different disciplines [216–218].

The growing use of complex systems terminology across disciplines has introduced a lot of concepts and key definitions. Information theory helps clarify these by providing a precise way to quantify complexity, emergence, self-organization, and homeostasis. When focusing on the information produced by a system, emergence and self-organization can be seen as opposing processes, while complexity results from their balance.

Information can be interpreted as a measurable pattern within data. Claude Shannon formalized this notion in the context of telecommunications, focusing on how to achieve reliable communication despite unreliable components [219]. Given a sequence X composed of elements x drawn from a probability distribution $P(x)$, the information content—also known as Shannon entropy—is defined as:

$$I = -\sum P(x) \log P(x), \tag{2.80}$$

where the logarithm is typically taken in base 2 for binary systems.

This measure quantifies the uncertainty in the system: if an outcome is fully predictable, such as $P(1) = 1$ and $P(0) = 0$, then $I = 0$ since no new information is gained. Conversely, in the case of maximal uncertainty—for example, a fair coin toss with $P(0) = P(1) = 0.5$—the information is maximal ($I = 1$ bit), as each outcome is entirely unpredictable and thus informative. Shannon information can therefore be understood as a measure of unpredictability. It reaches its maximum when all outcomes are equally probable and drops to zero when there is full certainty. Prokopenko et al. [218] provided a comprehensive review of how information theory underpins various measures of complexity, emergence, and self-organization, laying the groundwork for formalizing these concepts in the context of complex systems. The concept of complexity has been interpreted and measured in numerous ways [217, 220]. Etymologically, the term derives from the Latin *plexus*, meaning "interwoven," which conveys the idea that something complex is difficult to separate due to meaningful interactions among its components [215]. In the context of information theory, the central question becomes: *how complex is a given piece of information?* Here, complexity can be interpreted as the amount of information needed to describe a phenomenon at a given observational scale [215, 221]. The more information required for a description, the greater the complexity.

Some approaches to complexity suggest that highly ordered strings (where Shannon information $I \approx 0$) are simple, whereas those with maximal information content

($I \approx 1$) are complex. This view may be valid in certain contexts, but it leads to the paradox that purely random or chaotic systems would then have the highest complexity. To address this, alternative formulations, such as effective complexity [221], Langton's λ parameter [222], and Wuensche's Z measure [223], frame complexity as an intermediate condition between order and randomness. In this direction, the measure introduced by López-Ruiz et al. [224] is especially relevant. It defines complexity as the product of Shannon information and a notion of disequilibrium, reflecting the system's distance from thermodynamic equilibrium. In this formulation, ordered systems like crystals have low information and high disequilibrium, yielding low complexity. Chaotic systems, such as ideal gases, have high information but low disequilibrium, also resulting in low complexity. Maximum complexity emerges when both information and disequilibrium attain intermediate values, implying a balance between order and randomness. The study of complex systems often invokes a set of interrelated concepts, such as emergence, self-organization, homeostasis, and complexity, that describe the ways in which structure, coherence, and adaptive behavior arise from underlying dynamics. Although historically introduced in different disciplinary contexts, these notions can be jointly addressed within a unified framework grounded in information theory. This perspective enables the formalization and quantification of systemic organization by analyzing patterns of information production, transmission, and stability across scales. From this standpoint, emergence describes how macroscopic regularities arise from microscopic interactions; self-organization reflects the tendency of a system to develop order without external control; homeostasis captures the ability of a system to maintain functional stability amid internal or external perturbations; and complexity refers to the balance between order and randomness that characterizes systems capable of both flexibility and resilience. Emergence refers to the phenomenon by which properties arise at a macroscopic level that are not evident when examining the components individually. For instance, attributes such as color or malleability are not properties of individual gold atoms, but emerge at higher levels of structural organization [225]. Emergence may also be understood as a shift in the descriptive framework or as the manifestation of a higher-level pattern from lower-level interactions.

From an information-theoretic viewpoint, Shalizi in [226] proposed metrics to assess the causal architecture of complex systems, showing how coarse-grained descriptions may enhance predictive efficiency compared to micro-level models. In [226] has been introduced a metric that evaluates the efficiency of prediction across different scales, revealing that coarse-grained, macroscopic descriptions—such as those in thermodynamics—can be more predictive than micro-level descriptions involving all constituent particles. Similarly, other approaches to quantify emergence compare information at different system levels, indicating higher emergence when a system exhibits strong interdependence among its components [227].

Self-organization is characterized by the spontaneous formation of coherent structures or patterns due to local interactions among a system's components. Common examples include bird flocking, fish schooling, and insect swarming [228]. These systems evolve over time toward organized states through their internal dynamics, without centralized control. Ashby noted that any system moving toward an attractor could be considered self-organizing, provided that the attractor is interpreted as an organized state [229]. This highlights the contextual and observer-dependent nature of self-organization. Nevertheless, objective metrics based on information theory can be employed. For example, Gershenson and Heylighen proposed quantifying self-organization as the reduction in Shannon information over time: a decrease implies a gain in organization, while an increase suggests disorder [230]. Yet, this interpretation depends on the granularity and scale of observation, as the same system may be viewed as organizing or disorganizing depending on the observer's perspective.

Homeostasis describes the capacity of a system to maintain internal stability despite external or internal disturbances. Rather than implying stasis, homeostasis reflects adaptive constancy. The concept has evolved from its biological roots to describe the ability of a system to keep "essential variables" within a viability zone through internal regulation and adaptive feedback mechanisms [231]. In dynamical systems, homeostatic behavior is linked to the system's ability to remain near specific attractors in the presence of perturbations. When external changes occur, a homeostatic system adjusts its internal parameters to stay within acceptable operational boundaries. This trait is closely related to robustness, understood as the system's resilience to shocks or unexpected changes without losing functional integrity [231]. These foundational concepts—emergence, self-organization, and homeostasis—not only clarify the internal dynamics of complex systems but also provide a basis for understanding large-scale behaviors arising in real-world scenarios. Indeed, recent years have witnessed a growing interest in the study of complex networks and the collective dynamics that emerge in biological, social, and technological domains.

2.4.5.2 Modeling Collective Dynamics in Complex Systems

The quantitative study of complex systems has substantially improved our understanding of self-organization and adaptive behavior in biological, social, and technological domains. Agent-based modeling and network theory have proven particularly useful in revealing how collective dynamics emerge from local interactions among system components. Interdisciplinary approaches that draw on dynamical systems, graph theory, and statistical physics have enabled researchers to uncover general principles governing the evolution and robustness of natural and man-made systems [232]. These insights have found practical applications in the design of resilient infrastructures and distributed computing systems that mimic organic behaviors [233]. Furthermore, the increasing integration of social and technological systems requires modeling frameworks capable of capturing feedback loops between human behavior and technical platforms. This calls for dynamic representations of socio-technical systems, where interactions at different scales shape the collective outcomes of the whole [46].

2.4.5.3 Dynamical Influence and Node Centrality in Complex Networks

Complex networks have become an essential tool for studying the structure and behavior of heterogeneous systems in chemistry, biology, sociology, and technology [126, 136]. Their strength lies in their ability to model systems where a few key nodes play a critical role in the system's overall dynamics. While traditional centrality measures, such as degree, betweenness, eigenvector, are focused on topology [139, 234, 249], they often fail to fully capture the influence of nodes when dynamics are explicitly considered. In many real-world scenarios, node importance cannot be separated from the underlying processes occurring on the network. For example, choosing key nodes for immunization during an epidemic differs fundamentally from selecting optimal stimulation sites in neural therapies.

To bridge this gap, a new concept called *dynamical influence* has been introduced as a centrality measure that explicitly incorporates the interaction between network structure and dynamics. This measure evaluates how much the dynamical state of a node affects the collective behavior of the entire system. It has proven effective across a range of dynamical models, including epidemic spreading, diffusion processes, and nonlinear systems such as the Kuramoto and Rössler models [235]. Some efforts have also proposed Laplacian-based centralities, which account for system-specific dynamics and offer parallels to algorithms such as PageRank [236]. These dynamic-sensitive metrics are essential for applications in controllability, system optimization, and intervention strategies. Intermediate studies have explored Laplacian-based measures in the context of optimal control, highlighting the role of spectral properties in assessing control energy and node influence [237]. Similarly, Shimono and Tamura, recently in [238], proposed the Laplacian Eigenvector Centrality (LEC), leveraging spectral graph theory to assess node importance through the eigendecomposition of the Laplacian matrix.

2.4.5.4 Dynamical Influence in Networked Systems

To understand the role of individual nodes in shaping collective behavior, a dynamical system is considered, composed of N time-dependent variables $\mathbf{x} = (x_1, \ldots, x_N)$, whose evolution is governed by a coupled linear system $\dot{\mathbf{x}} = M\mathbf{x}$, where M is a real $N \times N$ matrix. The system's dynamics can be classified by the largest eigenvalue $\mu_{\max}$ of M: if $\mu_{\max} < 0$, the system asymptotically converges to a stable fixed point (i.e., $\mathbf{x}(t) \to 0$), whereas $\mu_{\max} > 0$ implies unbounded growth from most initial conditions. When M admits a non-degenerate zero eigenvalue ($\mu_{\max} = 0$), the system exhibits a conserved quantity. Specifically, the scalar product $w_c = \mathbf{c}^T\mathbf{x}$ remains constant, where $\mathbf{c}$ is the left eigenvector associated with $\mu_{\max}$. In this case, the asymptotic state $\mathbf{x}(\infty)$ is determined by:

$$\mathbf{x}(\infty) = \frac{\mathbf{c}^T\mathbf{x}(0)}{\mathbf{c}^T\mathbf{e}}\mathbf{e},$$

with $\mathbf{e}$ being the corresponding right eigenvector. This formulation reveals that $\mathbf{c}$ captures how the initial condition $\mathbf{x}(0)$ projects onto the long-term system behavior. Thus, the i-th component c_i measures the *dynamical influence* (DI) of node i.

An advantage of DI is its computational efficiency—it avoids extensive simulations and is interpretable through walk-counting arguments. Using the power method [239], the vector $\mathbf{c}$ can be approximated by iterative multiplication of a uniform vector $\mathbf{w}^{(0)} = (1, 1, \ldots, 1)$ with powers of M. After ℓ iterations:

$$\mathbf{w}^{(\ell)} = \mathbf{w}^{(0)} M^{\ell},$$

the i-th entry $w_i^{(\ell)}$ counts the number of walks of length ℓ starting at node i. For instance, at $\ell = 1$, $w_i^{(1)}$ equals the out-degree of node i, and at $\ell = 2$, it corresponds to the sum of the degrees of node i's neighbors. As $\ell \to \infty$, $\mathbf{w}^{(\ell)}$ converges (up to normalization) to $\mathbf{c}$ if $\mu_{\max}$ dominates all other eigenvalues. In summary, a node's dynamical influence reflects its capacity to start long walks across the network, an insight that bridges topology and dynamics in a principled manner.

2.4.5.5 Small-World Networks and Dynamical Systems

Networks composed of coupled dynamical elements serve as foundational models for a wide variety of self-organizing systems, including biological oscillators, Josephson junction arrays, excitable media, neural circuits, genetic regulatory networks, and spatial games [130, 240]. Traditionally, the network topology in these models has been considered either fully regular or entirely random. However, many real-world biological, technological, and social systems exhibit intermediate structures that lie between these two extremes [134]. To capture this hybrid nature, a class of networks has been introduced that interpolates between regular lattices and random graphs by means of a simple rewiring procedure. These structures, known as *small-world networks*, combine high local clustering, typical of regular lattices, with short average path lengths, characteristic of random graphs [130]. The term "small-world" draws from the well-known phenomenon that any two individuals are connected by a surprisingly short chain of acquaintances, often referred to as "six degrees of separation". Prominent real-world examples of small-world networks include the neural network of *Caenorhabditis elegans*, the U.S. Western power grid, and the co-actor collaboration graph in film [130]. Dynamical processes taking place on these networks, such as signal transmission, synchronization, and epidemic spreading, benefit from the small-world structure, which enhances propagation speed and collective coherence [240].

The construction of small-world networks begins with a regular ring lattice of n nodes, where each node is symmetrically connected to k neighbors. A fraction p of the edges is then randomly rewired, creating long-range links. This rewiring process allows a continuous transition from regular ($p = 0$) to random ($p = 1$) topologies, and offers a mechanism for analyzing the poorly understood intermediate regime

$(0 < p < 1)$ [130]. To quantify these networks, two metrics are central: the average shortest path length $L(p)$, which captures the global connectivity, and the clustering coefficient $C(p)$, which reflects the local interconnectedness of a node's neighborhood. As p increases from 0, $L(p)$ sharply decreases due to the introduction of shortcuts, while $C(p)$ remains significantly higher than in purely random graphs. This coexistence of short path lengths and strong clustering defines the small-world regime. Interestingly, even a small fraction of random long-range links is sufficient to drastically reduce $L(p)$, without significantly affecting $C(p)$. These shortcuts act as bridges that not only shorten the distance between individual nodes but also between entire neighborhoods, yielding profound and nonlinear reductions in path length across the network [130]. This structural property underlies the efficiency and robustness of many processes in small-world topologies.

2.4.5.6 Structural Insights Into the Small-World Phenomenon

In small-world networks, the introduction of shortcuts, edges that connect otherwise distant nodes, has a profound effect on the average shortest path length $L(p)$, dramatically reducing it even when only a small fraction of edges are rewired. In contrast, the clustering coefficient $C(p)$, which captures the local interconnectedness of nodes, is only marginally affected. This asymmetry arises because the removal of an edge from a clustered neighborhood typically has only a linear effect on C, whereas the addition of a shortcut can drastically alter global connectivity by reducing distances between large sections of the network [130]. As a result, at the local scale, reflected by $C(p)$, the transition to a small-world structure is almost imperceptible, despite the rapid global transformation indicated by $L(p)$. This suggests that local observations alone may be insufficient to detect the onset of small-world behavior.

The robustness of this phenomenon has been validated across various types of initial regular graphs and different rewiring algorithms. The key condition is that rewired edges should bridge nodes that are far apart in the original topology, thereby creating effective shortcuts. To further explore the prevalence of the small-world effect in real systems, several empirical networks have been analyzed. These include the collaboration network of film actors, constructed from publicly available data, which serves as a well-defined proxy for social interaction patterns; the electrical power grid of the western United States, pertinent to studies of infrastructure resilience and efficiency; and the fully mapped neural network of the nematode *Caenorhabditis elegans*, which provides a unique example of a complete biological connectivity map [130, 134]. All three cases exhibit characteristics consistent with the small-world regime, underscoring the ubiquity and relevance of this structural organization across diverse domains.

2.5 Multilayer, Pairwise and Higher Order Interactions: Structure and Dynamics

To provide a unified overview of the modeling frameworks discussed in this section, Fig. 2.1 summarizes the principal interaction units and the corresponding topological representations. From traditional pairwise links (edges) to 2-simplices and 2-hyperlinks, the diagram illustrates how real-world interactions can be modeled through graphs, multilayer networks, simplicial complexes, and hypergraphs. This visual taxonomy lays the groundwork for the structural and dynamical distinctions analyzed in the following subsections.

The intrinsic complexity of many biological, social, and technological systems arises from the richness and diversity of interactions among their constituent components. For decades, complex systems have been effectively described through network models, where nodes represent system units and links encode pairwise interactions between them [245]. However, in numerous empirical contexts, from collective human behavior to biochemical and ecological processes, interactions often involve more than two units simultaneously. These multi-node, or higher-order, interactions cannot be fully reduced to a set of dyads. For instance, communication within groups, enzymatic reactions among multiple proteins, or co-authorship in scientific collaborations often exhibit non-dyadic dependencies that shape system dynamics in fundamentally different ways. Until recently, the standard dyadic framework of networks largely neglected these higher-order structures. Yet, a growing body of work

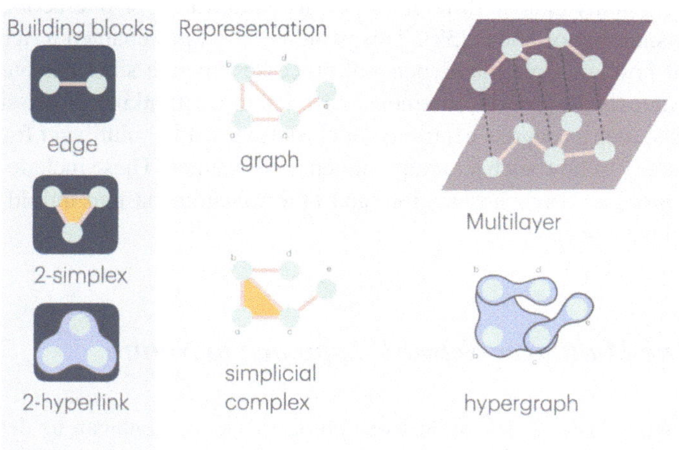

Fig. 2.1 Diagram illustrating different network representations. On the left, "Building blocks" are shown: an "edge" with two connected nodes, a "2-simplex" with three interconnected nodes forming a triangle, and a "2-hyperlink" with three nodes connected in a triangular shape. On the right, "Representation" includes: a "graph" with nodes labeled a to e connected by edges, a "simplicial complex" with nodes a to e and a highlighted 2-simplex, a "Multilayer" network with nodes on two layers connected by edges, and a "hypergraph" with nodes a to e connected by hyperedges

demonstrates that accounting for such complex architectures is essential to improve our understanding, modeling, and prediction of the dynamical behavior of real systems [245]. This shift has catalyzed the development of new mathematical models and computational tools, such as hypergraphs and simplicial complexes, which allow for the explicit representation of interactions among groups.

These higher-order representations not only provide richer structural insights but also reveal emergent dynamical behaviors absent in traditional models. For example, recent studies have shown that spreading processes, synchronization, and consensus dynamics are significantly affected by the presence of group interactions [246]. Such insights are particularly relevant in systems where collective coordination or higher-order dependencies play a central role.

In this section, the main elements of the modeling of pairwise and beyond-pairwise interactions are introduced by exploring the structural and dynamical consequences of incorporating higher-order representations in the study of complex systems. The discussion is also about the fundamental definitions of multilayer networks and what happens when different layers of interactions are taken into consideration. Over the past years, network science has provided powerful tools for investigating the interactions among the components of various complex systems, spanning domains such as biology, technology, and society [243]. Traditionally, these models have treated all elements and interactions uniformly, overlooking additional contextual dimensions such as the time or type of interaction.

With the advent of high-resolution datasets, researchers have begun to explore more realistic representations that capture the multilayer and temporal nature of real-world systems. These representations, often referred to as multiplex or multilayer networks, explicitly encode different types of relationships or interactions (layers) among the same set of nodes [37]. This paradigm shift has enabled a richer understanding of how multiple interdependent processes operate simultaneously across different contexts. Recent developments have led to a rethinking of classical structural metrics, such as degree, centrality, or clustering, within multilayer frameworks, and to the study of how such structure influences dynamics. These include diffusion, spreading processes, synchronization, and other phenomena that unfold across or within the layers [64].

2.5.1 The Multilayer Network Approach to Nature

A wide range of natural, social, and technological phenomena can be described as the emergent outcome of interactions among many individual units. These interactions often occur in complex and context-dependent ways, making network-based representations an essential tool for modeling and understanding such systems. Traditionally, complex systems have been represented as graphs where nodes denote entities and links represent pairwise interactions. However, this approach is sometimes insufficient when attempting to capture the rich, layered structure of real-world systems [37].

The *multilayer network theory* represents a framework that extends traditional network models by incorporating different layers of interaction. Each layer can represent a different type of relationship, activity, or context. This multilayer approach allows for a more accurate representation of systems in which elements interact across multiple types of connections or temporal scales [243]. For example, in social networks, individuals interact through various social contexts such as family, work, or hobbies. Modeling such a system as a single-layer network would ignore the layered structure of these interactions. Instead, a multilayer approach assigns each type of social relation to a separate layer, enabling a more realistic analysis of phenomena like information or rumor spreading.

Transportation systems also exemplify the necessity of multilayer modeling. Airports connected by flights from different airlines form distinct transportation layers. When simulating delay propagation or passenger rescheduling, it's essential to distinguish these layers, as airlines typically operate independently and incur costs for cross-company rescheduling [241]. In biology, the neural network of *C. elegans* provides another compelling example. Neurons in this organism are connected through chemical synapses and gap junctions—two interaction types with entirely different dynamics. Modeling such a system with a single layer would obscure these differences, whereas a multilayer model offers a clear framework to differentiate the functional role of each connection. These examples illustrate the limits of traditional network representations and motivate the growing interest in multilayer networks, which can better capture the diversity of real-world interactions. This paradigm shift has already led to significant theoretical advances and promises to open new directions in interdisciplinary research across physics, biology, sociology, and engineering. Complexity science focuses on systems composed of many interdependent elements, often interacting through different channels. These systems are central to understanding self-organization and emergent phenomena and are increasingly recognized as key to addressing challenges in the evolving global knowledge society. Across disciplines such as physics, biology, engineering, social sciences, and information technology, the study of complex systems has opened new perspectives on mechanisms that drive adaptation, transformation, and collective behavior in both natural and artificial environments. Most real-world complex systems are composed of multiple interconnected subsystems and layers of interaction. These systems are dynamic, multilevel within environments that are typically unstable and rapidly changing. Their structure reflects a composition of diverse components and subsystems, which co-evolve through internal feedbacks and external influences. As a result, they pose significant challenges for observation, modeling, and prediction—especially due to the multiscale and multi-component nature of their dynamics.

To address these challenges, there has been a growing need to move beyond traditional network representations. This has led to the development of a more general framework, multilayer network theory, which allows for the modeling of systems where nodes and links can span across multiple layers or contexts [37]. Several related concepts have emerged, related to this topic, including networks of networks, interdependent networks, multiplex and multidimensional networks [243].

Furthermore, new mathematical tools have been introduced to describe these systems, ranging from tensor-based formulations to generalized adjacency representations [242]. These approaches aim to extend fundamental ideas from single-layer network theory to the multilayer setting, enabling a comprehensive understanding of structure and dynamics across different layers. In this section, a synthesis of the main concepts and models proposed so far is provided, framing them within a unified multilayer network paradigm.

To support a unified understanding of multilayer networks, Table 2.2 summarizes the key concepts, structural parameters, and descriptive measures used in their characterization. Each entry includes the formal definition, symbol used in mathematical formulations, and a concise description of its role within the multilayer network framework.

2.5.1.1 Basic Formal Definitions

A *multilayer network* can be formally described as a pair $\mathcal{M} = (G, C)$, where $G = \{G^\alpha; \alpha \in \{1, ..., M\}\}$ represents a collection of layers, each modeled as a graph (directed or undirected, weighted or unweighted) $G^\alpha = (X^\alpha, E^\alpha)$, and $C = \{E^{\alpha\beta} \subseteq X^\alpha \times X^\beta; \alpha, \beta \in \{1, ..., M\}, \alpha \neq \beta\}$ is the set of interlayer connections between nodes belonging to different layers [37].

In this context:

- Elements of each E^α are referred to as *intralayer edges.*
- Elements of $E^{\alpha\beta}$ (with $\alpha \neq \beta$) are referred to as *interlayer edges.*

Let $X^\alpha = \{x_1^\alpha, ..., x_{N_\alpha}^\alpha\}$ be the set of nodes in layer α, and let $A^{[\alpha]} = (a_{ij}^\alpha) \in \mathbb{R}^{N_\alpha \times N_\alpha}$ be the adjacency matrix of G^α, defined as:

$$a_{ij}^\alpha = \begin{cases} 1 & \text{if } (x_i^\alpha, x_j^\alpha) \in E^\alpha \\ 0 & \text{otherwise} \end{cases}$$

The interlayer adjacency matrix $A^{[\alpha,\beta]} = (a_{ij}^{\alpha\beta}) \in \mathbb{R}^{N_\alpha \times N_\beta}$ is defined similarly for interconnections between different layers.

The *projection network* of $\mathcal{M}$ is the graph $\text{proj}(\mathcal{M}) = (X_\mathcal{M}, E_\mathcal{M})$ where $X_\mathcal{M} = \bigcup_{\alpha=1}^M X^\alpha$ and $E_\mathcal{M} = \left(\bigcup_{\alpha=1}^M E^\alpha\right) \cup \left(\bigcup_{\substack{\alpha,\beta=1 \\ \alpha\neq\beta}}^M E^{\alpha\beta}\right)$. The corresponding adjacency matrix is denoted by $A_\mathcal{M}$.

A particularly relevant subclass is the *multiplex network*, where all layers share the same node set $X^1 = X^2 = \cdots = X^M = X$, and each node is connected to its counterpart in other layers, i.e., $E^{\alpha\beta} = \{(x, x); x \in X\}$ for all $\alpha \neq \beta$ [243]. Multiplex networks are especially suited for modeling social systems, where different types of social relationships define different layers among the same individuals.

Table 2.2 Summary of multilayer network concepts and metrics

Concept	Definition (symbol)	Description
Multilayer Network	$\mathcal{M} = (G, C)$	General framework comprising multiple layers G^{α} with intralayer edges E^{α} and interlayer connections $C = \{E^{\alpha\beta}\}$
Multiplex Network	$\mathcal{M}_{\text{multiplex}}$	Special case of multilayer network where all layers share the same node set, with interlayer edges connecting identical nodes across layers
Projection Network	$\text{proj}(\mathcal{M})$	Flattened representation merging all layers and connections into a single graph
Supra-adjacency Matrix	$\widetilde{A}$	Block matrix aggregating all adjacency matrices and identity interlayer links for multiplex/unfolded representation
Vector Degree	$\mathbf{k}_i = (k_i^{[1]}, \ldots, k_i^{[M]})$	Degree of a node i in each layer, forming a vector in $\mathbb{R}^M$
Overlapping Degree	o_i	Total degree of a node across all layers: $o_i = \sum_{\alpha} k_i^{[\alpha]}$
Eigenvector Centrality (per layer)	$\mathbf{c}^{[\alpha]}$	Vector of eigenvector centralities for nodes in layer α
Global Heterogeneous Centrality	$c^{\otimes}$	Eigenvector centrality incorporating the influence matrix W among layers
Influence Matrix	$W = (w_{\alpha\beta})$	Matrix indicating directional influence between layers α and β
Clustering Coefficient	$C_M(i)$	Proportion of closed triplets among neighbors of node i in the multilayer projection
Distance	d_{uv}	Shortest path length between nodes u and v in multilayer topology
Diameter	$D(\mathcal{M})$	Longest shortest path between any two nodes in the multilayer network
Efficiency	$E(\mathcal{M})$	Inverse of average shortest path length: measures communication efficiency
Generalized Path Length	$\ell(\omega)$	Adjusted length of a walk penalizing interlayer transitions via a parameter β
Modularity (Multilayer)	Q	Quality function measuring density of intra-community edges across coupled slices
Motifs	–	Recurring subgraphs across or within layers, revealing functional microstructures
Multilinks	$\mathbf{m}$	Binary vector indicating presence of a link across M layers
Multidegree	$k_i^{\mathbf{m}}$	Number of neighbors of node i with multilink type $\mathbf{m}$

(continued)

Table 2.2 (continued)

Concept	Definition (symbol)	Description
Multistrength	$s^{\mathbf{m}}_{i,\alpha}$	Sum of link weights for a node i with multilink $\mathbf{m}$ in layer α
Inverse Multiparticipation Ratio	$Y^{\mathbf{m}}_{i,\alpha}$	Quantifies heterogeneity of multistrength distribution across neighbors
Activity Matrix	$b_{i\alpha}$	Binary indicator of node i's presence (activity) in layer α
Node Activity	B_i	Total number of layers in which node i is active
Layer Activity	N_α	Number of active nodes in layer α
Layer Pairwise Multiplexity	$Q_{\alpha\beta}$	Fraction of nodes active in both layers α and β
Node Pairwise Multiplexity	Q_{ij}	Fraction of layers in which nodes i and j are both active
Interlayer Degree Correlation	$r_{\alpha\beta}, \rho_{\alpha\beta}, \tau$	Pearson, Spearman, Kendall correlations between degree distributions across layers

A multiplex network $\mathcal{M}$ can also be represented by its projection proj$(\mathcal{M})$, whose adjacency matrix has elements:

$$a_{ij} = \begin{cases} 1 & \text{if } a_{ij}^{\alpha} = 1 \text{ for some } \alpha \in \{1, ..., M\} \\ 0 & \text{otherwise} \end{cases}$$

Alternatively, a *flattened* or *unfolded* monolayer network $\widetilde{\mathcal{M}} = (\widetilde{X}, \widetilde{E})$ can be constructed, where $\widetilde{X} = \bigcup_{\alpha=1}^{M} X^{\alpha}$ and $\widetilde{E}$ includes both intralayer and interlayer links. The corresponding adjacency matrix $\widetilde{A}$ is often referred to as the *supra-adjacency matrix* and can be represented as a block matrix of the form:

$$\widetilde{A} = \begin{pmatrix} A^{[1]} & I & \cdots & I \\ I & A^{[2]} & \cdots & I \\ \vdots & \vdots & \ddots & \vdots \\ I & I & \cdots & A^{[M]} \end{pmatrix} \in \mathbb{R}^{NM \times NM}$$

where I is the N-dimensional identity matrix. It is important to note that while the structures $\mathcal{M}$ and $\widetilde{\mathcal{M}}$ are related, their behaviors differ since a single node in $\mathcal{M}$ corresponds to multiple nodes in $\widetilde{\mathcal{M}}$. Thus, the dynamics and properties of a multilayer or multiplex network should be interpreted as a non-linear quotient of its corresponding flattened representation [242].

2.5.1.2 Multilayer Networks as a General Framework

The concept of multilayer networks generalizes a variety of mathematical objects that have been independently introduced to describe complex systems with multiple modes or dimensions of interaction. Below, briefly the summarizing of the main network types that can be encompassed within the multilayer network formalism.

- **Multiplex networks**: A multiplex network is composed of M layers $\{G^{\alpha}; \alpha \in \{1, ..., M\}\}$, where each layer $G^{\alpha} = (X^{\alpha}, E^{\alpha})$ shares the same node set $X^{\alpha} = X$ across layers. The only interlayer edges are self-edges, connecting each node to its counterpart in other layers, i.e., $E^{\alpha\beta} = \{(x, x); x \in X\}$ for $\alpha \neq \beta$ [243].
- **Temporal networks**: A temporal network $\{G(t)\}_{t=1}^{T}$ can be interpreted as a multilayer network, where each layer represents the network at time t: $G^{\alpha} = G(t)$. The interlayer connections only exist between consecutive time layers and are identity mappings: $E^{\alpha,\alpha+1} = \{(x, x); x \in X^{\alpha} \cap X^{\alpha+1}\}$, with $E^{\alpha\beta} = \emptyset$ otherwise [65].
- **Interconnected networks**: These consist of a family of interacting networks $\{G_1, ..., G_L\}$ modeled as layers of a multilayer system. The interlayer edges $E^{\alpha\beta}$ capture the interactions between nodes of different networks G^{α} and G^{β} [37].
- **Multidimensional networks**: Defined formally as edge-labeled multigraphs $G = (V, E, D)$, where D is the set of dimensions (labels), each edge is a triple (u, v, d) indicating a connection of type d. Such a network can be modeled as a multilayer system by mapping each label $d \in D$ to a separate layer $G^{d} = (V, E^{d})$, where $E^{d} = \{(u, v) \in V \times V \mid (u, v, d) \in E\}$.
- **Interdependent networks**: These consist of multiple networks (layers) with dependencies between their nodes. A node in one layer may depend on a node in another, with these dependencies modeled as interlayer edges. Such structures are particularly useful in modeling systems where failure in one layer affects another, such as critical infrastructures [244].
- **Multilevel networks**: A multilevel network is defined as a triple $\mathcal{M} = (X, E, \mathcal{S})$, where $G = (X, E)$ is the overall network and $\mathcal{S} = \{S_1, ..., S_p\}$ is a collection of subnetworks (or slices) such that $X = \bigcup X_j$ and $E = \bigcup E_j$. Each slice S_j becomes a layer in the multilayer formulation.
- **Hypergraphs and hypernetworks**: A hypergraph $H = (X, \mathcal{H})$ is composed of a node set X and a set $\mathcal{H}$ of hyperlinks, each connecting more than two nodes. These can be represented as multilayer networks by associating each hyperlink to a layer G^{h}—a complete graph over the nodes involved in the hyperlink—with interlayer edges linking nodes shared between different hyperlinks.

These different representations show how multilayer networks offer a unifying modeling paradigm that captures both the structure and the dynamics of systems with multi-dimensional, temporal, or interdependent interactions.

2.5.1.3 Centrality and Node Ranking in Multilayer Networks

A fundamental problem in complex network analysis is to identify the most central nodes within the structure. In traditional monolayer networks, several centrality measures have been developed to quantify the relevance of individual nodes. These include node degree, closeness centrality, betweenness centrality, eigenvector centrality as deeply discussed in Sect. 2.4.1.

In the multilayer context, these measures need to be generalized. A basic extension is the *vector degree*, which, for a node $i \in X$ in a multiplex network $\mathcal{M} = (G, C)$, is defined as:

$$\mathbf{k}_i = \left(k_i^{[1]}, \dots, k_i^{[M]}\right)$$

where $k_i^{[\alpha]} = \sum_j a_{ij}^{[\alpha]}$ denotes the degree of node i in layer α [37, 247].

Unlike in monolayer networks, the vectorial nature of $\mathbf{k}_i$ in $\mathbb{R}^M$ does not admit a unique natural ordering, making node ranking non-trivial. To address this, scalar aggregate metrics are employed. One such measure is the *overlapping degree*, defined as:

$$o_i = \sum_{\alpha=1}^{M} k_i^{[\alpha]} = \|\mathbf{k}_i\|_1,$$

which summarizes the total activity of a node across all layers. Alternatively, one might consider other aggregation functions, such as convex combinations or different norms of $\mathbf{k}_i$.

Other classical centrality measures, such as closeness and betweenness, rely on geodesic paths and metric properties of the network. These metrics can be extended to multilayer settings, provided that an appropriate definition of interlayer distances and multilayer geodesics is adopted.

2.5.1.4 Spectral Centrality in Multilayer Networks

An alternative framework for assessing node centrality in complex networks leverages the spectral properties of the adjacency matrix. Among these, *eigenvector centrality* stands out as it accounts not only for the quantity of a node's connections, but also for their quality, assigning more importance to nodes connected to other central nodes [249]. Extending this concept to multilayer networks can be approached in different ways [242, 247]. A straightforward strategy consists in computing the eigenvector centrality $\mathbf{c}^{[\alpha]} = (c_1^{[\alpha]}, \dots, c_N^{[\alpha]})$ independently for each layer $\alpha = 1, \dots, M$. As a result, the multilayer eigenvector centrality of a node $i \in X$ becomes a vector in $\mathbb{R}^M$:

$$\mathbf{c}_i = \left(c_i^{[1]}, \dots, c_i^{[M]}\right),$$

where each entry corresponds to the eigenvector centrality of node i in the respective layer. By assembling the layer-wise centralities, one obtains the matrix:

$$C = \left(\mathbf{c}^{[1]^T}, \mathbf{c}^{[2]^T}, \ldots, \mathbf{c}^{[M]^T}\right) \in \mathbb{R}^{N \times M},$$

which is column-stochastic under the usual normalization condition $\|\mathbf{c}^{[\alpha]}\|_1 = 1$ for all layers. To synthesize a scalar centrality score from the vector $\mathbf{c}_i$ for each node, various aggregation functions can be applied, such as the sum, maximum value, or an ℓ_p, norm. However, a limitation of this approach is that it fails to fully incorporate the influence of interlayer interactions on the overall centrality of the nodes.

2.5.1.5 Layer Interdependence and Influence Matrices

A more detailed strategy for evaluating node centrality in multilayer networks considers that layers may differ in their relative importance or influence. Instead of treating each layer as equally relevant, this approach incorporates the idea that certain layers may have a stronger effect on the others and that the position of a node in one layer can impact its perceived relevance across the whole network. For example, imagine a scenario where a company executive resides in the same building as one of their subordinates. Within the "residential layer" representing neighborhood relationships, the dynamic between these individuals differs greatly from their professional interaction in the "workplace layer". Still, the executive's influence may be elevated due to the overlap between personal and professional spheres. This suggests the necessity of considering asymmetric, heterogeneous interactions between layers when calculating node centrality.

To capture these effects, one can define a non-negative *influence matrix* $W = (w_{\alpha\beta}) \in \mathbb{R}^{M \times M}$, where $w_{\alpha\beta}$ quantifies the extent to which layer G_β influences layer G_α. By integrating W into centrality calculations, it's possible to reflect the varying levels of impact that layers exert on one another, yielding a more faithful representation of real-world multilayer structures [247, 248].

2.5.1.6 Global Heterogeneous Eigenvector Centrality

A further significant consideration in multilayer networks is that the centrality of a node x_i^α within a given layer α may not only depend on its intralayer connections, but also on its neighbors in other layers. This broader interdependence motivates the definition of a global heterogeneous eigenvector-like centrality [248].

Given a multiplex network M and an influence matrix $W = (w_{\alpha\beta})$, the global heterogeneous eigenvector centrality is defined as the positive, normalized eigenvector $c^{\otimes} \in \mathbb{R}^{NM}$, provided it exists, of the matrix:

$$A^{\otimes} = \begin{pmatrix} w_{11}(A^{[1]})^T & w_{12}(A^{[2]})^T & \cdots & w_{1M}(A^{[M]})^T \\ w_{21}(A^{[1]})^T & w_{22}(A^{[2]})^T & \cdots & w_{2M}(A^{[M]})^T \\ \vdots & \vdots & \ddots & \vdots \\ w_{M1}(A^{[1]})^T & w_{M2}(A^{[2]})^T & \cdots & w_{MM}(A^{[M]})^T \end{pmatrix} \in \mathbb{R}^{(NM)\times(NM)}$$

This matrix $A^{\otimes}$ is formed through the Khatri–Rao product between the influence matrix W and the transposed adjacency matrices $(A^{[\alpha]})^T$ of each layer [247]. Analogously to previous constructions, it is possible to write:

$$c^{\otimes} = \begin{pmatrix} c_1^{\otimes} \\ c_2^{\otimes} \\ \vdots \\ c_M^{\otimes} \end{pmatrix}, \quad \text{with } c_{\alpha}^{\otimes} \in \mathbb{R}^N,$$

and define the corresponding global heterogeneous centrality matrix as:

$$C^{\otimes} = \begin{pmatrix} c_1^{\otimes} & c_2^{\otimes} & \cdots & c_M^{\otimes} \end{pmatrix} \in \mathbb{R}^{N\times M}$$

Although $C^{\otimes}$ is not necessarily column-stochastic or row-stochastic, it is normalized such that the sum of all its entries equals 1. Other spectral centrality approaches also exist, such as variants of PageRank centrality, which incorporate random jumps in the analysis of stationary distributions of random walkers across multilayer networks.

2.5.1.7 Clustering in Multilayer Networks

The clustering coefficient introduced by Watts and Strogatz [130] is a key structural measure in complex network theory, commonly interpreted as a quantification of the tendency of nodes to form closed triplets or triangles—popularly stated as "the friend of your friend is also your friend." In a monolayer graph $G = (X, E)$, the local clustering coefficient of a node i is defined as:

$$c_G(i) = \frac{\text{Number of links between neighbors of } i}{\text{Maximum possible number of links between neighbors of } i}.$$

This metric captures the likelihood that any two neighbors of i are also directly connected, and the global clustering coefficient is typically computed as the average over all nodes. It serves as a measure of transitivity and neighborhood density.

An alternative, global definition—frequently used in social network analysis—is based on triads and triangles:

$$T = \frac{\text{Number of triangles in the network}}{\text{Number of triads in the network}},$$

which is commonly referred to as the network transitivity [143].

In the context of multilayer or multilevel networks, extending the notion of clustering requires accounting for both intralayer and interlayer connections. As described in [37], a meaningful generalization compares clustering coefficients across individual layers and the projection of the full multilayer system.

To formalize this, consider a node i in the projection network $\mathrm{proj}(M)$ of a multilayer network M. Let $N(i)$ denote the set of all neighbors of i in $\mathrm{proj}(M)$, and $N_\alpha(i)$ be its neighbors in layer α, i.e., $N_\alpha(i) = N(i) \cap X_\alpha$. Define $S_\alpha(i) = (N_\alpha(i), E_\alpha(i))$ as the subgraph induced by these neighbors in layer α, where:

$$E_\alpha(i) = \{(k, j) \in E_\alpha \mid k, j \in N_\alpha(i)\}.$$

The clustering coefficient of node i in M is then computed as:

$$C_M(i) = \frac{\sum_{\alpha=1}^{M} 2|E_\alpha(i)|}{\sum_{\alpha=1}^{M} |N_\alpha(i)|(|N_\alpha(i)| - 1)},$$

and the global clustering coefficient of the multilayer network M is the average of $C_M(i)$ over all nodes i.

2.5.1.8 Metric Structures: Shortest Paths and Distances

The metric structure of a complex network is fundamentally associated with the topological distances between nodes, typically defined through walks and paths. To extend these classical notions to the case of multilayer networks, the starting point is to formalize the definitions of walks, paths, and their lengths, following the framework introduced in [64].

Let $M = (G, C)$ be a multilayer network composed of a family of graphs $G = \{G_\alpha\}_{\alpha=1}^{M}$ and a set of interlayer connections C. Define the full set of edges as:

$$E(M) = \{E_1, ..., E_M\} \cup C.$$

A *walk* of length $q - 1$ is an alternating sequence of nodes and edges

$$\{x_1^{\alpha_1}, \ell_1, x_2^{\alpha_2}, \ell_2, ..., \ell_{q-1}, x_q^{\alpha_q}\},$$

such that for all $r < q$, there exists an edge $\ell_r \in E(M)$ satisfying $\ell_r = (x_r^{\alpha_r}, x_{r+1}^{\alpha_{r+1}})$. When the walk starts and ends at the same node, it is said to be closed. If all edges are weighted, the walk length can be computed as the sum of the inverses of the corresponding weights. A *path* is a walk where no node is repeated. A cycle is a closed path, and the network is said to be connected if a path exists between every pair of nodes. Importantly, various notions of connectivity can be introduced depending on whether only intralayer or both intra- and interlayer connections are considered [243]. The *length* of a path corresponds to the number of edges it includes. Given that multilayer networks contain both intralayer and interlayer edges, the choice of metric may vary depending on how these edge types are treated—equally or differently. A *geodesic* between two nodes u and v is defined as any shortest path connecting them. The *distance* d_{uv} is the length of such a geodesic. The *diameter* $D(M)$ is the maximum distance among all node pairs in M, and the *number of geodesics* between u and v is denoted by n_{uv}. If a node x or edge ℓ lies along a geodesic between u and v, their participations by $n_{uv}(x)$ and $n_{uv}(\ell)$ are defined, respectively. A *subnetwork* $N = (G', C')$ of M is defined such that each of its layer nodes X'_γ and edges E'_γ, $E'_{\gamma\delta}$ are subsets of those in the corresponding layers of M. A *connected component* is a maximal connected subnetwork.

Two paths are said to be *vertex-independent* if they share no nodes other than the endpoints. A k-component is a maximal subset of nodes such that every pair is connected by at least k vertex-independent paths. For $k = 2$ or 3, these are known as bi-components and tri-components, respectively.

Finally, the *characteristic path length* of a multilayer network is defined as:

$$\frac{1}{L(M)} = \frac{1}{N(N-1)} \sum_{\substack{u,v \in X_M \\ u \neq v}} d_{uv},$$

where $|X_M| = N$, and it serves as a standard measure of a network's overall navigability and efficiency.

2.5.1.9 Efficiency in Multilayer Networks

The concept of efficiency, originally introduced by Latora and Marchiori in the context of monolayer networks, can be naturally extended to the multilayer framework [138]. Using a similar notation to that presented in [37], the efficiency of a multilayer network M is defined as:

$$E(M) = \frac{1}{N(N-1)} \sum_{\substack{u,v \in X_M \\ u \neq v}} \frac{1}{d_{uv}}, \tag{2.81}$$

where $N = |X_M|$ is the number of nodes in the projection of M, and d_{uv} is the shortest path distance between nodes u and v in the multilayer structure. It is often useful to compare the efficiency $E(M)$ of the full multilayer network to that of its projection $E(\mathrm{proj}(M))$, or to the individual layers $E(G_\alpha)$. Analytical results presented in [37] provide insights into how these quantities relate to one another under different structural configurations.

When intra- and inter-layer edges are considered non-equivalent, one can adopt a generalized notion of path length. Let $M = (G, C)$ be a multilayer network, and consider a path:

$$\omega = \{x_1^{\alpha_1}, \ell_1, x_2^{\alpha_2}, \ell_2, ..., \ell_q, x_{q+1}^{\alpha_{q+1}}\}. \tag{2.82}$$

The length of ω can then be defined as:

$$\ell(\omega) = q + \beta \sum_{j=2}^{q} \Delta(j), \tag{2.83}$$

where

$$\Delta(j) = \begin{cases} 1 & \text{if } \ell_j \in C \quad \text{(i.e., if the edge is inter-layer)} \\ 0 & \text{otherwise} \end{cases} \tag{2.84}$$

and β is a non-negative parameter that penalizes layer transitions. When $\beta = 0$, this definition coincides with the natural metric of the projection network $\mathrm{proj}(M)$. For $\beta > 0$, it accounts for the extra cost of switching between layers, thus providing a more nuanced view of navigation efficiency across the multilayer structure.

2.5.1.10 Mesoscales: Motifs and Modular Structures

One of the core methodologies in network analysis is the identification of mesoscopic structures, commonly referred to as communities or cohesive groups. These communities are subsets of nodes that exhibit a higher density of internal connections relative to their links with the rest of the network [150]. A widely used approach for community detection is based on the optimization of modularity, a scalar quality function that compares the observed number of intra-community edges to the expected number under a random null model. In the context of multilayer networks, modularity has been extended through tensorial frameworks that capture the null model across layers [250]. Such generalizations allow the analysis of multislice networks, in which multiple adjacency matrices (corresponding to different layers or time steps) are coupled together, facilitating community detection across temporal and multiplex dimensions.

While communities are typically defined at a global mesoscale level, more localized mesoscopic structures also play an important role in understanding network function. Among them, *motifs*—small subgraphs that recur more frequently than expected in randomized networks—are particularly significant. These structures are often interpreted as fundamental building blocks of networks, each potentially associated with specific functionalities [147]. By separating positive and negative interactions into different layers, researchers found that certain triadic configurations occur with frequencies either substantially higher or lower than those predicted by random models, highlighting the relevance of multilayer motifs in uncovering underlying structural patterns [251].

2.5.1.11 Correlations in Multiplex Networks

Multiplex networks provide a richer and more specific representation of complex systems than single-layer networks, primarily due to the interdependencies and correlations among layers. Understanding the statistical correlations across layers, both in node roles and structural properties, is a critical frontier in network science.

Several types of correlations have been identified and analyzed so far:

- **Interlayer Degree Correlations**: These correlations assess whether nodes that serve as hubs in one layer are also prominent in another. A positive correlation indicates that centrality is preserved across layers, whereas a negative correlation suggests a role, shift between layers [252].
- **Overlap and Multidegree**: When nodes participate in multiple layers, their connectivity patterns may overlap. This is often observed in social networks where communication with the same individuals occurs via various platforms (e.g., phone, email, social media). Such overlap can be quantified globally or locally, or captured through the multidegree vector of nodes [247].
- **Multistrengths and Inverse Multiparticipation Ratio**: In weighted multiplex networks, link weights may correlate with structural characteristics such as node strength or layer activity. For instance, stronger collaboration ties in co-authorship networks often occur in specific research fields and are not uniformly distributed. These patterns are captured through metrics like multistrengths and the inverse multiparticipation ratio [247].
- **Node Pairwise Multiplexity**: When nodes are not active in all layers, it becomes relevant to study whether pairs of nodes are co-active in the same subset of layers. This kind of correlation highlights shared participation patterns among node pairs.
- **Layer Pairwise Multiplexity**: Similarly, correlations can be analyzed at the layer level, particularly when layers share a subset of active nodes. These correlations inform us about functional or structural dependencies between layers.

Analyzing these correlations offers key insights into the organization of real-world systems and reveals interdependencies that cannot be captured by traditional single-layer network models.

2.5.1.12 Degree Correlations in Multiplex Networks

In a multiplex network, each node $i = 1, 2, \ldots, N$ has a degree $k_i^{[\alpha]}$ in each layer $\alpha = 1, 2, \ldots, M$. These degree values can be statistically correlated across layers. For instance, a node that is a hub in a scientific collaboration layer is often also highly connected in the corresponding citation layer. This reflects a positive interlayer correlation. Conversely, in some cases, hubs in one layer may have low degrees in another, indicating negative correlations [252].

Several methods can be employed to quantify these correlations:

- **Joint Degree Distribution**: A complete description can be provided by computing the matrix $P(k^{[\alpha]}, k^{[\beta]})$, where each entry reflects the fraction of nodes with degree $k^{[\alpha]}$ in layer α and $k^{[\beta]}$ in layer β:

$$P(k^{[\alpha]}, k^{[\beta]}) = \frac{N(k^{[\alpha]}, k^{[\beta]})}{N},$$

  where $N(k^{[\alpha]}, k^{[\beta]})$ is the number of nodes with the given degree combination, and N is the total number of nodes.
- **Conditional Average Degree**: A more coarse-grained approach computes the average degree in one layer given the degree in another:

$$\bar{k}^{[\alpha]}(k^{[\beta]}) = \sum_{k^{[\alpha]}} k^{[\alpha]} P(k^{[\alpha]}|k^{[\beta]}) = \frac{\sum_{k^{[\alpha]}} k^{[\alpha]} P(k^{[\alpha]}, k^{[\beta]})}{\sum_{k^{[\alpha]}} P(k^{[\alpha]}, k^{[\beta]})}.$$

  If this function increases (or decreases) with $k^{[\beta]}$, it suggests positive (or negative) degree correlations.
- **Statistical Correlation Coefficients**: measures include:
  - *Pearson coefficient* $r_{\alpha\beta}$, based on covariance and standard deviations of the degree distributions:

$$r_{\alpha\beta} = \frac{\langle k_i^{[\alpha]} k_i^{[\beta]} \rangle - \langle k_i^{[\alpha]} \rangle \langle k_i^{[\beta]} \rangle}{\sigma_\alpha \sigma_\beta}.$$

  - *Spearman coefficient* $\rho_{\alpha\beta}$, using ranks of degrees instead of actual values. It is less sensitive to outliers but not unique due to rank degeneracy.
  - *Kendall's τ coefficient*, which compares the concordance of ranked degree pairs:

$$\tau = \frac{n_c - n_d}{\sqrt{(n_0 - n_1)(n_0 - n_2)}},$$

    where n_c and n_d count the concordant and discordant pairs, and the other terms correct for tied ranks.

- **Controlling Degree Correlations**. Degree correlations across layers can also be tuned by relabeling nodes according to their degree ranks. When nodes are ranked increasingly in both layers, the result is a maximally positive correlation (MP). Conversely, assigning increasing rank in one layer and decreasing rank in the other results in a maximally negative correlation (MN), as explored in [247].

These approaches are crucial to understand how node roles are maintained, transformed, or diversified across the different layers of a multiplex network.

2.5.1.13 Overlap and Multidegree in Multiplex Networks

Many empirical datasets—ranging from airport networks to online social platforms, collaborations, and citation networks—exhibit significant link overlap between layers [247, 251]. This means that the same pair of nodes is often connected in multiple layers of the network.

The *total overlap* $O_{\alpha\beta}$ between two layers α and β is defined as the number of common links:

$$O_{\alpha\beta} = \sum_{i<j} a_{ij}^{[\alpha]} a_{ij}^{[\beta]},$$

where $a_{ij}^{[\alpha]}$ and $a_{ij}^{[\beta]}$ are the adjacency matrix elements of layers α and β, respectively.

A *local version* of this measure captures the overlap at the level of individual nodes:

$$o_i^{\alpha\beta} = \sum_{j=1}^{N} a_{ij}^{[\alpha]} a_{ij}^{[\beta]},$$

quantifying how many of node i's neighbors are shared across both layers.

To systematically account for the presence or absence of links across all layers, the notion of *multilinks* is introduced. For M layers, a multilink is described by a binary vector $\mathbf{m} = (m_1, m_2, \ldots, m_M)$, where $m_\alpha = 1$ indicates a connection in layer α. Using this, the *multiadjacency matrix* $A^{\mathbf{m}}$ is defined as:

$$A_{ij}^{\mathbf{m}} = \prod_{\alpha=1}^{M} \left[a_{ij}^{[\alpha]} m_\alpha + (1 - a_{ij}^{[\alpha]})(1 - m_\alpha) \right],$$

which equals 1 if the link pattern between i and j matches $\mathbf{m}$ and 0 otherwise.

Each node i is then associated with a *multidegree* $k_i^{\mathbf{m}}$, denoting the number of links of type $\mathbf{m}$ incident on i:

$$k_i^{\mathbf{m}} = \sum_{j=1}^{N} A_{ij}^{\mathbf{m}}.$$

For a duplex (i.e., two layers), one can explicitly compute the multidegrees:

$$k_i^{(1,0)} = \sum_{j=1}^{N} a_{ij}^{[1]}(1 - a_{ij}^{[2]}), \quad k_i^{(0,1)} = \sum_{j=1}^{N} (1 - a_{ij}^{[1]}) a_{ij}^{[2]}, \quad k_i^{(1,1)} = \sum_{j=1}^{N} a_{ij}^{[1]} a_{ij}^{[2]}.$$

These count, respectively, the number of neighbors connected only in the first layer, only in the second, and in both.

When M is large, it becomes unfeasible to track every type of multilink. In such cases, the *multiplicity of overlap* ν_{ij} provides a simplified measure:

$$\nu_{ij} = \sum_{\alpha=1}^{M} a_{ij}^{[\alpha]},$$

counting the number of layers in which node pair (i, j) is connected.

These tools enable the characterization of the structural redundancy and interdependence across layers in multiplex systems.

2.5.1.14 Multistrength and Inverse Multiparticipation Ratio in Weighted Multiplex Networks

In weighted multiplex networks, link weights can exhibit non-trivial correlations with the structure of the multiplex itself. To explore this aspect, two structural descriptors have been proposed: the *multistrength* and the *inverse multiparticipation ratio* [247].

Given a multilink type $\mathbf{m}$ and a node i in layer α such that $m_\alpha > 0$, the multistrength $s_{i,\alpha}^{\mathbf{m}}$ of node i is defined as:

$$s_{i,\alpha}^{\mathbf{m}} = \sum_{j=1}^{N} a_{ij}^{[\alpha]} A_{ij}^{\mathbf{m}},$$

where $a_{ij}^{[\alpha]}$ is the weighted adjacency element in layer α, and $A_{ij}^{\mathbf{m}}$ encodes the presence of a multilink of type $\mathbf{m}$ between nodes i and j.

The corresponding *inverse multiparticipation ratio* $Y_{i,\alpha}^{\mathbf{m}}$ measures the heterogeneity of those weights:

$$Y_{i,\alpha}^{\mathbf{m}} = \sum_{j=1}^{N} \left(\frac{a_{ij}^{[\alpha]} A_{ij}^{\mathbf{m}}}{\sum_r a_{ir}^{[\alpha]} A_{ir}^{\mathbf{m}}} \right)^2.$$

For example, in a social multiplex network, where one layer represents virtual interaction and another represents email interactions, the multistrength $s_{i,1}^{(1,1)}$ captures the total interactions strength of node i with those it also emails, while $s_{i,1}^{(1,0)}$ counts virtual-only contacts.

Since only multilinks for which $m_\alpha = 1$ are relevant for a given layer α, the number of non-trivial multistrengths per node and per layer is $2^M - 1$, leading to a total of $M(2^M - 1)$ multistrength values per node in a multiplex with M layers.

Just as in monolayer networks [144], these quantities can be analyzed statistically. In particular, the average multistrength and inverse multiparticipation ratio for nodes with a given multidegree $\mathbf{k}_m$ are defined as:

$$s_{\mathbf{m},\alpha}(\mathbf{k}_m) = \langle s_{i,\alpha}^{\mathbf{m}} \delta(\mathbf{k}_i^{\mathbf{m}}, \mathbf{k}_m) \rangle, \quad Y_{\mathbf{m},\alpha}(\mathbf{k}_m) = \langle Y_{i,\alpha}^{\mathbf{m}} \delta(\mathbf{k}_i^{\mathbf{m}}, \mathbf{k}_m) \rangle,$$

and are expected to scale as:

$$s_{\mathbf{m},\alpha}(\mathbf{k}_m) \propto (\mathbf{k}_m)^{\beta_{\mathbf{m},\alpha}}, \quad Y_{\mathbf{m},\alpha}(\mathbf{k}_m) \propto \frac{1}{(\mathbf{k}_m)^{\lambda_{\mathbf{m},\alpha}}},$$

with exponents $\beta_{\mathbf{m},\alpha} \geq 1$ and $\lambda_{\mathbf{m},\alpha} \leq 1$.

Empirical analysis of weighted multiplex networks, such as the APS dataset of scientific collaborations and citations, confirms these scaling laws and shows that the values of the exponents differ depending on whether the multilinks are overlapping or not [252]. These findings highlight the importance of structural-weight correlations in understanding multiplex architectures.

2.5.1.15 Node Activities and Pairwise Multiplexity

In multiplex networks, it is possible to define the *activity* of nodes across layers using the layer-specific adjacency matrices $a^{[\alpha]}$ for $\alpha = 1, 2, \ldots, M$. This leads to the construction of an $N \times M$ *activity matrix* b whose elements $b_{i\alpha}$ indicate whether node i is active in layer α.

For undirected multiplex networks, a node is considered active in a given layer if it has a non-zero degree in that layer:

$$b_{i\alpha} = 1 - \delta_{0,k_i^{[\alpha]}} = 1 - \delta_{0,\sum_{j=1}^{N} a_{ij}^{[\alpha]}},$$

where δ is the Kronecker delta function. In the directed case, the node is considered inactive if both its in-degree and out-degree are zero:

$$b_{i\alpha} = 1 - \delta_{0,\sum_{j=1}^{N} a_{ij}^{[\alpha]}} \cdot \delta_{0,\sum_{j=1}^{N} a_{ji}^{[\alpha]}}.$$

The *activity* B_i of node i is defined as the total number of layers in which it is active:

$$B_i = \sum_{\alpha=1}^{M} b_{i\alpha},$$

while the *activity* N_α of a layer α is given by:

$$N_\alpha = \sum_{i=1}^{N} b_{i\alpha}.$$

Empirical studies show that the distribution $P(B_i)$ of node activities is broad and often follows a power-law, i.e., $P(B_i) \sim B_i^{-\delta}$ with δ typically in the range $[1.5, 3.0]$ [252]. This implies that in many real-world systems, node activity is highly heterogeneous, and some nodes are present in many layers while others participate in very few.

To capture correlations between activities across layers, one can define the *layer pairwise multiplexity* $Q_{\alpha\beta}$:

$$Q_{\alpha\beta} = \frac{1}{N} \sum_{i=1}^{N} b_{i\alpha} b_{i\beta},$$

which measures the fraction of nodes active in both layers α and β.

Similarly, the *node pairwise multiplexity* Q_{ij} quantifies the extent to which two nodes are simultaneously active across layers:

$$Q_{ij} = \frac{1}{M} \sum_{\alpha=1}^{M} b_{i\alpha} b_{j\alpha}.$$

These quantities help to reveal non-trivial correlations in the structure of multiplex networks. For instance, in a study of European airline networks, connections between companies were weighted by their layer pairwise multiplexity, revealing clusters of airlines with overlapping operational activity [37].

2.5.2 Spreading Processes and Games

As in the case of single-layer networks, understanding the structure of multilayer networks must be complemented by analyzing how their topology influences dynamic processes. Given the prevalence of multilayer structures in biological, social, and technological systems, it becomes essential to revisit and extend many classical dynamical frameworks within this new context. This section aims to explore how structural characteristics of multilayer networks impact their function. The most studied processes include stochastic dynamics such as random walks, epidemic spreading, and evolutionary game dynamics. Importantly, each layer may support a distinct dynamical process, and these processes can influence one another. For instance, the multilayer formalism enables the investigation of interactions among co-evolving processes, such as competing strains of a virus spreading through a population. Here, it has been addressed diffusion processes, starting from linear diffusion and random walks. As a generalization, taking into consideration congestion phenomena in communication networks, where nodes are limited in their capacity to send and store packets. In this setting, a key question is how multiple inter-layer routing paths affect the onset of congestion or jamming. The following subsections discuss on the generalization of epidemic models to the multilayer framework. Central to this discussion is the computation of the epidemic threshold, but additional aspects such as the effect of control strategies, the co-evolution of disease and awareness, and the transmission of information about the outbreak are also considered. Opinion dynamics is another form of contact process discussed here, where the objective is to understand how consensus emerges and how it may be influenced by inter-layer interactions among agents holding different opinions. Then, the discussion addresses the study of cooperation and its persistence under natural selection. Although selfish strategies typically yield individual benefits, thereby threatening the emergence of cooperation, recent studies on spatial evolutionary games have revealed that cooperation can flourish through the formation of tightly-knit clusters—a mechanism known as *network reciprocity*. This mechanism is already enhanced in heterogeneous networks, and a natural question is whether multilayer coupling can further promote cooperative behavior. Various models coupling evolutionary dynamics across network layers have been proposed, and this section reviews the principal results in that direction.

2.5.2.1 Diffusion on Multilayer Networks

A fundamental topic in network science is the analysis of transport and navigation processes over network topologies [78]. These processes serve as realistic proxies for a wide range of systems, including data flow in computer networks, resource distribution in technological infrastructures, and the movement of goods or people in transportation systems. The nature of such dynamics ranges from basic diffusive behavior and random walks [253] to sophisticated protocols tailored for optimal routing and navigation in real networks [254]. These approaches have made it possible to characterize network-level properties that cannot be inferred from purely local node-level metrics and to design strategies optimized for the underlying structure. Among the most relevant dynamical processes studied in complex networks, disease spreading and the design of containment strategies have received particular attention in the literature [28, 255]. This interest stems from the successful adaptation of classical compartmental models from mathematical epidemiology, such as SIS and SIR models, into network-based frameworks that better reflect the heterogeneous structure of real-world interactions [190]. Furthermore, from a physics standpoint, epidemic models on networks have expanded the study of non-equilibrium phase transitions beyond traditional lattice models [256]. In addition to epidemiology, these models are also useful for understanding information diffusion and rumor dynamics in social systems.

The multilayer network formalism has recently provided a more comprehensive framework to model the spreading of contagions, accounting for the co-existence and interplay of multiple types of interactions across different layers.

To briefly review the two most common compartmental models:

- In the **SIS (Susceptible–Infected–Susceptible)** model, individuals can transition from susceptible to infected upon contact with an infected individual. After some time, infected individuals recover and become susceptible again. This model is relevant for diseases that do not confer long-term immunity, such as many sexually transmitted infections.

- In the **SIR (Susceptible–Infected–Recovered)** model, infected individuals eventually recover and gain immunity, meaning they no longer participate in disease transmission. This framework suits infections such as influenza or measles.

In both models, contagion occurs with a given probability β, and recovery happens with probability μ per time unit. A critical point of analysis is whether the disease spreads significantly enough to trigger an epidemic outbreak, which is typically controlled by the ratio β/μ.

The behavior of the system differs depending on the model:

- In the **SIR model**, the disease naturally dies out when no infected individuals remain. The total number of recovered individuals at the end of the process is the key metric of impact.
- In the **SIS model**, reinfections can sustain the epidemic over time. The relevant observable here is the steady-state fraction of infected individuals, which serves as an order parameter to characterize the transition between healthy and epidemic regimes.

The application of these models to multilayer networks allows for a more accurate and flexible representation of spreading dynamics in complex systems, where individuals may interact in multiple contexts or platforms.

2.5.2.2 Disease Spreading on Multiplex Networks

Multiplex networks, a specific case of multilayer networks, consist of identical nodes replicated across multiple layers, each connected to its counterparts in the other layers through interlayer links. This structural property has significant implications for modeling contagion processes, particularly disease spreading. A fundamental characteristic of spreading in multiplex systems is that once a node becomes infected in one layer, it is considered infected in all layers, due to the tight interdependence among layers.

From a mathematical standpoint, disease spreading in such networks can be studied using spectral properties of the supra-adjacency matrix $\hat{A}$ of the system, which can be decomposed as:

$$\hat{A} = A + C = \begin{bmatrix} 0 & A_2 & 0 & \cdots & 0 \\ A_1 & 0 & \cdots & \cdots & 0 \\ \vdots & \vdots & \ddots & \vdots & \vdots \\ 0 & \cdots & \cdots & 0 & A_M \end{bmatrix} + \begin{bmatrix} I & 0 & I & \cdots & I \\ 0 & I & \cdots & \cdots & I \\ \vdots & \vdots & \vdots & \ddots & \vdots \\ I & \cdots & \cdots & I & 0 \end{bmatrix}.$$

In this configuration, disease transmission across layers is immediate and complete. Unlike interdependent networks, where coupling can be tuned, the multiplex setting imposes full synchronization of node states across layers. The SIS model on multiplex networks has been analyzed using a microscopic approach, where the probability that a node i is not infected by any of its neighbors across all layers is given by:

$$q_i(t) = \prod_{\alpha=1}^{M} \prod_{j=1}^{N} \left(1 - \beta_\alpha (A^\alpha)_{ij} \cdot p_j(t)\right),$$

with A^α denoting the adjacency matrix of layer α, and $p_j(t)$ the infection probability of node j at time t [257]. Unlike general multilayer models, here the vector $\mathbf{p}$ has N components only, since the infection state of each node is shared across all layers. The epidemic threshold β_c is related to the largest eigenvalue of the matrix R:

$$R_{ij} = 1 - \prod_{\alpha=1}^{M} \left(1 - \beta_\alpha (A^\alpha)_{ij}\right),$$

and it has been shown that β_c is always lower than that of any individual layer in isolation.

The SIR model has also been explored within the multiplex setting. One analysis introduced a cost for switching layers during disease transmission and studied the impact of this cost on the epidemic threshold. For strong interlayer coupling, the epidemic threshold again turns out to be lower than in the individual layers, suggesting an enhancement of spreading due to multiplexity [258]. A particularly insightful result relates the epidemic threshold β_c to the difference in transmission rates between intra- and inter-layer links ($\delta\lambda$), and the difference in mean degrees between the layers (δz). When δz is small, β_c increases with $\delta\lambda$, but for large δz, the trend reverses and β_c decreases, highlighting a complex dependency on the network's structural heterogeneity [259]. Furthermore, the SIR process has been studied in multiplex networks where a second layer contains only a subset of the nodes of the first (physical) layer. Varying the size of this subset reveals conditions under which disease spreading is enhanced or suppressed, providing a finer understanding of multilayer resilience and control.

2.5.2.3 Competing Epidemics on Multiplex Networks

Multiplex networks offer an ideal framework for modeling the dynamical interplay between multiple contagion processes occurring over the same population. This structural setup allows one to assign a different network layer to each spreading process, which is particularly suitable for studying the competition between two distinct epidemic outbreaks. In a seminal model proposed by Newman [260], two successive SIR diseases propagate on a single-layer network. The first disease spreads following standard dynamics, while individuals recovered from it are assumed to gain immunity against the second one. As a result, the second disease encounters a reduced susceptible population. Even in scale-free networks, this setting yields a non-zero epidemic threshold for the second disease.

Subsequent works extended this model to the case of simultaneous infections and to SIS dynamics [261, 263]. The natural generalization to multiplex networks has provided additional insights, especially when considering distinct transmission channels for each disease. In such cases, each disease propagates in a separate network layer. In particular, Funk and Jansen [263] considered a multiplex structure with two network layers and SIR dynamics, where the first disease spreads in one layer and the recovered individuals are marked as immune in the second layer before the second contagion takes place. They found that positive correlations between node degrees across the two layers increase the effective epidemic threshold for the second disease, thereby enhancing overall network immunization.

This framework was later generalized to account for simultaneous spreading of two SIR diseases with possible delays and partial immunity [264]. Moreover, interacting contagions have also been modeled using SIS dynamics by introducing mutual exclusivity, resulting in the so-called SI_1I_2S model [262]. In this setting, an individual infected by one disease becomes temporarily immune to the other.

The SI_1I_2S dynamics can be captured using a set of coupled equations. Let A_1 and A_2 be the adjacency matrices of layers 1 and 2, and let β_1, β_2 and μ_1, μ_2 be the infection and recovery rates for diseases 1 and 2, respectively. The evolution of infection probabilities for node i in each layer is given by:

$$p_1^i(t+1) = p_1^i(t) - \mu_1 p_1^i(t) + \beta_1(1 - p_1^i(t) - p_2^i(t)) \sum_{j=1}^{N} (A_1)_{ij} p_1^j(t),$$

$$p_2^i(t+1) = p_2^i(t) - \mu_2 p_2^i(t) + \beta_2(1 - p_1^i(t) - p_2^i(t)) \sum_{j=1}^{N} (A_2)_{ij} p_2^j(t).$$

Linearizing these equations around the disease-free equilibrium ($p_i^\alpha = 0$ for all i, α), one can derive epidemic thresholds as a function of the control parameters $\tau_1 = \beta_1/\mu_1$ and $\tau_2 = \beta_2/\mu_2$. If both $\tau_1 < 1/\Lambda(A_1)$ and $\tau_2 < 1/\Lambda(A_2)$, the infection dies out in both layers. However, if either τ_1 or τ_2 exceeds the corresponding threshold, complex competitive dynamics emerge. An analytical and numerical study in [262] revealed four possible regimes: (N) both diseases die out, (I) disease 1 spreads and disease 2 does not, (II) disease 2 spreads and disease 1 does not, and (III) both diseases coexist. Interestingly, if one disease has a sufficiently high transmission rate, it can suppress the other—even if the latter would surpass its own threshold in isolation.

A key finding is that when the network layers are structurally identical, the coexistence phase (III) disappears, as the dominant disease consistently outcompetes the other.

2.5.2.4 Interaction Between Social Dynamics and Epidemic Spreading

The multiplex network paradigm enables the coupling of distinct dynamical processes operating over the same population but evolving on structurally different layers. One particularly relevant application is the interplay between disease spreading and social awareness, where the dynamics of awareness can mitigate the epidemic impact [259].

To explore this interaction, the authors of [259] coupled a classical SIS process on a physical network with an awareness diffusion model operating on a separate informational layer. In this setting, individuals are either Aware (A) or Unaware (U), and the transitions between these states follow an SIS-like mechanism. Specifically, A agents convert U neighbors into A with probability λ, and A nodes revert to U with probability δ. Importantly, awareness reduces the susceptibility of individuals to the disease by a factor γ, where $\gamma = 0$ implies perfect immunization. Moreover, unaware individuals cannot be infected unless they simultaneously become aware. Independently, each dynamical process has a well-defined critical point: $\beta_c = \mu/\Lambda(A_1)$ for the SIS dynamics and $\lambda_c = \delta/\Lambda(A_2)$ for awareness spreading, where A_1 and A_2 are the adjacency matrices of the respective layers. Once the two processes are coupled, feedback emerges: infected nodes raise awareness, which in turn reduces disease transmission. As shown in [259], this feedback increases the epidemic threshold, particularly when δ/μ is small, meaning that the awareness process strongly influences the spread of infection.

2.5.2.5 Impact of Interlayer Topological Properties

Beyond the interplay of dynamics, the structural properties of interlayer connections significantly influence disease propagation. One essential factor is the coupling density, which accounts for partial node overlap between layers. In [265], the authors considered a fraction q of nodes shared across layers in a multiplex SIR model. Their analysis revealed that the epidemic threshold decreases with increasing q—reaching a minimum when $q \to 1$, i.e., full overlap. In this case, the disease can spread synchronously across layers. Conversely, when q is small, the threshold increases, approaching that of the layer with higher spreading capability. Interestingly, even minimal overlap ($q \ll 1$) can substantially alter the epidemic threshold of the stronger layer, highlighting how sparse interconnections serve as an efficient control lever in non-pharmaceutical interventions.

Another structural property of interest is how nodes are interconnected across layers. In [266], the authors explored two new measures to assess interlayer influence: (1) degree–degree correlation (DDC), indicating whether high-degree nodes in one layer tend to be connected to high-degree nodes in the other; and (2) average similarity of neighbors (ASN), measuring the extent of shared neighbors across layers. They found that higher DDC values facilitate disease propagation by lowering the epidemic threshold and reducing the outbreak size, even if each layer is below its individual epidemic threshold. In contrast, ASN had minimal effect on the overall dynamics.

These findings underscore the critical role of specific interlayer configurations in shaping contagion outcomes.

2.5.2.6 Opinion Formation Dynamics in Multilayer Systems

The emergence of large-scale consensus in systems composed of interacting agents is a fundamental question in both natural and social sciences. In the last decades, numerous models have been proposed to study the dynamics of opinion formation, drawing from statistical physics, economics, and sociology [267]. Among the various frameworks, the voter model stands out for its efficiency and ability to reproduce non-trivial behaviors [268]. In its basic version, the voter model assigns to each node a binary opinion variable $s_i = \pm 1$. At every time step, an agent i randomly adopts the state of a randomly chosen neighbor j, i.e., $s_i \leftarrow s_j$. This imitation process continues until the system reaches consensus. Despite its simplicity, the model exhibits rich dynamical behavior, particularly in relation to the network topology. On regular lattices, it shows different coarsening dynamics depending on the dimension, while on scale-free networks, the consensus time scales differently depending on the exponent of the degree distribution [269]. Several extensions of the voter model have been introduced to capture more realistic phenomena. These include the nonlinear voter model, the vacillating voter model, and the introduction of heterogeneity in agents' behavior. Particularly interesting are models where the agents' ability to change their opinion depends on their memory (i.e., persistence time), showing that such mechanisms can accelerate consensus formation [270].

In coevolving networks, where agents are allowed to rewire their connections based on disagreement, a competition arises between the imitation of opinions and the adaptation of the network structure. This often leads to a fragmentation transition, where the network breaks into disconnected components each with local consensus. The interface density ρ, which measures the fraction of links connecting agents with opposing opinions, serves as an order parameter: $\rho > 0$ corresponds to active states, while $\rho = 0$ marks the frozen phase. Another relevant modification involves the presence of zealots—agents that never change their opinion. Zealots can drastically alter the dynamics, biasing the global outcome toward their own fixed opinion in low-dimensional systems, while in higher-dimensional networks they mostly influence their local neighborhood without triggering global consensus. Finally, the voter model has been successfully applied beyond opinion dynamics, such as in ecological systems and evolutionary biology, where it models competition and fixation phenomena among species.

2.5.2.7 Voter Model on Multilayer Networks

With the rapid advancement of network science, opinion dynamics have been naturally extended to multilayer network frameworks. One common approach involves establishing interlayer connections that allow opinion exchange across different network layers. In a recent study, the dynamics were investigated on a two-clique graph where two identical cliques, each of size N, are interconnected through m interlayer links [271]. Each node is, on average, involved in m/N interlayer links, either arranged regularly or distributed randomly. At each step, a link is selected uniformly at random and an opinion imitation event occurs between the two connected nodes. The analysis revealed that the mean consensus time $\langle T \rangle$ scales linearly, $O(N)$, in the regime of dense interlayer connectivity, but grows quadratically, $O(N^2)$, in the sparse regime. A crossover between these two regimes is observed when each node has approximately one interlayer link.

Further insights emerged from the study of coevolving multilayer voter dynamics [272]. Here, a fraction q of nodes shares interlayer links between two network layers. Within each layer $l = 1, 2$, agents update their opinions and rewire their links according to internal time scales governed by a rewiring probability p_l. In the symmetric case ($p_1 = p_2$), interlayer coupling plays a decisive role in controlling the fragmentation transition of the system. When $p_1 \neq p_2$, the asymmetry induces a "shattered fragmentation" phase, where one of the layers fragments into many disconnected components, each characterized by different opinions. Notably, a critical threshold of interlayer connectivity was identified, above which fragmentation in one layer could be suppressed due to its connection with a more cohesive counterpart.

In the context of political dynamics, a multiplex formulation of the voter model has been proposed to simulate the competition between two political parties during elections [273]. Each party is modeled as a separate layer of a social network, where opinion contagion occurs. On election day, each agent can vote for one party (i.e., be active on one network) or abstain (inactive in both). Using simulated annealing techniques, the model exhibits a rich phase diagram. In broad parameter regions, pluralism emerges where both parties capture significant fractions of votes. The denser network often dominates the election, though small committed minorities can drastically affect the outcome, aligning with real-world political observations.

2.5.2.8 Evolutionary Games on Multilayer Networks

The study of evolutionary games on networks provides a framework to explore the emergence of cooperation within structured populations. In classical formulations such as the Prisoner's Dilemma, agents can adopt one of two strategies—cooperation (C) or defection (D)—with payoffs determined by mutual interactions. While well-mixed populations typically evolve toward full defection, the presence of a network structure enables cooperation to emerge through mechanisms like network reciprocity [274, 275].

Multilayer networks extend this framework by allowing agents to interact in multiple layers, each representing a distinct context (e.g., social, biological, or technological). In this setting, the strategies and payoffs may vary across layers, and the interplay between layers can influence the global dynamics of cooperation. For example, an agent might cooperate in one layer and defect in another, depending on the local incentives.

Recent studies have also considered the Public Goods Game (PGG) in multilayer environments. In the PGG, each cooperator contributes to a common pool, which is enhanced and redistributed among all participants, including defectors. This setup captures the tension between individual benefit and collective welfare. When played on multilayer networks, cooperation can be promoted by coupling layers with different topological or dynamical properties [276, 277]. Overall, multilayer network models provide a richer and more realistic representation of social dilemmas and evolutionary processes, capturing the complexity of interactions observed in natural and artificial systems.

2.5.2.9 Games on Multiplex Networks

A relevant extension of evolutionary dynamics considers multiplex networks, where all M layers share the same set of N nodes, each representing an agent active in multiple contexts. In this setting, each node may adopt different strategies in different layers. For example, an individual might cooperate in some layers while defecting in others, depending on local incentives or environmental conditions.

The total payoff of an agent i in a given layer α at time t is computed by aggregating interactions with its neighbors according to the standard payoff matrix. The global fitness of the agent is then obtained by summing its payoffs across all layers:

$$f_i(t) = \sum_{\alpha=1}^{M} f_i^{\alpha}(t).$$

Crucially, the strategy update in any layer depends on the cumulative payoff across the entire multiplex, which introduces interlayer dynamical correlations [278].

Interestingly, the presence of multiple layers enhances the robustness of cooperation, even though it may slightly lower the overall level of cooperation for low temptation to defect. This behavior reflects the stabilizing role of multiplexity in evolutionary games.

Another modeling approach considers separating the interaction layer (where agents collect payoffs) from the updating layer (where strategies are revised). This configuration, often motivated by the distinct roles social environments play in real-life scenarios, shows that asymmetries in network mixing (e.g., assortative vs. disassortative) can suppress the formation of cooperative clusters and hinder the emergence of cooperation [279].

2.5.3 Beyond Pairwise Interactions

Any meaningful understanding of a complex system must be grounded in system-level descriptions. Consider the following scenario: take an ecosystem and disassemble it into its component species. Regardless of how precise and thorough our knowledge of each species may be, our ability to comprehend population dynamics, such as how species abundances evolve over time, will likely remain superficial. The same holds when trying to explain phenomena such as epileptic seizures by focusing on individual neurons, or the viral spread of rumors through societies by analyzing isolated psychological traits. These approaches are often inadequate because they fail to account for a crucial element of complex systems: the intricate pattern of nonlinear interactions among their components.

After many years of reductionist perspectives, science has increasingly acknowledged that the collective behaviors of complex systems cannot be understood by examining their individual units in isolation. Instead, a shift towards embracing complexity as a fundamental principle of our world has taken place [225]. Within this conceptual framework, networks have emerged as an essential modeling tool for representing complex systems [134]. Networks serve as structured maps that describe the physical or virtual spaces where interactions take place. Building on earlier developments in mathematics, sociology, and ecology, a series of landmark papers around the turn of the millennium sparked widespread interest and led to the rise of a vibrant multidisciplinary field known as Network Science. This community has combined elements of graph theory and statistical mechanics into a unified framework capable of modeling a vast array of phenomena, from physics and biology to economics and social behavior.

As investigations of real-world systems deepened, researchers have realized that the traditional notion of a network, as a set of nodes and pairwise edges, needs to be extended to more accurately reflect the diversity and richness of real interactions. Over the years, significant efforts have been made to incorporate additional structure into networks, such as edge directionality, weights, and signs indicating cooperative or antagonistic relationships. Furthermore, temporal networks have been introduced to account for interactions that evolve over time, and multilayer or multiplex networks have been proposed to represent systems in which nodes are connected through different kinds of relationships or interactions [243]. Despite these developments, all such network representations still fundamentally rely on pairwise interactions. Yet, empirical evidence from various domains—including social systems, neuroscience, ecology, and biology—shows that many interactions occur at the level of groups rather than individual pairs. Neuronal activity often reflects mesoscopic dynamics that depend on simultaneous input from multiple neurons. In ecological systems, groups of three or more species frequently interact in competition for resources, where the presence of one species can modulate the relationship between the other two. Social mechanisms such as peer influence, collective decision-making, or collaboration naturally extend beyond the scope of dyadic interactions.

Although early models in network science attempted to approximate group interactions using tools such as bipartite graphs or probabilistic contagion processes, they were still embedded within a pairwise framework. The increasing recognition that such models are insufficient has led to a growing interest in mathematical formalisms that explicitly describe higher-order interactions. Simplicial complexes and hypergraphs have emerged as powerful and natural frameworks capable of capturing the collective interactions among multiple units. In recent years, the adoption of these higher-order representations has fundamentally transformed our approach to complex systems. While their importance had been acknowledged in the past, it is only with the current wave of enthusiasm that a deeper and more comprehensive understanding has become possible. It is now widely accepted that moving beyond dyadic interactions is essential to predict and explain collective behaviors that could not be accounted for with traditional network models.

2.6 Higher-Order Representations of Networks

The discussion begins by clarifying again what it is intended with the term *interaction*, and subsequently, what qualifies as a higher-order interaction. An interaction is defined as a set $I = [p_0, p_1, \ldots, p_{k-1}]$ consisting of an arbitrary number k of fundamental elements of the system under study—commonly referred to as nodes or vertices. These interactions may describe a variety of real-world phenomena, such as co-authorship in a scientific paper, a set of genes cooperating to fulfill a biological function, or the co-activation of a neuronal ensemble during a specific task. About the *order* (or dimension) of an interaction involving k nodes to be $k - 1$, an interaction involving only a single node is of order zero, while a pairwise interaction has order one, a triadic interaction has order two, and so on. Accordingly, higher-order interactions refer to cases where $k \geq 3$, representing group interactions beyond pairs. Conversely, interactions of order $k \leq 1$ (i.e., self and pairwise interactions) are classified as low-order.

This distinction between low- and high-order systems is conceptually important for two main reasons. Firstly, it emphasizes the contrast between traditional graph-based models, which have long dominated the study of complex systems, and more recent approaches that are explicitly designed to capture group interactions. Secondly, this classification allows us to explore the relationships, overlaps, and possible mappings between these modeling paradigms. It is important to note that our definition of higher-order interactions excludes other kinds of structural dependencies, such as those found in multilayer networks with multiple link types [243], or those based on non-Markovian dynamics in time-stamped interaction data [280, 281]. These cases involve higher-order dependencies of a different nature and fall outside the scope of this particular discussion.

Let us illustrate these concepts with a concrete example. Consider a node set $V = \{a, b, c, d, e\}$ and an interaction set $I = \{[a, b, c], [a, d], [d, c], [c, e]\}$. This system comprises one 2-order interaction and three 1-order interactions. While this list fully captures the interaction structure, extracting meaningful system-level properties typically requires the use of a representation that facilitates comparison and analysis across different orders of interaction. For instance, assessing how densely connected the system is, or whether one node is reachable from another, depends on having a unified representation that accounts for the complexity and heterogeneity of interaction orders.

The Table 2.3 provides a summary of the main concepts and measures used in higher-order representations of complex systems. It distinguishes between modeling frameworks, such as simplicial complexes, hypergraphs, and bipartite graphs, and analytical tools including matrices, tensors, and centrality measures.

Table 2.3 Summary of key higher-order representations and measures

Concept	Definition (Symbol)	Description
Higher-order interaction	$I = [p_0, \ldots, p_{k-1}]$	A general interaction involving k nodes; order defined as $k - 1$
Simplex	$\sigma = [p_0, \ldots, p_k]$	A set of $k + 1$ nodes representing a k-order interaction in a simplicial complex
Simplicial complex	K	A set of simplices closed under inclusion of all subfaces. Models nested group interactions
Hypergraph	(V, H)	A generalized graph where each hyperedge is a subset of nodes; does not require subface inclusion
Bipartite graph	(U, W, E)	Two-layer graph where U are system nodes and W are interactions; preserves group membership
Hasse diagram	$HD(K)$	A DAG representing simplices as nodes and inclusion relations as edges
Facet representation	–	A representation of a simplicial complex through its maximal simplices
Incidence matrix	$I_{i\alpha}$	Binary matrix encoding node–interaction membership in hypergraphs or simplicial complexes
Adjacency matrix (higher-order)	A_{ij}	Number of shared interactions (e.g., hyperedges or simplices) between nodes i and j
Intersection profile	$P = I^T I$	Matrix indicating number of shared nodes between interaction sets
Adjacency tensor	$A^{[d]}$	Multidimensional tensor encoding d-order interactions
k-walk	–	A sequence of higher-order interactions sharing at least k nodes; generalizes paths
Generalized degree	$k_{d,\delta}(\alpha)$	Number of d-simplices that include a δ-simplex α
Eigenvector centrality	x_v	Recursive centrality based on influence of connected elements. Extended to incidence and adjacency tensors
Katz centrality	$K_{k,\alpha}$	Centrality of a k-simplex computed from a power series of its adjacency matrix
Sub-hypergraph centrality	$C_{\text{sh}}(v_i)$	Spectral measure of walk-based self-return centrality over higher-order structures

2.6.1 Graph-Based Representations

Graphs represent the most prevalent method for modeling families of interactions in complex systems. A graph $G = (V, E)$ is constructed from a node set V and an edge set E, where each edge connects a pair of nodes. As such, graphs intrinsically represent 1-interactions, i.e., interactions of the form $I = [i, j]$. To accommodate higher-order interactions within this framework, it is customary to decompose any k-interaction into its constituent 1-interactions. For instance, given the set of interactions $I = \{[a, b, c], [a, d], [d, c], [c, e]\}$, the graph-based representation I_G becomes $\{[a, b], [b, c], [c, a], [a, d], [d, c], [c, e]\}$.

This reduction to pairwise interactions enables the use of well-established tools from graph theory and network science [134]. It has underpinned major advances across a wide range of disciplines, including biology, ecology, social science, engineering, and neuroscience. However, such simplification introduces significant limitations. Once interactions are projected to pairwise edges, the original group structure is lost: there exists no unambiguous way to reconstruct the original k-interactions from the pairwise representation. For instance, using only I_G, one cannot discern whether the triplet $[a, d, c]$ was part of the original interaction set.

In some cases, high-order relationships are inferred from dense patterns of low-order interactions, employing methods like community detection, clique analysis, or stochastic block models. Still, these inferences are indirect and often suffer from incompleteness or ambiguity. An alternative low-order framework that retains information on higher-order interactions is the bipartite graph. A bipartite graph is defined by two disjoint node sets (U, W) and an edge set E such that all edges connect a node in U to a node in W. In the context of higher-order modeling, it can be considered the following: $U = V$ (the original nodes), and $W = I$ (the set of interactions). In this formulation, each node is linked to the interactions it participates in, thus preserving full information about the group structures. This representation is widely used in social network analysis, recommendation systems, and ecological modeling [282].

Nevertheless, in a bipartite graph, original system nodes no longer interact directly, but rather via the interaction nodes. This structural asymmetry adds complexity to the analysis. To mitigate this, unipartite projections are often used, in which each high-order interaction is mapped to a fully connected subgraph. However, this again sacrifices the integrity of group interactions.

Motifs offer an additional tool to characterize interaction patterns. A motif is a small, recurring subgraph configuration that appears significantly often in a network. Motifs are seen as structural signatures tied to the function or optimization of a system. They allow one to capture fine-grained structural differences between networks, such as across functional brain states or evolutionary pathways [146]. However, due to their exponential growth in number with respect to size, motifs can become computationally intractable for large graphs. A special class of motifs, cliques, has been widely adopted as a canonical representation of group interaction. A k-clique is a fully connected subgraph of k nodes and encodes an interaction of order $k - 1$. While cliques capture dense group structures and are essential to community detection, they can also introduce artifacts. For instance, in the earlier example, both sets $\{a, b, c\}$ and $\{a, d, c\}$ form 3-cliques in the projected graph, but only the former corresponds to a true group interaction. Thus, clique-based representations can falsely infer nonexistent higher-order relations, in contrast to edge-only graphs, which tend to underestimate them.

2.6.2 Explicit Higher-Order Representations

To model systems with higher-order interactions, it is crucial to represent these interactions explicitly rather than reducing them to pairwise approximations. A powerful mathematical structure for this purpose is the *simplex*. A k-simplex σ is formally defined as a set of $k + 1$ nodes: $\sigma = [p_0, p_1, \ldots, p_k]$. This formulation aligns naturally with the previously introduced concept of a k-interaction and borrows notation from algebraic topology, where nodes often represent points in a topological space. In the context of combinatorial or abstract systems, these nodes are typically relabeled as $v_0, v_1, \ldots$ to emphasize their role as vertices within a system of interactions, rather than as points in a metric space. As before, a k-simplex represents an interaction of order k, with the convention that the order equals the number of nodes involved minus one.

Analogous to how a graph is a set of edges, a *simplicial complex* is a set of simplices. However, simplicial complexes come with an added constraint: for every k-simplex $\sigma \in K$, all of its subfaces (i.e., simplices of lower order formed from subsets of nodes in σ) must also be present in K. For example, if $[a, b, c]$ is a simplex in K, then all its constituent edges $[a, b]$, $[a, c]$, $[b, c]$ and nodes $[a]$, $[b]$, $[c]$ must be part of K as well. This constraint ensures consistency across dimensions and allows for a precise encoding of the topological structure of the system. Unlike clique complexes, where higher-order interactions are inferred from densely connected subgraphs, simplicial complexes make it possible to distinguish between interactions that are truly present and those that emerge artifactually through projection.

Simplicial complexes offer a rich mathematical framework equipped with tools such as boundary operators and higher-dimensional Laplacians. These tools enable the study of topological features like cycles, cavities, and homological holes, which are essential in applications ranging from geometry to topological data analysis [286]. However, the requirement that all lower-dimensional faces of a simplex be present in the complex can be overly restrictive in some scenarios. For example, in studying scientific collaborations, it might be observed that three researchers coauthor a paper without any pairwise collaborations among them. In such cases, representing the group as a full simplex would be misleading.

To address this limitation, *hypergraphs* are introduced. A hypergraph consists of a node set V and a set of hyperedges H, where each hyperedge is a non-empty subset of V. Hypergraphs allow for arbitrary interactions without enforcing subface inclusion. For instance, the hyperedge $[a, b, c]$ can exist independently, without requiring the presence of $[a, b]$, $[b, c]$, or $[a, c]$ as separate interactions. This flexibility makes hypergraphs ideal for representing systems where the group interaction itself has unique functional or structural meaning. Nevertheless, their generality introduces analytical challenges. Extending classical graph-theoretic concepts such as Laplacians to hypergraphs is nontrivial and has only recently seen significant development [283, 284]. Moreover, applications of hypergraph-based models are still emerging, despite their promising theoretical foundation.

2.6.3 Relations and Links Between Representations

When analyzing multiple representations of the same interacting system, several natural questions arise: To what extent do these representations overlap? Can they be mapped onto one another in a canonical way? What kind of information is preserved, or lost, when transitioning between them?

A classic example is the comparison between graphs and simplicial complexes composed exclusively of 1-dimensional simplices (i.e., edges). At first glance, such a simplicial complex may appear equivalent to a standard graph. However, this equivalence is not absolute and depends heavily on the chosen representation [285].

Let us begin by considering simplicial complexes, focusing on two prominent representations: the *Hasse diagram* and the *facet representation*. The **Hasse diagram** of a simplicial complex K is a directed acyclic graph $HD(K) = (V_{HD}, E_{HD})$, where the nodes V_{HD} correspond to all simplices in K, and the edges E_{HD} denote inclusion relations between simplices that differ in dimension by exactly one. Specifically, an edge $(\sigma, \tau) \in E_{HD}$ exists if $\sigma \subset \tau$ and $\dim(\tau) = \dim(\sigma) + 1$.

This diagram reveals the internal organization of the complex through a layered, multipartite structure. It serves not only as a visual representation but also as a blueprint for defining walks and diffusions on the complex. The structure of the Hasse diagram directly informs the construction of higher-order Laplacians, which are central to generalizing classical graph operators such as the Laplacian to higher dimensions. These constructions are based on (co)boundary operators that link simplices of adjacent dimensions, mirroring differential operators in geometry. For instance, applying a boundary operator to a 2-simplex (a triangle) yields a sum of its three edges. Applying the operator again returns zero, since a boundary has no boundary. Such rules underlie the construction of combinatorial Laplacians, which thus inherently describe transitions only between simplices of co-dimension one [286].

The second useful representation is the **facet representation**, which encodes a simplicial complex via its maximal simplices (facets). A facet is a simplex that is not a face of any higher-dimensional simplex in the complex. In the Hasse diagram, facets correspond to nodes with zero out-degree. This representation is particularly efficient because it implies the existence of all lower-dimensional subsimplices without explicitly listing them.

Structurally, the facet representation can be interpreted as a directed bipartite graph: one layer contains the nodes (vertices) of the system, the other contains the facets, and directed edges link a node to the facets it belongs to. This bipartite graph can further be interpreted as a hypergraph, where each facet corresponds to a hyperedge. However, not every bipartite or hypergraph representation gives rise to a valid simplicial complex in the facet representation. This only occurs when no set of vertices associated with a facet (hyperedge) is a proper subset of another—ensuring compliance with the non-inclusion condition that characterizes simplicial facets [287].

2.7 Higher-Order Measures

Many classical graph-based metrics have been extended to apply to higher-order systems such as hypergraphs, simplicial complexes, and clique complexes. In this section, the discussion is focused on how interactions can be represented and analyzed using matrices and tensors.

2.7.1 Matrix Representations of Higher-Order Systems

2.7.1.1 Incidence Matrix

The incidence matrix provides a foundational way to encode the relationships between nodes and interactions. Originally introduced by Kirchhoff in 1847 for graphs, the incidence matrix $I = \{I_{i\alpha}\}$ of a graph $G = (V, E)$ is an $n \times m$ binary matrix, where n is the number of nodes and m is the number of edges. The entry $I_{i\alpha} = 1$ if node i is incident to edge α, and 0 otherwise. This concept generalizes naturally to hypergraphs and simplicial complexes. In a hypergraph, the incidence matrix represents the membership of each node in each hyperedge. The matrix can also be weighted to reflect multiple instances of a node in a single hyperedge. Structurally, the incidence matrix of a hypergraph is equivalent to the adjacency matrix of a bipartite graph composed of nodes and hyperedges [287]. From the incidence matrix, two immediate measures can be defined: the degree of a node i (sum of row i) and the size of a hyperedge α (sum of column α). These local metrics serve as primary indicators of the structure in HOrSs.

2.7.1.2 Adjacency Matrix and Intersection Profile

The adjacency matrix A of a graph can be derived from its incidence matrix via the product $A = II^T - D$, where D is the diagonal matrix of node degrees. In higher-order systems, this formulation is extended directly. Here, A_{ij} counts the number of hyperedges (or simplices) shared by nodes i and j. When hyperedges carry weights, $A = IWI^T - D$, where W is a diagonal matrix containing the weights, is defined.

An additional construction is the intersection profile $P = I^T I$, which yields an $m \times m$ matrix counting the number of common nodes between hyperedges α and β. This profile is especially useful in statistical studies of edge intersections [288].

2.7.1.3 Adjacency Tensors and Dimensional Interactions

To represent adjacency at arbitrary dimensions, adjacency tensors are introduced. For each dimension d, one can define an $n \times n \times \cdots \times n$ tensor $A^{[d]}$, where each entry $a_{i_1,\ldots,i_d} = 1$ if the d-set $\{i_1, \ldots, i_d\}$ forms a higher-order interaction, and 0 otherwise. This representation captures interactions of order $d - 1$ and generalizes classical adjacency matrices to arbitrary higher-order interactions [289].

Additionally, adjacency matrices can be defined specifically for k-simplices, focusing on their relationships via *lower* and *upper* adjacency. Two k-simplices are lower adjacent if they share a common $(k-1)$-face, and upper adjacent if they are both faces of the same $(k+1)$-simplex. These relationships define the matrices $A_L^{[k]}$ and $A_U^{[k]}$, respectively. One may also define an adjacency matrix $A^{[k]}$ that encodes only lower adjacencies not involved in upper ones. These constructs are key to defining centralities and diffusion processes over higher-dimensional walks.

2.7.2 *Walks, Paths, and Centrality Measures*

Centrality measures evaluate the importance of nodes in a network. They can be based on local properties, like the number of connections a node has, or on global structural aspects such as distances and paths. In this section, generalizations of centrality measures to higher-order systems (HOrSs), including hypergraphs and simplicial complexes are presented.

2.7.2.1 Degree Centralities

The most basic centrality metric is the degree of a node, defined as the number of interactions it participates in. In graph-based systems, this corresponds to the number of adjacent vertices. For higher-order systems, the degree of a node i can be defined from the adjacency matrix A as:

$$\deg(i) = \sum_{j=1}^{n} a_{ij}.$$

This notion generalizes to adjacency tensors. Let α be a δ-dimensional simplex; its generalized degree with respect to d-simplices is given by:

$$k_{d,\delta}(\alpha) = \sum_{\alpha' \supset \alpha} a_{\alpha'},$$

where $a_{\alpha'}$ indicates the presence of a d-simplex that contains α [289]. When $\delta = 0$, this reduces to the classical node degree.

In weighted hypergraphs, the definition of degree depends on whether one counts the number of hyperedges or the number of adjacent nodes. Weights can be derived from co-occurrence frequency, collaboration strengths, or probabilities of joint interactions. Degrees can also be assigned to hyperedges or simplices. For simplicial complexes, one may define upper, lower, or mixed adjacency degrees. The maximal simplicial degree is often written as $\deg(\alpha) = \deg_U(\alpha) + \deg_L(\alpha)$, where the terms represent counts of adjacent simplices above and below α in dimension [288].

2.7.2.2 Paths and Path-Based Centralities

To define global centrality measures such as closeness or betweenness, the notion of walks to higher-order systems is extended. A walk in a hypergraph is typically defined as a sequence of hyperedges where consecutive ones share at least one node. A path is a walk that does not repeat vertices. In this context, centralities are defined analogously to the graph case. For instance:

- **Betweenness centrality**: number of shortest paths passing through a node or simplex.
- **Closeness centrality**: reciprocal of the average distance from the node (or simplex) to all others.
- **Sub-hypergraph centrality**: number of closed walks of different lengths starting and ending at the same vertex.

The sub-hypergraph centrality of a node v_i can be written in spectral form as:

$$C_{\text{sh}}(v_i) = \sum_j u_{ij}^2 e^{\lambda_j},$$

where u_{ij} is the ith component of the jth eigenvector of the adjacency matrix and λ_j its corresponding eigenvalue [288].

More generally, k-walks between hyperedges or simplices require that consecutive elements intersect in at least k nodes. For simplicial complexes, a k-walk often involves only k-simplices that are lower adjacent (i.e., share a k-dimensional face). The simplicial closeness centrality of a k-simplex is then the reciprocal of the total k-walk distance to all others. Topological constraints significantly influence connectivity. For example, imposing a higher threshold k for shared elements between interactions reduces the number of valid walks, potentially fragmenting the system.

Moreover, whether a system is modeled as a hypergraph or a simplicial complex affects walkability, since simplicial complexes imply the existence of all subfaces of included simplices.

2.7.2.3 Eigenvector Centralities

In many contexts, it is essential to evaluate the influence of a node not only based on local or path-based considerations but also on its global impact across the network. Eigenvector centrality captures this idea by assigning high scores to nodes that are connected to other high-scoring nodes. Originally proposed in sociology by Bonacich [249], the centrality x_v of a node v satisfies the eigenvalue equation:

$$x_v = \frac{1}{\lambda} \sum_{t \sim v} a_{vt} x_t,$$

where a_{vt} is an entry of the adjacency matrix, and λ is the associated eigenvalue.

This recursive definition lends itself to several generalizations. For example, degree centrality counts walks of length one, while eigenvector centrality aggregates walks of infinite length. Alpha centrality introduces exogenous sources of influence, and subgraph centrality restricts the count to closed walks such as triangles or cycles [290].

In bipartite and higher-order systems such as hypergraphs or simplicial complexes, eigenvector centrality can be derived from the spectrum of their incidence matrices. Bonacich's extension to bipartite graphs enables computing centrality scores for both node sets using a shared eigenvalue. In this way, one can obtain eigenvector centralities for both nodes and hyperedges via the singular vectors of the incidence matrix [291]. When extending these ideas to motifs, one defines a motif adjacency matrix A_M, and the motif-based eigenvector centrality of node i corresponds to the i-th entry of the leading eigenvector of a weighted motif adjacency matrix. This construction becomes particularly relevant when motifs (such as cliques) play a functional role in the network [292].

For tensor-based generalizations to k-dimensional interactions, one considers the adjacency tensor A_k of order m, where eigenvector centralities are derived as solutions to:

$$Tc^{m-1} = \lambda c^m, \quad \|c\|_1 = 1,$$

in the case of Z-eigenvectors, or more generally,

$$Tc^{m-1} = \lambda c,$$

for H-eigenvectors. These nonlinear tensor eigenvector formulations allow capturing the multiplicative influence across nodes in higher-order hyperedges or simplices [292].

Finally, the eigenvector centrality of a k-simplex α is defined as the corresponding entry in the principal eigenvector of A_k, the adjacency matrix (or tensor) at that dimension. The Katz centrality of α is defined via a power series:

$$K_{k,\alpha} = \left[\sum_{p=0}^{\infty} x^p (A_k)^p \right]_\alpha ,$$

with $0 < x < \frac{1}{\lambda_1(A_k)}$ ensuring convergence.

2.8 Spreading and Social Dynamics

Dynamical processes that emulate human behaviors have been extensively studied, particularly in connection with the underlying structure of social interactions [240]. Social settings provide a natural testbed for higher-order representations, as individuals often interact not only in pairs but also within groups. Consequently, these dynamics must account for the potential nonlinear effects arising from such group-level interactions. In this section, a variety of models, originally developed on graphs, that have recently been extended to higher-order systems (HOrSs) is reviewed. The starting point are the spreading processes, traditionally rooted in epidemic modeling, and later adapted to capture mechanisms of social contagion. As follows, broader classes of models are focused on consensus formation and opinion dynamics.

2.8.1 Spreading in Higher-Order Networks

The study of spreading dynamics has long represented a central area of research within network science. Historically, this line of inquiry builds upon classical compartmental epidemiological models such as SIS (Susceptible–Infected–Susceptible) and SIR (Susceptible–Infected–Recovered). In both cases, susceptible individuals may become infected via interactions with already infected individuals. The key distinction lies in the presence of immunity: while in the SIS model individuals can return to the susceptible state and thus sustain a persistent endemic phase, the SIR model includes an absorbing state (Recovered), reflecting acquired immunity and halting further reinfection.

Advancements in this field have been propelled by the increasing availability of real-world data, which has enabled more accurate parameter estimation and more realistic modeling of disease propagation [28]. These developments prompted a more refined modeling approach, gradually moving from mean-field (MF) approximations and heterogeneous mean-field (HMF) models to sophisticated microscopic Markov chain-based formulations [68]. While these epidemic models were originally conceived for infectious diseases like Ebola or influenza, their applicability has expanded to social processes such as rumor spreading, viral content diffusion, or the adoption of technologies. However, social contagions involve mechanisms—e.g., social influence, peer pressure—that are not typically present in biological contagions. This gave rise to the framework of *complex contagion* [293], wherein multiple simultaneous exposures are often necessary to trigger the transition from a neutral to an active state.

With these dynamics in mind, the representation of the underlying contact structure becomes crucial. Transitioning from dyadic to higher-order structures, such as simplicial complexes or hypergraphs, allows a better encoding of collective interactions. For example, simplicial complexes are particularly suited when it is reasonable to assume that group interactions also imply all corresponding pairwise interactions. In contrast, hypergraphs may be more appropriate when only group-level effects matter, without necessitating all subsets to be present. Here it is possible to distinguish between these two modeling approaches—simplicial-based and hypergraph-based—and highlight recent models and results developed for both. Our aim is not to establish a definitive standard for modeling higher-order spreading dynamics, but rather to present a broad overview of recent research directions and challenges in extending contagion dynamics to HOrSs.

2.8.2 Evolutionary Games

Imitation is a fundamental mechanism underlying many social dynamics, yet individuals often do not base their decisions solely on peer pressure or social influence. In various settings, strategic reasoning comes into play, particularly when individuals aim to maximize a benefit or mitigate a cost. This logic is well represented in biology, where certain traits prevail due to their evolutionary advantage, and in social systems where decision-making hinges on evaluating outcomes across different scenarios.

In traditional models, interactions are commonly dyadic, with agents choosing between strategies such as cooperation (C) or defection (D). The payoffs from these interactions are captured in a standard matrix, where R represents the reward for mutual cooperation, T the temptation to defect, S the sucker's payoff, and P the punishment for mutual defection. The relative ordering of these payoffs defines the nature of the game. One of the most well-known dilemmas is the Prisoner's Dilemma, characterized by the ordering $T > R > P > S$, where mutual defection becomes the Nash equilibrium despite being suboptimal for both players. This game illustrates the challenges of promoting cooperation in self-interested populations. Alternative scenarios include the Stag Hunt game, in which cooperation yields the highest payoff but requires mutual commitment. This game has two pure Nash equilibria—one cooperative and one risk-dominant. Another notable example is the Snowdrift (or Chicken) game, where cooperation by at least one individual is necessary for a task to be completed. In this case, the best response depends on the opponent's strategy, leading to anti-coordination and the emergence of two equilibria where each player adopts opposing strategies.

Originally conceived as static, these games were extended into dynamic frameworks by evolutionary game theory, a field pioneered by Maynard Smith. In a well-mixed population, the temporal evolution of cooperative behavior can be modeled via replicator dynamics:

$$\dot{x}_c = x_c(1 - x_c)[\bar{\pi}_C - \bar{\pi}_D],$$

where $\bar{\pi}_C$ and $\bar{\pi}_D$ denote the average payoffs of cooperators and defectors, respectively.

Simulations play a crucial role in exploring these dynamics, especially when interactions are constrained by network structures. In agent-based models, strategy updates can occur through best-response dynamics or imitation mechanisms influenced by payoff differences. A landmark result by Nowak and May showed that spatial structure—modeled using a square lattice—could foster cooperation through network reciprocity: the reinforcement of cooperative ties due to repeated interactions [294]. Further studies revealed that scale-free or clustered topologies enhance this effect by concentrating influence in cooperative hubs or cohesive groups. Nonetheless, experimental investigations involving real human subjects often fail to reproduce such cooperation levels [295], highlighting the complexity of translating theoretical models into empirical results.

In contrast to the Prisoner's Dilemma, the Snowdrift game's payoff structure allows for stable coexistence between cooperators and defectors. Early computational studies, however, suggested that spatial structure might hinder cooperation in this setting. Despite this, the integration of evolutionary game theory with network science has flourished in recent years, leading to a wealth of new insights into the dynamics of strategic behavior in higher-order systems [105].

2.9 Conclusion

This chapter has explored the structural and dynamic foundations of complex networks, offering a theoretical and methodological overview of the models that describe both natural and engineered systems. Beginning with classical frameworks such as random graphs and small-world networks, and progressing through scale-free, multilayer, and higher-order representations, we have traced how network topologies influence the behavior, efficiency, and robustness of complex systems. By emphasizing the importance of metrics such as degree distribution, clustering coefficient, path length, betweenness centrality, and modularity, the chapter has illustrated how the architecture of a network shapes the dynamics that unfold upon it—whether it be the diffusion of information, the spread of contagion, or the emergence of coordination and cooperation.

A key insight is that structure and dynamics are inherently linked: topology not only constrains but also enables certain behaviors, while dynamic processes can in turn reshape the structure through adaptive feedback, rewiring, or co-evolutionary mechanisms. This duality is particularly evident in multilayer and higher-order networks, where interdependencies across layers and group interactions extend beyond the capabilities of pairwise models. Throughout the discussion, the focus has remained on the relevance of these models for communication networks. As mobile and wireless infrastructures become increasingly complex, layered, and adaptive, the use of complexity-aware models becomes not only beneficial but necessary. The application of network science in this domain allows for more robust design, more effective resource allocation, and a deeper understanding of system-level vulnerabilities and opportunities.

The theoretical framework established here serves as a basis for the following chapters, where we will move from abstract network properties to applied contexts. In particular, Chap. 3 will explore how the principles of complexity can impact and inspire the design of communication architectures such as network-of-networks, digital twins, and AI-native 6G systems, and how they can be harnessed to design more adaptive, resilient, and intelligent infrastructures.

References

1. Bianconi, G., Arenas, A., Biamonte, J., Carr, L.D., Kahng, B., Kertész, J., Kurths, J., Lü, L., Masoller, C., Motter, A.E., et al.: Complex systems in the spotlight: next steps after the 2021 Nobel Prize in Physics. J. Phys. Complex. **4**(1), 010201 (2023)
2. Parisi, G.: Order parameter for spin-glasses. Phys. Rev. Lett. **50**(24), 1946 (1983)
3. Carlson, J.M., Doyle, J.: Complexity and robustness. Proc. Natl. Acad. Sci. U.S.A. **99**(Suppl 1), 2538–2545 (2002)
4. Radicchi, F., Fortunato, S., Markines, B., et al.: Universality of citation distributions: toward an objective measure of scientific impact. Proc. Natl. Acad. Sci. U.S.A. **105**(45), 17268–17272 (2008)
5. Bjerknes, V.: The problem of weather prediction, considered from the viewpoint of mechanics and physics. Meteorol. Z. **21**, 1–7 (1904)
6. Richardson, L.F.: Weather Prediction by Numerical Process. Cambridge University Press (1922)
7. Biamonte, J., Faccin, M., De Domenico, M.: Complex networks from classical to quantum. Commun. Phys. **1**, 81 (2017)
8. Scatà, M., Attanasio, B., Aiosa, G.V., La Corte, A.: The dynamical interplay of collective attention, awareness and epidemics spreading in the multiplex social networks during COVID-19. IEEE Access **8**, 189203–189223 (2020)
9. Batty, M.: The New Science of Cities. MIT Press (2013)
10. Sergiou, C., Lestas, M., Antoniou, P., Liaskos, C., Pitsillides, A.: Complex systems: a communication networks perspective towards 6G. IEEE Access **8**, 89007–89030 (2020)
11. Newman, M.: Resource letter CS-1: complex systems. Am. J. Phys. **79**, 800–810 (2011)
12. Thurner, S., Hanel, R., Klimek, P.: Introduction to the Theory of Complex Systems. Oxford University Press, Oxford (2018)
13. Schueler, G.J.: The unpredictability of complex systems. J. Wash. Acad. Sci. **84**, 3–12 (1996)
14. Jiang, W., Han, B., Habibi, M.A., Schotten, H.D.: The road towards 6G: a comprehensive survey. IEEE Open J. Commun. Soc. **2**, 334–366 (2021)

15. Artime, O., et al.: The origin of bursts and heavy tails in human dynamics. Nat. Hum. Behav. **6**, 1188–1199 (2022)
16. Batty, M.: Cities as complex systems: scaling, interaction, networks, dynamics and urban morphologies. In: Meyers, R.A. (ed.) Encyclopedia of Complexity and Systems Science, pp. 1041–1071. Springer, New York (2009)
17. Mitchell, M.: Complex systems: network thinking. Artif. Intell. **170**(18), 1194–1212 (2006)
18. Górriz, J.M., Álvarez-Illán, I., Álvarez-Marquina, A., et al.: Computational approaches to explainable artificial intelligence: advances in theory, applications and trends. Inf. Fusion **100**, 101945 (2023)
19. Ye, J.C.: Geometry of Deep Learning. Springer, Berlin (2022)
20. Menche, J., Sharma, A., Kitsak, M., et al.: Uncovering disease–disease relationships through the incomplete interactome. Science **347**(6224), 1257601 (2015)
21. Chinazzi, M., Davis, J.T., Ajelli, M., et al.: The effect of travel restrictions on the spread of the 2019 novel coronavirus (COVID-19) outbreak. Science **368**(6489), 395–400 (2020)
22. Lom, M., Pribyl, O.: Smart city model based on systems theory. Int. J. Inf. Manag. **56**, 102092 (2021)
23. Oughton, E.J., Usher, W., Tyler, P., Hall, J.W.: Infrastructure as a complex adaptive system. Complexity **1**, 3427826 (2018)
24. Mitchell, M.: Artificial Intelligence: A Guide for Thinking Humans. Farrar, Straus and Giroux, New York (2019)
25. Berry, B.J.L., Kiel, L.D., Elliott, E.: Adaptive agents, intelligence, and emergent human organization: capturing complexity through agent-based modeling. Proc. Natl. Acad. Sci. U.S.A. **99**(Suppl 3), 7187–7188 (2002)
26. Estrada, E.: Complex Networks: An Algorithmic Perspective. Springer, Cham (2024)
27. Nowzari, C., Preciado, V.M., Pappas, G.J.: Analysis and control of epidemics: a survey of spreading processes on complex networks. IEEE Control Syst. Mag. **36**(1), 26–46 (2016)
28. Pastor-Satorras, R., Vespignani, A.: Epidemic spreading in scale-free networks. Phys. Rev. Lett. **86**(14), 3200–3203 (2001)
29. Railsback, S.F., Grimm, V.: Agent-Based and Individual-Based Modeling: A Practical Introduction. Princeton University Press, Princeton (2019)
30. Tang, Y., Kurths, J., Lin, W., Ott, E., Kocarev, L.: Introduction to focus issue: when machine learning meets complex systems: networks, chaos, and nonlinear dynamics. Chaos **30**(6) (2020)
31. Mitchell, M.: Complexity: A Guided Tour. Oxford University Press, Oxford (2009)
32. Boccara, N.: Modeling Complex Systems. Springer, New York (2004)
33. Miller, J.H., Page, S.E.: Complex Adaptive Systems: An Introduction to Computational Models of Social Life. Princeton University Press, Princeton (2009)
34. Holland, J.H.: Signals and Boundaries: Building Blocks for Complex Adaptive Systems. MIT Press, Cambridge (2012)
35. Barabási, A.L.: Network Science. Cambridge University Press, Cambridge (2016)
36. Newman, M.E.J.: The structure and function of complex networks. SIAM Rev. **45**(2), 167–256 (2003)
37. Boccaletti, S., Bianconi, G., Criado, R., et al.: The structure and dynamics of multilayer networks. Phys. Rep. **544**(1), 1–122 (2014)
38. Scatà, M., Attanasio, B., La Corte, A.: Cognitive profiling of nodes in 6G through multiplex social network and evolutionary collective dynamics. Future Internet **13**(5), 135 (2021)
39. De Domenico, M.: Multilayer modeling and analysis of complex systems. Nat. Phys. **18**, 769–779 (2022)
40. Strogatz, S.H.: Sync: the emerging science of spontaneous order. Hyperion. New York (2003)
41. Peitgen, H.O., Jürgens, H., Saupe, D.: Chaos and Fractals: New Frontiers of Science. Springer, New York (2004)
42. Ott, E., Grebogi, C., Yorke, J.A.: Controlling chaos. Phys. Rev. Lett. **64**(11), 1196–1199 (1990)

43. May, R.M.: Simple mathematical models with very complicated dynamics. Nature **261**, 459–467 (1976)
44. Wolfram, S.: A New Kind of Science. Wolfram Media, Champaign (2002)
45. Langton, C.G.: Studying artificial life with cellular automata. Physica D **22**(1–3), 120–149 (1986)
46. Barabási, A.L., Albert, R.: Emergence of scaling in random networks. Science **286**(5439), 509–512 (1999)
47. Bak, P., Tang, C., Wiesenfeld, K.: Self-organized criticality: an explanation of 1/f noise. Phys. Rev. Lett. **59**(4), 381–384 (1987)
48. Newman, M.E.J.: Power laws, Pareto distributions and Zipf's law. Contemp. Phys. **46**(5), 323–351 (2005)
49. Koza, J.R.: Genetic Programming: On the Programming of Computers by Means of Natural Selection. MIT Press, Cambridge (1992)
50. Axelrod, R.: The Evolution of Cooperation. Basic Books, New York (2006)
51. Gintis, H.: The Bounds of Reason: Game Theory and the Unification of the Behavioral Sciences. Princeton University Press, Princeton (2009)
52. Di Stefano, A., Scatà, M., Attanasio, B., La Corte, A., Lió, P., Das, S.K.: A novel methodology for designing policies in mobile crowdsensing systems. Pervasive Mob. Comput. **67**, 101230 (2020)
53. Attanasio, B., La Corte, A., Scatà, M.: Evolutionary dynamics of MEC's organization in a 6G scenario through EGT and temporal multiplex social network. ICT Express **7**(2), 138–142 (2021)
54. Cover, T.M., Thomas, J.A.: Elements of Information Theory. Wiley, New York (1991)
55. Bennett, C.H.: How to define complexity in physics, and why. In: Zurek, W.H. (ed.) Complexity, Entropy and the Physics of Information, pp. 137–148. Addison-Wesley, Redwood City (1990)
56. Moore, C., Mertens, S.: The Nature of Computation. Oxford University Press, Oxford (2011)
57. Sipser, M.: Introduction to the Theory of Computation, 2nd edn. Thomson, Boston (2006)
58. Schelling, T.C.: Dynamic models of segregation. J. Math. Sociol. **1**(2), 143–186 (1971)
59. Palmer, T.N., Anderson, D.L.T., Brunet, G., et al.: The ECMWF 'ensemble prediction' system. ECMWF Newsletter **65**, 2–10 (1994)
60. Gilbert, N.: Agent-Based Models. Sage Publications, London (2019)
61. Gilbert, N.: Emergence in social simulation. In: Artificial Societies, pp. 134–143. Routledge, London (2006)
62. Bianconi, G., Barabási, A.L.: Bose-Einstein condensation in complex networks. Phys. Rev. Lett. **86**(24), 5632–5635 (2001)
63. Bianconi, G.: Entropy of network ensembles. Phys. Rev. E **66**(5), 056123 (2002)
64. De Domenico, M., Solé-Ribalta, A., Cozzo, E., et al.: Mathematical formulation of multilayer networks. Phys. Rev. X **3**(4), 041022 (2013)
65. Holme, P., Saramäki, J.: Temporal networks. Phys. Rep. **519**(3), 97–125 (2012)
66. Bianconi, G.: Higher-Order Networks. Cambridge University Press (2021)
67. Haldane, A.G., May, R.M.: Systemic risk in banking ecosystems. Nature **469**, 351–355 (2011)
68. Pastor-Satorras, R., Castellano, C., Van Mieghem, P., Vespignani, A.: Epidemic processes in complex networks. Rev. Mod. Phys. **87**(3), 925–979 (2015)
69. Sporns, O.: Networks of the Brain. MIT Press, Cambridge (2016)
70. Boers, N.: Observation-based early-warning signals for a collapse of the Atlantic Meridional Overturning Circulation. Nat. Clim. Change **11**, 680–688 (2021)
71. Amani, A.M., Jalili, M.: Power grids as complex networks: resilience and reliability analysis. IEEE Access **9**, 119010–119031 (2021)
72. Masoller, C.: Noise-induced resonance in delayed feedback systems. Phys. Rev. Lett. **88**(3), 034102 (2002)
73. Motter, A.E., Zhou, C., Kurths, J.: Network synchronization, diffusion, and the paradox of heterogeneity. Phys. Rev. E **71**(1), 016116 (2005)
74. Helbing, D.: Globally networked risks and how to respond. Nature **497**(7447), 51–59 (2013)

75. Schadt, E.E.: Molecular networks as sensors and drivers of common human diseases. Nature **461**(7261), 218–223 (2009)
76. Alber, M., Buganza Tepole, A., Cannon, W., De, S., Dura-Bernal, S., Garikipati, K., Karniadakis, G., Lytton, W.W., Perdikaris, P., Petzold, L., Kuhl, E.: Integrating machine learning and multiscale modeling: perspectives, challenges, and opportunities in the biological, biomedical, and behavioral sciences. NPJ Digit. Med. **2**, 115 (2019)
77. Dura-Bernal, S., Suter, B.A., Gleeson, P., Cantarelli, M., Quintana, A., Rodriguez, F., Kedziora, D.J., Lytton, W.W.: NetPyNE, a tool for data-driven multiscale modeling of brain circuits. eLife **8**, e44494 (2019)
78. Arenas, A., Díaz-Guilera, A., Kurths, J., Moreno, Y., Zhou, C.: Synchronization in complex networks. Phys. Rep. **469**(3), 93–153 (2008)
79. Perseguers, S., Lewenstein, M., Acín, A., Cirac, J.I.: Quantum random networks. Nat. Phys. **6**(7), 539–543 (2010)
80. Ablowitz, M.J., Been, J.B., Carr, L.D.: Fractional integrable nonlinear soliton equations. Phys. Rev. Lett. **128**(18), 184101 (2022)
81. Lee, Y., Lee, J., Oh, S.M., Lee, D., Kahng, B.: Homological percolation transitions in growing simplicial complexes. Chaos **31**(4) (2021)
82. Evans, D.J., Searles, D.J.: The fluctuation theorem. Adv. Phys. **51**(7), 1529–1585 (2002)
83. Allen, R.J., Valeriani, C., ten Wolde, P.R.: Forward flux sampling for rare event simulations. J. Phys.: Condens. Matter **21**(46), 463102 (2009)
84. Vasilyeva, E., Kozlov, A., Alfaro-Bittner, K., Musatov, D., Raigorodskii, A.M., Perc, M., Boccaletti, S.: Multilayer representation of collaboration networks with higher-order interactions. Sci. Rep. **11**, 5666 (2021)
85. Kertész, J., Török, J., Murase, Y., Jo, H.H., Kaski, K.: Modeling the complex network of social interactions. In: Pathways Between Social Science and Computational Social Science: Theories, Methods, and Interpretations. Springer, pp. 3–19 (2021)
86. Danziger, M.M., Bonamassa, I., Boccaletti, S., Havlin, S.: Dynamic interdependence and competition in multilayer networks. Nat. Phys. **15**(2), 178–185 (2019)
87. Runge, J., Nowack, P., Kretschmer, M., Flaxman, S., Sejdinovic, D.: Detecting and quantifying causal associations in large nonlinear time series datasets. Sci. Adv. **5**(11), eaau4996 (2019)
88. Lü, L., Zhang, Y.C., Yeung, C.H., Zhou, T.: Leaders in social networks, the delicious case. PLoS ONE **6**(6), e21202 (2011)
89. Törnberg, P.: Echo chambers and viral misinformation: modeling fake news as complex contagion. PLoS ONE **13**(9), e0203958 (2018)
90. Motter, A.E., Myers, S.A., Anghel, M., Nishikawa, T.: Spontaneous synchrony in power-grid networks. Nat. Phys. **9**, 191–197 (2013)
91. De Vincenzo, I., Massari, G.F., Giannoccaro, I., Carbone, G., Grigolini, P.: Mimicking the collective intelligence of human groups as an optimization tool for complex problems. Chaos, Solitons Fractals **110**, 259–266 (2018)
92. Radicchi, F.: Percolation in real interdependent networks. Nat. Phys. **11**(7), 597–602 (2015)
93. Guimerà, R., Reichardt, I., Aguilar-Mogas, A., Massucci, F.A., Miranda, M., Pallarès, J., Sales-Pardo, M.: A Bayesian machine scientist to aid in the solution of challenging scientific problems. Sci. Adv. **6**(5), eaav6971 (2020)
94. Gatrell, A.C.: Complexity theory and geographies of health: a critical assessment. Soc. Sci. Med. **72**(12), 1929–1930 (2011)
95. Brambilla, M., Ferrante, E., Birattari, M., Dorigo, M.: Swarm robotics: a review from the swarm engineering perspective. Swarm Intell. **7**(1), 1–41 (2013)
96. Scatà, M., Di Stefano, A., La Corte, A., Liò, P.: A multiplex social contagion dynamics model to shape and discriminate D2D content dissemination. IEEE Trans. Cogn. Commun. Netw. (2020)
97. Mathews, L.G., Jones, A.: Using systems thinking to improve interdisciplinary learning outcomes: reflections on a pilot study in land economics. Issues Integr. Stud. **26**, 73–104 (2008)
98. Byrne, D., Callaghan, G.: Complexity Theory and the Social Sciences: The State of the Art. Routledge, London (2022)

99. Artime, O., Grassia, M., Domenico, M., Gleeson, J.P., Makse, H.A., Mangioni, G., Perc, M., Radicchi, F.: Robustness and resilience of complex networks. Nat. Rev. Phys. **6**(2), 114–131 (2024)
100. Gómez, S., Arenas, A., Borge-Holthoefer, J., Meloni, S., Moreno, Y.: Discrete-time Markov chain approach to contact-based disease spreading in complex networks. EPL **89**, 38009 (2010)
101. Walschaers, M., Sundar, B., Treps, N., Carr, L.D., Parigi, V.: Emergent complex quantum networks in continuous-variables non-Gaussian states. Quantum Sci. Technol. **8**(3), 035009 (2023)
102. Luna-Pla, I., Nicolás-Carlock, J.R.: Corruption and complexity: a scientific framework for the analysis of corruption networks. Appl. Netw. Sci. **5**(1), 13 (2020)
103. Sastry, S.: Nonlinear Systems: Analysis, Stability, and Control, vol. 10. Springer (2013)
104. Motter, A.E.: Cascade control and defense in complex networks. Phys. Rev. Lett. **93**, 098701 (2004)
105. Perc, M., Jordan, J.J., Rand, D.G., et al.: Statistical physics of human cooperation. Phys. Rep. **687**, 1–51 (2017)
106. Radicchi, F., Castellano, C.: Breaking of the site-bond percolation universality in networks. Nat. Commun. **6**, 10196 (2015)
107. Landi, P., Minoarivelo, H.O., Brännström, Å., Hui, C., Dieckmann, U.: Complexity and stability of ecological networks: a review of the theory. Popul. Ecol. **60**(4), 319–345 (2018)
108. Artime, O., De Domenico, M.: From the origin of life to pandemics: emergent phenomena in complex physical and socio-technical systems. Phil. Trans. R. Soc. A **380**(2214), 20210365 (2022)
109. Arenas, A., et al.: Modeling the spatiotemporal epidemic spreading of COVID-19 and the impact of mobility and social distancing interventions. Phys. Rev. X **10**(4), 041055 (2020)
110. Lee, D., Jo, H.-H., Kahng, B.: Homological percolation transitions in growing simplicial complexes. Chaos: Interdiscip. J. Nonlinear Sci. **31**(4), 043109 (2021)
111. Reisch, T., Heiler, G., Diem, C., Klimek, P., Thurner, S.: Monitoring supply networks from mobile phone data for estimating the systemic risk of an economy. Sci. Rep. **12**(1), 18206 (2022)
112. Hasselmann, K.: Stochastic climate models Part I: theory. Tellus **28**, 473–485 (1976)
113. Manabe, S., Wetherald, R.T.: Thermal equilibrium of the atmosphere with a given distribution of relative humidity. J. Atmos. Sci. **24**(3), 241–259 (1967)
114. Watts, D.J.: Everything is Obvious: How Common Sense Fails Us. Crown Publishing Group (2011)
115. Deza, J.I., Barreiro, M., Masoller, C.: Inferring interdependencies in climate networks constructed at inter-annual, intra-season and longer time scales. Eur. Phys. J. Spec. Top. **222** (2013)
116. Nishikawa, T., Menck, P.J., Motter, A.E.: Comparative analysis of existing models for power-grid synchronization. New J. Phys. **19**, 015012 (2017)
117. Radicchi, F., Fortunato, S.: Explosive percolation in scale-free networks. Phys. Rev. Lett. **103**, 168701 (2009)
118. Makridakis, S., Hogarth, R.M., Gaba, A.: Forecasting and uncertainty in the economic and business world. Int. J. Forecast. **25**(4), 794–812 (2009)
119. San Miguel, M., Johnson, J.H., Kertész, J., Kaski, K., Díaz-Guilera, A., MacKay, R.S., Loreto, V., Érdi, P., Helbing, D.: Challenges in complex systems science. Eur. Phys. J. Spec. Top. **214**, 245–271 (2012)
120. Margetts, H., John, P., Hale, S.A., Yasseri, T.: Political Turbulence: How Social Media Shape Collective Action. Princeton University Press, Princeton (2016)
121. Scatà, M., Di Stefano, A., La Corte, A., Liò, P.: Quantifying the propagation of distress and mental disorders in social networks. Sci. Rep. **8**(1), 5005 (2018)
122. Bollobás, B.: Modern Graph Theory. Springer, New York (2001)
123. Broder, A., et al.: Graph structure in the web. Comput. Netw. **33**(1–6), 309–320 (2000)

124. Pavlopoulos, G.A., Secrier, M., Moschopoulos, C.N., Soldatos, T.G., Kossida, S., Aerts, J., Schneider, R., Bagos, P.G.: Using graph theory to analyze biological networks. BioData Min. **4**, 10 (2011)
125. Van Steen, M.: Graph Theory and Complex Networks. An Introduction, vol. 144(1) (2010)
126. Boccaletti, S., Latora, V., Moreno, Y., Chavez, M., Hwang, D.-U.: Complex networks: structure and dynamics. Phys. Rep. **424**(4–5), 175–308 (2006)
127. Alexanderson, G.: About the cover: Euler and Königsberg's Bridges: a historical view. Bull. Am. Math. Soc. **43**(4), 567–573 (2006)
128. Freeman, L.: The development of social network analysis. A Study Sociol. Sci. **1**(687), 159–167 (2004)
129. Barabási, A.-L.: Emergence of scaling in complex networks. In Handbook of Graphs and Networks: From the Genome to the Internet. Wiley-VCH, Berlin (2003)
130. Watts, D.J., Strogatz, S.H.: Collective dynamics of 'small-world' networks. Nature **393**, 440–442 (1998)
131. Barabási, A.-L., Bonabeau, E.: Scale-free networks. Sci. Am. **288**(5), 50–59 (2003)
132. Cohen, R., Erez, K., Ben-Avraham, D., Havlin, S.: Resilience of the Internet to random breakdowns. Phys. Rev. Lett. **85**, 4626 (2000)
133. Bollobás, B.: Random Graphs. Springer, Berlin (1998)
134. Newman, M.: Networks. Oxford University Press, Oxford (2018)
135. West, D.B.: Introduction to Graph Theory, 2nd edn. Prentice Hall, Upper Saddle River (2001)
136. Albert, R., Barabási, A.L.: Statistical mechanics of complex networks. Rev. Mod. Phys. **74**, 47–97 (2002)
137. Kitsak, M., Ganin, A., Elmokashfi, A., et al.: Finding shortest and nearly shortest path nodes in large substantially incomplete networks by hyperbolic mapping. Nat. Commun. **14**, 186 (2023)
138. Latora, V., Marchiori, M.: Efficient behavior of small-world networks. Phys. Rev. Lett. **87**, 198701 (2001)
139. Freeman, L.C.: A set of measures of centrality based on betweenness. Sociometry **40**(1), 35–41 (1977)
140. Girvan, M., Newman, M.E.J.: Community structure in social and biological networks. Proc. Natl. Acad. Sci. U.S.A. **99**, 7821–7826 (2002)
141. Cormen, T.H., Leiserson, C.E., Rivest, R.L., Stein, C.: Introduction to Algorithms, 4th edn. MIT Press, Cambridge (2022)
142. Brandes, U.: A faster algorithm for betweenness centrality. J. Math. Sociol. **25**, 163–177 (2001)
143. Wasserman, S., Faust, K.: Social Network Analysis: Methods and Applications. Cambridge University Press, Cambridge (1994)
144. Barrat, A., Barthélemy, M., Pastor-Satorras, R., Vespignani, A.: The architecture of complex weighted networks. Proc. Natl. Acad. Sci. U.S.A. **101**, 3747–3752 (2004)
145. Palla, G., Derényi, I., Farkas, I., Vicsek, T.: Uncovering the overlapping community structure of complex networks in nature and society. Nature **435**, 814–818 (2005)
146. Alon, U.: Network motifs: theory and experimental approaches. Nat. Rev. Genet. **8**, 450–461 (2007)
147. Milo, R., Shen-Orr, S., Itzkovitz, S., et al.: Network motifs: simple building blocks of complex networks. Science **298**, 824–827 (2002)
148. Ingram, P.J., Stumpf, M.P.H., Stark, J.: Network motifs: structure does not determine function. BMC Genomics **7**, 108 (2006)
149. Girvan, M., Newman, M.E.J.: Community structure in social and biological networks. Proc. Natl. Acad. Sci. U.S.A. **99**, 7821–7826 (2002)
150. Fortunato, S.: Community detection in graphs. Phys. Rep. **486**, 75–174 (2010)
151. Chung, F.R.K.: Spectral Graph Theory. American Mathematical Society, Providence (1997)
152. Farkas, I.J., Derényi, I., Barabási, A.L., Vicsek, T.: Spectra of "real-world" graphs: beyond the semicircle law. Phys. Rev. E **64**, 026704 (2001)

153. Jeong, H., Tombor, B., Albert, R., Oltvai, Z.N., Barabási, A.L.: The large-scale organization of metabolic networks. Nature **407**, 651–654 (2000)
154. Erdős, P., Rényi, A.: On random graphs. Publ. Math. **6**, 290–297 (1959)
155. Bender, E., Canfield, E.: The asymptotic number of labeled graphs with given degree sequences. J. Comb. Theory Ser. A **24**, 296–307 (1978)
156. Molloy, M., Reed, B.: A critical point for random graphs with a given degree sequence. Random Struct. Algorithms **6**, 161–180 (1995)
157. Newman, M., Strogatz, S., Watts, D.: Random graphs with arbitrary degree distributions and their applications. Phys. Rev. E **64**, 026118 (2001)
158. Milgram, S.: The small-world problem. Psychol. Today **2**, 60–67 (1967)
159. Dodds, P.S., Muhamad, R., Watts, D.J.: An experimental study of search in global social networks. Science **301**, 827–829 (2003)
160. Barrat, A., Weigt, M.: On the properties of small-world network models. Eur. Phys. J. B **13**(3), 547–560 (2000)
161. Monasson, R.: Diffusion, localization and dispersion relations on small-world lattices. Eur. Phys. J. B **12**, 555–567 (1999)
162. Pastor-Satorras, R., Vázquez, A., Vespignani, A.: Dynamical and correlation properties of the internet. Phys. Rev. Lett. **87**, 258701 (2001)
163. Clauset, A., Shalizi, C.R., Newman, M.E.J.: Power-law distributions in empirical data. SIAM Rev. **51**, 661–703 (2009)
164. Aiello, W., Chung, F., Lu, L.: A random graph model for massive graphs. In: Proceedings of the 32nd Annual ACM Symposium on Theory of Computing, pp. 171–180 (2000)
165. Goh, K.I., Kahng, B., Kim, D.: Universal behavior of load distribution in scale-free networks. Phys. Rev. Lett. **87**, 278701 (2001)
166. Caldarelli, G., Capocci, A., De Los, R.P., Muñoz, M.A.: Scale-free networks from varying vertex intrinsic fitness. Phys. Rev. Lett. **89**, 258702 (2002)
167. Toroczkai, Z.: Why do scale-free networks emerge in nature? From gradient networks to transport efficiency. APS March Meet. Abstr. **2004**, V18–007 (2004)
168. Dorogovtsev, S.N., Mendes, J.F.F., Samukhin, A.N.: Structure of growing networks with preferential linking. Phys. Rev. Lett. **85**, 4633 (2000)
169. Krapivsky, P.L., Redner, S., Leyvraz, F.: Connectivity of growing random networks. Phys. Rev. Lett. **85**, 4629 (2000)
170. Ravasz, E., Barabási, A.L.: Hierarchical organization in complex networks. Phys. Rev. E **67**, 026112 (2003)
171. Klemm, K., Eguiluz, V.M.: Highly clustered scale-free networks. Phys. Rev. E **65**, 036123 (2002)
172. Solé, R.V., Pastor-Satorras, R., Smith, E., Kepler, T.: A model of large-scale proteome evolution. Adv. Complex Syst. **5**, 43–54 (2002)
173. Vázquez, A., Pastor-Satorras, R., Vespignani, A.: Large-scale topological and dynamical properties of the Internet. Phys. Rev. E **67**, 046111 (2003)
174. Onnela, J.P., Saramäki, J., Kertész, J., Kaski, K.: Intensity and coherence of motifs in weighted complex networks. Phys. Rev. E **71**, 065103(R) (2005)
175. Menichetti, G., Remondini, D., Panzarasa, P., Mondragón, R.J., Bianconi, G.: Weighted multiplex networks. PLoS ONE **9**(6), e97857 (2014)
176. Gastner, M.T., Newman, M.E.J.: The spatial structure of networks. Eur. Phys. J. B **49**, 247–252 (2006)
177. Yook, S.H., Jeong, H., Barabási, A.L.: Modeling the Internet's large-scale topology. Proc. Natl. Acad. Sci. U.S.A. **99**, 13382–13386 (2002)
178. Barthélemy, M.: Spatial networks. Phys. Rep. **499**, 1–101 (2011)
179. Albert, R., Jeong, H., Barabási, A.L.: Error and attack tolerance of complex networks. Nature **406**, 378–382 (2000)
180. Callaway, D.S., Newman, M.E.J., Strogatz, S.H., Watts, D.J.: Network robustness and fragility: percolation on random graphs. Phys. Rev. Lett. **85**, 5468 (2000)

181. Cohen, R., Erez, K., ben-Avraham D, Havlin, S.: Breakdown of the Internet under intentional attack. Phys. Rev. Lett. **86**, 3682 (2001)
182. Stauffer, D., Aharony, A.: Introduction to Percolation Theory. Taylor & Francis, London (2018)
183. Cohen, R., ben-Avraham, D., Havlin, S.: Percolation critical exponents in scale-free networks. Phys. Rev. E **66**, 036113 (2002)
184. Molloy, M., Reed, B.: The size of the giant component of a random graph with a given degree sequence. Comb. Probab. Comput. **7**, 295–305 (1998)
185. Vázquez, A., Moreno, Y.: Resilience to damage of graphs with degree correlations. Phys. Rev. E **67**, 015101 (2003)
186. Boguñá, M., Pastor-Satorras, R., Vespignani, A.: Epidemic spreading in complex networks with degree correlations. Lect. Notes Phys. **625**, 127–147 (2003)
187. Dorogovtsev, S.N., Mendes, J.F.F.: Evolution of Networks: From Biological Nets to the Internet and WWW. OUP Oxford, Oxford (2013)
188. Xulvi-Brunet, R., Sokolov, I.M.: Reshuffling scale-free networks: from random to assortative. Phys. Rev. E **70**, 066102 (2004)
189. Hobbs, W.R., Burke, M.K.: Connective recovery in social networks after the death of a friend. Nat. Hum. Behav. **1**(5), 0092 (2017)
190. Anderson, R.M., May, R.M.: Infectious Diseases of Humans: Dynamics and Control. Oxford University Press, Oxford (1991)
191. Vespignani, A.: Modelling dynamical processes in complex socio-technical systems. Nat. Phys. **8**, 32–39 (2012)
192. Kampen, N.G.: Stochastic Processes in Physics and Chemistry, 3rd edn. Elsevier, Amsterdam (2007)
193. Eubank, S., Guclu, H., Kumar, V.S., Marathe, M.V., Srinivasan, A., Toroczkai, Z., Wang, N.: Modelling disease outbreaks in realistic urban social networks. Nature **429**, 180–184 (2004)
194. Ferguson, N.M., Cummings, D.A., Fraser, C., Cajka, J.C., Cooley, P.C., Burke, D.S.: Strategies for mitigating an influenza pandemic. Nature **442**, 448–452 (2006)
195. Balcan, D., Colizza, V., Gonçalves, B., Hu, H., Ramasco, J.J., Vespignani, A.: Multiscale mobility networks and the spatial spreading of infectious diseases. Proc. Natl. Acad. Sci. U.S.A. **106**, 21484–21489 (2009)
196. Keeling, M.J., Rohani, P.: Modeling Infectious Diseases in Humans and Animals. Princeton University Press, Princeton (2008)
197. Riley, S.: Large-scale spatial-transmission models of infectious disease. Science **316**, 1298–1301 (2007)
198. Vespignani, A.: Predicting the behavior of techno-social systems. Science **325**, 425–428 (2009)
199. Butts, C.T.: Revisiting the foundations of network analysis. Science **325**, 414–416 (2009)
200. Albert, R., Barabási, A.L.: Statistical mechanics of complex networks. Rev. Mod. Phys. **74**, 47–97 (2002)
201. Caldarelli, G.: Scale-Free Networks: Complex Webs in Nature and Technology. Oxford University Press, Oxford (2007)
202. Danon, L., Ford, A.P., House, T., Jewell, C.P., Keeling, M.J., Roberts, G.O., Ross, J.V., Vernon, M.C.: Networks and the epidemiology of infectious disease. Interdiscip. Perspect. Infect. Dis. **2011**, 284909 (2011)
203. Keeling, M.J., Eames, K.T.D.: Networks and epidemic models. J. R. Soc. Interface **2**, 295–307 (2005)
204. Andersson, H., Britton, T.: Stochastic Epidemic Models and Their Statistical Analysis. Springer, New York (2012)
205. Brauer, F., Castillo-Chavez, C.: Mathematical Models in Population Biology and Epidemiology. Springer, New York (2012)
206. Diekmann, O., Heesterbeek, H., Britton, T.: Mathematical Tools for Understanding Infectious Disease Dynamics. Princeton University Press, Princeton (2013)
207. Ross, S.M.: Stochastic Processes, 2nd edn. Wiley, New York (1995)

208. Hethcote, H.W.: The mathematics of infectious diseases. SIAM Rev. **42**, 599–653 (2000)
209. Domb, C.: Phase Transitions and Critical Phenomena, vol. 19. Elsevier (2000)
210. Marro, J., Dickman, R.: Nonequilibrium Phase Transitions in Lattice Models. Cambridge University Press, Cambridge (2005)
211. Henkel, M., Hinrichsen, H., Lübeck, S.: Non-equilibrium Phase Transitions, vol. 1. Springer, Dordrecht (2008)
212. Yeomans, J.M.: Statistical Mechanics of Phase Transitions. Oxford University Press, Oxford (1992)
213. Harris, T.E.: Contact interactions on a lattice. Ann. Probab. **2**, 969–988 (1974)
214. Grassberger, P.: On the critical behavior of the general epidemic process and dynamical percolation. Math. Biosci. **63**, 157–172 (1983)
215. Bar-Yam, Y.: Multiscale variety in complex systems. Complexity **9**(4), 37–45 (2004)
216. Gershenson, C., Fernández, N.: Complexity and information: measuring emergence, self-organization, and homeostasis at multiple scales. Complexity **18**(2), 29–44 (2012)
217. Lloyd, S.: Measures of complexity: a nonexhaustive list. IEEE Control Syst. Mag. **21**(4), 7–8 (2001)
218. Prokopenko, M., Boschetti, F., Ryan, A.: An information-theoretic primer on complexity, self-organization and emergence. Complexity **15**(1), 11–28 (2009)
219. Shannon, C.E.: A mathematical theory of communication. Bell Syst. Tech. J. **27**, 379–423 (1948)
220. Edmonds, B.: Syntactic measures of complexity. PhD Thesis, University of Manchester (1999)
221. Gell-Mann, M., Lloyd, S.: Information measures, effective complexity, and total information. Complexity **2**(1), 44–52 (1996)
222. Langton, C.G.: Computation at the edge of chaos: phase transitions and emergent computation. Physica D **42**(1–3), 12–37 (1990)
223. Wuensche, A.: Classifying cellular automata automatically: finding gliders, filtering, and relating space-time patterns, attractor basins, and the Z parameter. Complexity **4**(3), 47–66 (1999)
224. Lopez-Ruiz, R., Mancini, H., Calbet, X.: A statistical measure of complexity. In: Concepts and Recent Advances in Generalized Information Measures and Statistics, pp. 147–168. Bentham Science Publishers (2013)
225. Anderson, P.W.: More is different: broken symmetry and the nature of the hierarchical structure of science. Science **177**(4047), 393–396 (1972)
226. Shalizi, C.R., Crutchfield, J.P.: Causal architecture, complexity and self-organization in time series and cellular automata. J. Stat. Phys. **104**(3–4), 817–879 (2001)
227. Hoel, E.P., Albantakis, L., Tononi, G.: Quantifying causal emergence shows that macro can beat micro. Proc. Natl. Acad. Sci. U.S.A. **110**(49), 19790–19795 (2013)
228. Camazine, S., Deneubourg, J.L., Franks, N.R., Sneyd, J., Theraulaz, G., Bonabeau, E.: Self-Organization in Biological Systems. Princeton University Press, Princeton (2020)
229. Ashby, W.R.: Principles of the self-organizing system. In: Pickering, A. (ed.) Systems Research for Behavioral Science. Routledge, London, pp. 108–118 (2017)
230. Gershenson, C., Heylighen, F.: When Can We Call a System Self-Organizing? In: Advances in Artificial Life. Lecture Notes in Computer Science, vol. 2801, pp. 606–614. Springer, Berlin (2003)
231. Kitano, H.: Biological robustness. Nat. Rev. Genet. **5**, 826–837 (2004)
232. Helbing, D.: Social Self-Organization. Springer, Berlin (2012)
233. Vespignani, A.: Predicting the behavior of techno-social systems. Science **325**(5939), 425–428 (2009)
234. Kitsak, M., Gallos, L.K., Havlin, S., et al.: Identification of influential spreaders in complex networks. Nat. Phys. **6**, 888–893 (2010)
235. Nicosia, V., Skardal, P.S., Arenas, A., Latora, V.: Collective phenomena emerging from the interactions between dynamical processes in multiplex networks. Phys. Rev. Lett. **118**(13), 138302 (2017)

236. Brin, S., Page, L.: The anatomy of a large-scale hypertextual Web search engine. Comput. Netw. ISDN **30**(1–7), 107–117 (1998)
237. Klickstein, I., Sorrentino, F., Díaz-Guilera, A.: Energy scaling of targeted optimal control of complex networks. Nat. Commun. **8**, 15423 (2017)
238. Shimono, M., Tamura, K.: Laplacian Eigenvector Centrality: A flexible spectral-based centrality measure for complex networks. arXiv preprint, arXiv:2501.11024 (2025)
239. Golub, G.H., Loan, C.F.: Matrix Computations, 4th edn. Johns Hopkins University Press, Baltimore (2013)
240. Barrat, A., Barthélemy, M., Vespignani, A.: Dynamical Processes on Complex Networks. Cambridge University Press, Cambridge (2008)
241. Cardillo, A., Gómez-Gardeñes, J., Zanin, M., et al.: Emergence of network features from multiplexity. Sci. Rep. **3**, 1344 (2013)
242. De Domenico, M., Solé-Ribalta, A., Cozzo, E., et al.: Mathematical formulation of multilayer networks. Phys. Rev. X **3**(4), 041022 (2013)
243. Kivelä, M., Arenas, A., Barthelemy, M., et al.: Multilayer networks. J. Complex Netw. **2**(3), 203–271 (2014)
244. Buldyrev, S.V., Parshani, R., Paul, G., Stanley, H.E., Havlin, S.: Catastrophic cascade of failures in interdependent networks. Nature **464**, 1025–1028 (2010)
245. Battiston, F., Cencetti, G., Iacopini, I., Latora, V., Lucas, M., Patania, A., Young, J.-G., Petri, G.: The physics of higher-order interactions in complex systems. Nat. Phys. **17**(10), 1093–1098 (2021)
246. Iacopini, I., Petri, G., Barrat, A., Latora, V.: Simplicial models of social contagion. Nat. Commun. **10**(1), 1–9 (2019)
247. Battiston, F., Nicosia, V., Latora, V.: Structural measures for multiplex networks. Phys. Rev. E **89**(3), 032804 (2014)
248. De Domenico, M., Solé-Ribalta, A., Omodei, E., et al.: Ranking in interconnected multilayer networks reveals versatile nodes. Nat. Commun. **6**, 6868 (2015)
249. Bonacich, P.: Power and centrality: a family of measures. Am. J. Sociol. **92**(5), 1170–1182 (1987)
250. Mucha, P.J., Richardson, T., Macon, K., Porter, M.A., Onnela, J.-P.: Community structure in time-dependent, multiscale, and multiplex networks. Science **328**(5980), 876–878 (2010)
251. Szell, M., Lambiotte, R., Thurner, S.: Multirelational organization of large-scale social networks. Proc. Natl. Acad. Sci. U.S.A. **107**(31), 13636–13641 (2010)
252. Nicosia, V., Latora, V.: Measuring and modeling correlations in multiplex networks. Phys. Rev. E **92**(3), 032805 (2015)
253. Noh, J.D., Rieger, H.: Random walks on complex networks. Phys. Rev. Lett. **92**(11), 118701 (2004)
254. Gómez, S., Díaz-Guilera, A., Gómez-Gardeñes, J., et al.: Diffusion dynamics on multiplex networks. Phys. Rev. Lett. **110**(2), 028701 (2013)
255. Newman, M.E.J.: Spread of epidemic disease on networks. Phys. Rev. E **66**(1), 016128 (2002)
256. Dorogovtsev, S.N., Goltsev, A.V., Mendes, J.F.F.: Critical phenomena in complex networks. Rev. Mod. Phys. **80**(4), 1275–1335 (2008)
257. de Arruda, H.F., Cozzo, E., Moreno, Y.: Fundamentals of spreading processes in multilayer networks. Phys. Rep. **756**, 1–59 (2018)
258. Sanz, J., Xia, C.Y., Meloni, S., Moreno, Y.: Dynamics of interacting diseases. Phys. Rev. X **4**(4), 041005 (2014)
259. Granell, C., Gómez, S., Arenas, A.: Dynamical interplay between awareness and epidemic spreading in multiplex networks. Phys. Rev. Lett. **111**(12), 128701 (2013)
260. Newman, M.E.J.: Threshold effects for two pathogens spreading on a network. Phys. Rev. Lett. **95**, 108701 (2005)
261. Sahneh, F.D., Chowdhury, F.N., Scoglio, C.: On the existence of a threshold for preventive behavioral responses to suppress epidemic spreading. Sci. Rep. **2**, 632 (2012)
262. Marceau, V., Noël, P.-A., Hébert-Dufresne, L., Allard, A., Dubé, L.J.: Modeling the dynamical interaction between epidemics on overlay networks. Phys. Rev. E **84**, 026105 (2011)

263. Funk, S., Salathé, M., Jansen, V.A.: Modelling the influence of human behaviour on the spread of infectious diseases: a review. J. R. Soc. Interface **7**(50), 1247–1256 (2010)
264. Saumell-Mendiola, A., Serrano, M.Á., Boguñá, M.: Epidemic spreading on interconnected networks. Phys. Rev. E—Statistical, Nonlinear, Soft Matter Phys. **86**(2), 026106 (2012)
265. Buono, C., Alvarez Zuzek, L.G., Macri, P.A., Braunstein, L.A.: Epidemics in partially overlapped multiplex networks. Phys. A **410**, 482–488 (2014)
266. Min, B., Yi, S.D., Lee, K.M., Goh, K.I.: Network robustness of multiplex networks with interlayer degree correlations. Phys. A **436**, 855–864 (2015)
267. Castellano, C., Fortunato, S., Loreto, V.: Statistical physics of social dynamics. Rev. Mod. Phys. **81**(2), 591–646 (2009)
268. Sood, V., Redner, S.: Voter model on heterogeneous graphs. Phys. Rev. Lett. **94**, 178701 (2005)
269. Suchecki, K., Eguíluz, V.M., San Miguel, M.: Voter model dynamics in complex networks: Role of dimensionality, disorder, and degree distribution. Phys. Rev. E—Statistical, Nonlinear, Soft Matter Phys. **72**(3), 036132 (2005)
270. Stark, H.-U., Tessone, C.J., Schweitzer, F.: Slower is faster: fostering consensus formation by heterogeneous inertia. Adv. Complex Syst. **11**(4), 551–563 (2008)
271. Masuda, N.: Voter model on the two-clique graph. Phys. Rev. E **90**(1), 012802 (2014)
272. Diakonova, M., San Miguel, M., Eguíluz, V.M.: Absorbing and shattered fragmentation transitions in multilayer coevolution. Phys. Rev. E **89**(6), 062818 (2014)
273. Battiston, F., Cairoli, A., Nicosia, V., Baule, A., Latora, V.: Interplay between consensus and coherence in a model of interacting opinions. Physica D **323**, 12–19 (2016)
274. Nowak, M.A., May, R.M.: Evolutionary games and spatial chaos. Nature **359**, 826–829 (1992)
275. Santos, F.C., Pacheco, J.M.: Scale-free networks provide a unifying framework for the emergence of cooperation. Phys. Rev. Lett. **95**(9), 098104 (2005)
276. Szolnoki, A., Perc, M.: Information sharing promotes prosocial behaviour. New J. Phys. **15**, 053010 (2013)
277. Wang, Z., Wang, L., Szolnoki, A., Perc, M.: Evolutionary games on multilayer networks: a colloquium. Eur. Phys. J. B **88**, 124 (2015)
278. Gómez-Gardeñes, J., Reinares, I., Arenas, A., Floría, L.M.: Evolution of cooperation in multiplex networks. Sci. Rep. **2**, 620 (2012)
279. Wang, Z., Szolnoki, A., Perc, M.: Interdependent network reciprocity in evolutionary games. Sci. Rep. **3**, 1183 (2013)
280. Rosvall, M., Esquivel, A.V., Lancichinetti, A., West, J.D., Lambiotte, R.: Memory in network flows and its effects on spreading dynamics and community detection. Nat. Commun. **5**, 4630 (2014)
281. Scholtes, I., Wider, N., Pfitzner, R., Garas, A., Tessone, C.J., Schweitzer, F.: Causality-driven slow-down and speed-up of diffusion in non-Markovian temporal networks. Nat. Commun. **5**, 5024 (2014)
282. Lambiotte, R., Rosvall, M., Scholtes, I.: From networks to optimal higher-order models of complex systems. Nat. Phys. **15**, 313–320 (2019)
283. Zhou, D., Huang, J., Schölkopf, B.: Learning with hypergraphs: Clustering, classification, and embedding. Adv. Neural Inf. Process. Syst. **19** (2006)
284. Rodriguez, J.A.: On the Laplacian eigenvalues and metric parameters of hypergraphs. Linear Multilinear Algebr. **50**(1), 1–14 (2002)
285. Otter, R.: The number of trees. Ann. Math. **49**(3), 583–599 (1948)
286. Horak, D., Maletić, S., Rajković, M.: Persistent homology of complex networks. J. Stat. Mech: Theory Exp. **03**, P03034 (2009)
287. Berge, C.: Hypergraphs: Combinatorics of Finite Sets, vol. 45. Elsevier (1984)
288. Chodrow, P.S.: Configuration models of random hypergraphs. J. Complex Netw. **8**(3), cnaa018 (2020)
289. Courtney, O.T., Bianconi, G.: Generalized network structures: the configuration model and the canonical ensemble of simplicial complexes. Phys. Rev. E **93**(6), 062311 (2016)

290. Estrada, E., Rodriguez-Velazquez, J.A.: Subgraph centrality in complex networks. Phys. Rev. E **71**(5), 056103 (2005)
291. Tudisco, F., Higham, D.J.: Node and edge nonlinear eigenvector centrality for hypergraphs. Commun. Phys. **4**, 201 (2021)
292. Benson, A.R., Gleich, D.F., Leskovec, J.: Higher-order organization of complex networks. Science **353**(6295), 163–166 (2016)
293. Centola, D.: The spread of behavior in an online social network experiment. Science **329**(5996), 1194–1197 (2010)
294. Nowak, M.A., May, R.M.: Evolutionary games and spatial chaos. Nature **359**, 826–829 (1992)
295. Grujić, J., Fosco, C., Moro, L.A., Sánchez, A., Cuesta, J.A.: Social experiments in the mesoscale: humans playing a spatial prisoner's dilemma. PLoS ONE **5**(11), e13749 (2010)

Chapter 3
A Complex-Based Rethinking of Communications

Abstract This chapter explores how the approach that follows the complex systems theory can offer a transformative lens and a shifting action for understanding and designing future communication infrastructures, not only in the transition toward 6G but also beyond. By integrating insights from network science and data science, evolutionary game theory, machine learning, collective dynamics, and multiplex and hypergraph modeling, it will be discussed a novel conceptual and analytical roadmap for managing the ever-increasing dynamism, heterogeneity, and intelligence of modern communication ecosystems.

3.1 Introduction

Building upon the fundamental concepts introduced in Chap. 2, where the principles and dynamics of complex systems were thoroughly explored, in this chapter, the aim is to discuss how these ideas intersect with the design and evolution of modern communication networks. Our world is increasingly recognized as a deeply interconnected environment, where systems, whether technological, natural, or social-based structured as vast networks of interactions. From communication networks [1–3] to transportation infrastructures [4, 5], from ecosystems [6] to financial markets [7], and from neural networks [8] to social communities [9–11], it is possible to find out a recurring pattern: highly dynamic systems composed of numerous interacting elements. These systems are what has been discussed in Chap. 2 and what has been identified as complex systems.

Complex systems science seeks to understand how local interactions among individual components give rise to collective, often unpredictable behaviors, and how such systems respond to environmental changes. Understanding a complex system, therefore, requires a dual point: both on the features of individual components and on the emergent properties that arise from their interactions. These systems typically exhibit features such as emergence, self-organization, adaptability, robustness, resilience, decentralization, and non-linearity. Structurally, they are often described as non-hierarchical, distributed, modular, and flat networks. As a result, network theory provides a natural framework for capturing their organization and behavior. The

M. Scatà, *The Power of Complex Systems*, Studies in Systems, Decision and Control 630,
https://doi.org/10.1007/978-3-032-04243-9_3

proliferation of data across scientific disciplines has empowered empirical studies of complex systems. From molecular biology and neuroscience to social sciences and information technology. The convergence between these topics under a common language has pushed the growth of interdisciplinary field of complexity science into a robust field joining mathematics, computer science, engineering, physics, sociology, and beyond. Crucially, complex systems demand a shift in how we approach problem-solving and system design. In these systems, causality is often indirect and non-obvious–problems may be hard to solve precisely because their causes and effects are deeply entangled. Complexity theory studies the emergence of patterns through interaction, yet these patterns are often perceivable but not predictable. Even when regular patterns appear, the mechanisms that generate them are typically hidden or also sensitive to perturbation, sometimes even disrupted by the observation.

In the context of communication networks, the challenge becomes even more urgent. As infrastructures evolve from 5G to 6G and beyond, they grow increasingly dense, heterogeneous, and dynamic. Their components, ranging from physical base stations to AI-based control planes, interact across multiple scales and modalities. Thus, information flow and architectural complexity risks of overcoming the capabilities of classical modeling and control approaches. To manage this complexity, new paradigms are required, based on complex systems science, that could be paradigms capable of modeling systems that are open, distributed, adaptive, and evolving. We must move beyond static architectures toward networked systems that self-organize and optimize autonomously in response to shifting demands. Complex systems theory thus offers both conceptual tools, to think rigorously about emergence, control, and feedback, and computational instruments, to simulate, monitor, and guide the behavior of systems too intricate to manage by traditional tools. From artificial societies and biological ecosystems to the smart networks of the future, this perspective paves the way for a new generation of robust and intelligent communication infrastructures [11–13].

This chapter sets the stage for a rethinking, a novel approach for the future of communication networks by establishing the key role and relevance of complex systems theory. The concepts introduced in the previous chapter, Chap. 2, such as emergence, self-organization, adaptability, multilayer interactions, collective dynamics and so on, serve as the conceptual pillars for various interdisciplinary works. Building upon the solid theoretical foundation laid out in Chap. 2, where the principles and dynamics of complex systems were thoroughly examined, and following the technological background introduced in Chap. 1, which motivated the need for complexity-inspired methodologies in next-generation communication networks, Chap. 3 aims to bridge these theoretical insights with practical applications. In particular, this chapter explores how the complex systems paradigm has inspired innovations in communication networks, delving into both current applications and future potentials. Through an in-depth analysis of Networks of Networks (NoN), Digital Twins, Federated Learning and so on, it is possible to how the intrinsic properties of complex systems can enhance resilience, optimize resource distribution, and foster adaptive synchronization in highly interconnected infrastructures.

Furthermore, the investigation continues not only in the sectors where complex systems have already demonstrated substantial impact, such as smart cities, IoT ecosystems, and edge computing, but also where they hold the promise for transformative contributions. The final sections of this chapter will be dedicated to a forward-looking perspective, highlighting how the principles of complexity can serve as a blueprint for the design and optimization of paradigmatic application scenarios. These include resilient urban infrastructures, scalable edge intelligence, and autonomous vehicular networks, where real-time adaptation, self-healing, and cooperative strategies are not merely advantageous but essential for sustainable and intelligent communication ecosystems. Each of the following topics, from Network-of-Networks (NoN) structures to Digital Twins and Complexity-Inspired Machine Learning, will explore how these principles are not only applicable but essential to the design and orchestration of future 6G infrastructures. These explorations will demonstrate that complex systems theory is not just a lens for understanding communication networks, it is a necessary framework for building resilient, adaptive, and intelligently orchestrated digital ecosystems.

To provide a structured overview of the paradigms discussed throughout this chapter, Table 3.1 summarizes the key concepts at the intersection of communication networks and complexity science. For each concept, we include a brief definition and describe how the adoption of a complexity-inspired approach can enhance, extend, or rethink the corresponding technological framework. This synthesis aims to highlight the systemic and dynamic nature of future communication infrastructures, where interaction, emergence, and adaptivity play a central role.

Thus, the following sections are intended as a conceptual and practical bridge, translating the theoretical underpinnings of complex systems into actionable strategies for the evolution of communication networks. It sets the stage for understanding not only how these principles currently shape technological advancements, but also how they can be harnessed to address the unprecedented challenges posed by the ever-growing complexity and interdependence of global communication infrastructures.

3.2 A New Roadmap

Many modern systems, from communication infrastructures like the Internet and wireless networks to ecosystems, traffic flows, financial markets, and biological or social interactions, can be modeled as networks of massively interconnected and interacting components. These systems can be viewed for a better understanding through the lens of complexity [12].

Complex systems science investigates how the interaction among system components leads to collective behavior and how the system, as a whole, responds to its environment. To understand these systems, it is crucial not only to analyze the individual entities but also how their interactions shape emergent, often unpredictable global patterns. These systems are typically characterized by features such as emergence,

Table 3.1 Summary of 6G-related topics and benefits from applying a complexity-based approach

Topic	Definition	Contribution of a complexity-based approach
Network-of-Networks	Interconnected and interdependent network infrastructures	Improves adaptability and resilience via decentralized control
Digital twins	Digital replicas of physical systems for prediction and monitoring	Enables real-time co-evolution and feedback mechanisms
Edge intelligence	AI applied at the network edge for local data processing	Supports emergent behavior from local device interactions
Federated learning	Decentralized learning preserving data privacy	Adapts to node heterogeneity and dynamic topologies
Complexity-inspired machine learning	AI inspired by adaptive, emergent, and non-linear dynamics	Enhances learning robustness and multi-scale behavior
Floating content architectures	Local content sharing in dynamic environments	Boosts relevance using mobility-aware diffusion models
Society 5.0	A smart society integrating human and digital systems	Models societal needs via dynamic system feedback
Gossip protocols	Probabilistic, local data dissemination protocols	Ensures scalable propagation through local interactions
AI-native communication	AI built into the core of communication systems	Allows adaptive decisions across protocol layers
Digital social twins	Digital models of collective social behaviors	Captures social dynamics to guide network adaptation
Resilience and robustness	Ability to maintain functionality under disruption	Designs systems resistant to cascading failures
Edge AI	AI that adapts to local edge conditions	Drives optimization through emergent edge behaviors
Communication-efficient learning	Learning methods minimizing communication overhead	Improves learning with local, adaptive data use
Intelligent digital twin networks (I-DTN)	AI-based digital twins for smart environments	Enables prediction in complex urban systems
AI-driven trust and security	AI-enhanced security and trust mechanisms	Activates responsive security via distributed learning
Complexity-based optimization	Resource strategies based on system adaptivity	Balances demands with multi-objective strategies
Quantum slicing	Partitioning of quantum networks for services	Adapts resources to quantum state variability
Artificial Intelligence Of Things (AIoT)	IoT systems enhanced with embedded intelligence	Fosters autonomy in massive distributed environments
Self-organization	Systems that self-structure via local rules	Triggers recovery via decentralized reconfiguration
Game-theoretic adaptation	Adaptation modeled via strategic agent interactions	Guides adaptive equilibria in strategic settings
Cross-layer complexity	Coordination across layers for optimal performance	Optimizes flow via inter-layer coordination

adaptability, self-organization, robustness, resilience, decentralization, and responsiveness. Structurally, they tend to be flat, non-hierarchical, distributed, networks in the most general sense [14]. The empirical study of complex systems has expanded significantly thanks to the explosion of available data across diverse domains. This has revealed shared theoretical challenges among disciplines, encouraging the rise of complexity science as a mature and interdisciplinary scientific field.

The study of complex systems centers around understanding indirect effects, situations where causality is not easily discernible. In many real-world systems, causes and consequences are distributed across multiple agents and interactions, making the dynamics difficult to trace and even harder to predict. Complexity aims to uncover how macroscopic patterns emerge from the local interactions of numerous heterogeneous components. These emergent patterns are often unpredictable and may change as the system evolves, especially when the system's hidden patterns are not fully observable or when observation itself perturbs the dynamics [15]. As Newman outlined in [16], the study of networked complex systems can follow three directions: (a) identifying statistical properties such as path lengths, clustering, and degree distributions that characterize their structure and dynamics; (b) constructing generative models that explain how such networks form and evolve; and (c) using structural and local information to predict the future behavior of the system and the evolution of emerging patterns, as properly explained and discussed in Chap. 2.

Today's technological ecosystems, particularly mobile communication networks, are rapidly growing in both size and complexity [3, 12]. The transition from 5G to 6G technologies introduces a greater number of interconnected components, high-density infrastructure, and a massive surge in information flow. This increasing heterogeneity and interdependence pose significant challenges for traditional paradigms. Managing, modeling, and optimizing such systems requires a significant shift from centralized, static architectures to dynamic, decentralized ones, where autonomy, adaptability, and real-time responsiveness are key considerations. In this context, complex systems science has the potential to provide essential tools and theoretical insights [3, 12, 13]. By viewing communication networks as evolving systems composed of distributed agents, which represent the device or users, base stations, etc., it becomes possible to incorporate concepts like self-organization, robustness, and collective behavior into the design and control strategies [13, 17]. This paradigm enables the development of communication networks that are not only scalable but also capable of learning, adapting, and optimizing their functions in real time to meet changing user demands and environmental conditions. This is important as 6G networks are expected to integrate cutting-edge technologies such as AI/ML, Terahertz communications, semantic-aware protocols, and softwarized, service-oriented architectures [18]. Traditional information theory, while foundational, is no longer sufficient to capture the full range of dynamics in such complex, multi-layered environments. Thus, complexity theory stands out as a promising approach for modeling, monitoring, and designing the next generation of communication systems [12].

Over the past two decades, cellular network technologies have undergone a remarkable transformation, evolving from the first generation (1G), centered on analog voice transmission, to the fifth generation (5G), which introduces a sophisticated digital infrastructure supporting a wide array of services. The rollout of 5G around 2020 marked a significant milestone, bringing forth novel innovations such as millimeter-wave (mmWave) transmission, ultra-dense network deployments, and support for massive device heterogeneity [19, 20]. However, as 5G deployment expands globally, the academic and industrial research communities have already shifted their focus toward the next evolutionary leap, 6G mobile networks. A growing body of literature is exploring this paradigm, generally converging on the idea that 6G will not merely be a linear extension of 5G, but rather a disruptive transformation introducing groundbreaking technologies and revolutionary architectural frameworks [18, 21, 22]. A key innovation introduced by 5G is the softwarization of network functionalities, exemplified by the rise of the cloud-native architecture. This architectural model centralizes network functions and service applications within cloud data centers, enabling logical network slicing over a shared infrastructure. Each slice can be dynamically customized to meet the stringent and diverse requirements of various service categories, from ultra-reliable low-latency communications to massive machine-type communications [18].

Looking forward, the vision for 6G introduces an even more ambitious shift: the evolution from the Internet of Things (IoT), or the concept of "everything connected," toward the era of Connected Intelligence [23]. In this paradigm, networks will no longer merely interconnect devices but will embed Artificial Intelligence (AI) and Machine Learning (ML) as core capabilities, enabling self-optimization, adaptive control, and intelligent orchestration of network resources and services. Such a shift will impose far more stringent performance demands than ever before. To support Connected Intelligence, 6G systems will need to achieve extremely high data rates and ultra-reliable transmission, with aggregate speeds reaching 1 Tbps in some scenarios, and individual user rates of 100 Gbps becoming a realistic target [21, 24]. These requirements will inevitably reshape the main features of mobile network design, pushing the boundaries of communication theory, signal processing, and distributed intelligence, while significantly increasing system complexity.

In addition to the extreme data rate and reliability requirements previously discussed, 6G networks are also expected to deliver ultra-low end-to-end latency, exceptionally high energy efficiency, and seamless support for cloud-native applications under the Network-as-a-Service (NaaS) paradigm. This model will enable flexible orchestration of services and network functions via cloud computing platforms, favoring dynamic scalability and cost-efficiency. Furthermore, 6G will leverage a vastly extended frequency spectrum, including frequencies in the terahertz (THz) range, opening up new frontiers in bandwidth availability and communication speed [21]. Beyond the terrestrial domain, the integration of non-terrestrial networks, including satellites, high-altitude platforms, unmanned aerial vehicles (UAVs), and connected autonomous vehicles, will significantly augment the infrastructure, yet concurrently increase the complexity, heterogeneity, and dynamism of the overall system [24]. According to the International Telecommunication Union (ITU), 5G networks have been designed around three fundamental service categories, which are expected to characterize also the 6G era: enhanced Mobile Broadband (eMBB), Ultra-Reliable Low-Latency Communications (URLLC), and massive Machine-Type Communications (mMTC) [25]. Each of these service classes addresses a specific class of application needs. eMBB aims to provide high data rates, especially in stable network conditions, while maintaining acceptable performance at the cell edge. The mMTC is optimized for supporting massive numbers of low-power, intermittently active devices, such as sensors and Internet-of-Things (IoT) nodes, that transmit small packets. Meanwhile, URLLC addresses mission-critical applications by ensuring minimal latency and maximal reliability for small data payloads, important for scenarios like industrial automation or remote control of critical systems. However, the 6G vision goes further, proposing the introduction of three new advanced service types [23]: Computation-Oriented Communications (COC), Contextually Agile eMBB Communications (CAeC), and Event-Defined URLLC (EDuRLLC). The COC paradigm introduces a fundamental shift in the Quality of Service (QoS) model. Instead of static, pre-defined thresholds, service guarantees will be dynamically adapted based on real-time availability of computational and communication resources, thus achieving a targeted level of computational accuracy. The CAeC model will enhance traditional eMBB by making it contextually aware and responsive to network conditions, including metrics like congestion levels, topology changes, user location, and reliability requirements. This will lead to more intelligent, content-aware, and adaptive broadband services. The EDuRLLC service class is designed to respond to extreme or emergency events, such as natural disasters or critical infrastructure failures, by adapting to highly variable traffic loads, device densities, and unpredictable conditions. Unlike conventional URLLC, EDuRLLC emphasizes situational adaptability, enabling 6G networks to maintain reliability and responsiveness under stress.

Collectively, these emerging service types illustrate the growing functional complexity of 6G networks. These networks are envisioned to serve as the foundational infrastructure for autonomous systems, space-air-ground integrated networks, smart vehicular ecosystems, and immersive technologies such as virtual/augmented reality (VR/AR) and holographic communication [26, 27]. Applications such as remote surgery, multi-sensory telepresence, autonomous driving, Internet of Vehicles, networked UAV coordination, and VAR-based gaming are all expected to become integral components of the 6G landscape, turning long-imagined concepts into everyday experiences. The development of 6G mobile communication systems is expected to introduce a range of novel architectural and operational concepts that will significantly increase network complexity. These innovations mark a substantial shift away from the relatively static configurations of earlier generations, ushering in dynamic, heterogeneous, and context-aware infrastructures [12, 23].

In 6G, network topology is expected to become fully dynamic. Devices, ranging from smartphones to autonomous vehicles and Internet-of-Things (IoT) sensors, will continuously connect to the subnetwork that provides the best real-time Quality of Service (QoS). This dynamic association will redefine network behavior, driven by constantly changing link conditions, node mobility, and service requirements. The proliferation of high-mobility nodes such as unmanned aerial vehicles (UAVs), drones, and low-orbit satellites will increase topological variability. To ensure reliable service under these conditions, advanced models for interference dynamics and intelligent handover mechanisms will be required, rooted in complex systems theory and adaptive mathematics. To meet the extremely high throughput and spectral efficiency goals of 6G, future systems will exploit terahertz (THz) frequencies, beyond the conventional millimeter-wave bands. This will lead to the creation of so-called "tiny cells" with coverage radio of only a few meters. These ultra-dense deployments will necessitate new strategies for traffic and mobility management, as well as congestion control. The propagation characteristics of THz signals–marked by high path loss, environmental sensitivity, and power constraints–will demand a deep rethinking of the physical layer. These conditions call for the design of new MAC, link, and network protocols, potentially leveraging technologies such as programmable electromagnetic surfaces and advanced wave control [12]. As deeply discussed in [12], the anticipated exponential increase in traffic volume by 2030 will put severe pressure on existing backhaul infrastructures. The ITU's Focus Group on Network 2030 has highlighted the risk that traditional access networks may become bottlenecks under 6G performance demands. To overcome this, new high-frequency bands–such as the D-band around 60 GHz–are being explored. Additionally, future backhaul solutions may include free-space optical communications and even quantum communication channels. Drone-based access points have also been proposed to enhance connectivity in underserved areas and hotspots, supported by satellite links through LEO constellations and CubeSats. These architectures point toward a unified, multi-layered wireless infrastructure that merges terrestrial, airborne, and satellite networks into a coherent and responsive system. As mobile communication technology evolves, an increasing number of users demand uninterrupted high-quality connectivity even while traveling in high-speed environments–trains, aircraft, autonomous vehicles, and more. To address this rising challenge, the concept of Moving Networks has emerged. These are a special class of ad hoc networks composed of mobile nodes and mobile relay units embedded within moving platforms. Such nodes often operate as mobile small cells, designed to maintain consistent connectivity for passengers in motion.

However, ensuring service continuity under these conditions is far from trivial. One of the primary obstacles is Vehicular Penetration Loss (VPL), which refers to the degradation of signal strength due to attenuation through metal frames and insulation materials of vehicles. Studies show that VPL can reach up to 25 dB in common minivans at 2.4 GHz, and is likely to be even more severe at higher frequencies used in future communication systems. This degradation not only reduces received signal power but also increases interference levels, posing a major threat to service quality. A promising strategy to mitigate VPL involves network densification, although it introduces its own challenges, particularly inter-cell interference. Nevertheless, Mobile Relay Nodes (MRNs) are inherently more robust than user equipment (UE), as they are less limited by size, power, and computational constraints. This flexibility allows the integration of advanced multi-antenna techniques, sophisticated signal processing algorithms, and interference cancellation schemes, thereby enhancing performance even in dynamic vehicular contexts. Moreover, as data demands increase, millimeter-wave (mmWave) technology is being explored to enable high-throughput links for

Moving Networks. While mmWave can help meet bandwidth requirements, it further exacerbates penetration losses, reinforcing the need for innovative architectures and materials.

One of the key innovations in 6G communication networks is the integration of aerial-assisted architectures, supported by Unmanned Aerial Vehicles (UAVs). These airborne platforms enable on-demand, low-latency connectivity for both ground and aerial users. Advanced designs incorporate massive MIMO-enabled base stations with spatial multiplexing capabilities, allowing simultaneous service for multiple users across diverse environments [12]. To enhance the performance of these mobile and aerial networks, 6G will leverage emerging technologies such as HyperSurfaces (HSFs), Reconfigurable Intelligent Surfaces (RIS), and Ultra-Massive MIMO arrays. These innovations allow for programmable manipulation of electromagnetic waves, dynamically reshaping the wireless medium for optimized propagation. Ultra-Massive MIMO systems use large antenna arrays to adapt radiation patterns in real-time, while metasurfaces–engineered ultra-thin materials–introduce advanced functionalities like wave steering, filtering, and absorption, even in challenging scenarios. HyperSurfaces represent an evolution of metasurfaces by integrating software-defined capabilities directly into the physical environment. This enables real-time, programmable control of wave propagation through APIs, extending coverage and enhancing performance in both indoor and outdoor settings. These surfaces are conceptualized as a dynamic multigraph, where nodes represent programmable surfaces and edges model electromagnetic connectivity. Pathfinding and optimization within this framework require complex algorithms inspired by systems theory, reflecting the adaptive nature of 6G environments. Applications include Non-Line-of-Sight (NLOS) communication, wireless power transfer, interference mitigation, and enhanced security, positioning HyperSurfaces as a cornerstone of adaptive, intelligent 6G infrastructures.

The design and orchestration of 6G network functions will demand a fundamental shift in how intelligence, architecture, and system dynamics are managed across a multi-domain, multi-layered infrastructure. As 6G is envisioned as a network of networks, encompassing heterogeneous physical and virtual environments, intelligence-driven automation will become crucial to ensure scalable and responsive management across all levels of the architecture [18, 28, 29].

This architecture will span a continuum from the cloud to the edge, enabling real-time data collection and decision-making based on inputs from the radio access network (e.g., via O-RAN), the transport network, and the application layer. Resource-aware and cross-layer orchestration mechanisms will support intelligent service function placement, adaptive radio resource management, and end-to-end Quality of Service (QoS) [30, 31]. Notably, the placement of Network Functions (NFs) must become more flexible and context-sensitive than in previous generations, accommodating new service types such as digital twins, immersive applications, and ubiquitous AI services [32].

To this end, Machine Learning (ML) techniques are playing an increasingly central role in enabling adaptive, data-driven orchestration. ML facilitates dynamic NF placement decisions based on evolving network states, resource availability, and application demands, outperforming static or rule-based systems in complex and volatile scenarios [33–35]. At the same time, the rise of federated learning and game-theoretic approaches supports decentralized and privacy-preserving strategies for distributed network optimization [36, 37].

In highly dynamic systems like 6G, Machine Learning alone may not capture strategic, long-term interactions among autonomous network agents. This is where Evolutionary Game Theory (EGT) becomes a complementary tool. EGT excels in modeling behavioral adaptation, strategy evolution, and cooperation-competition tradeoffs, especially in settings involving heterogeneous agents competing for shared resources [38, 39]. A hybrid EGT-ML framework offers the ability to model adaptive learning and strategic decision-making under uncertainty, enabling more intelligent and resilient orchestration of virtual network functions. The complexity of these decisions demands adequate system modeling. In this regard, complex network theory has gained traction as a modeling paradigm capable of capturing the multiscale structure and dynamics of 6G networks [13, 40, 41]. A complex system is not merely a large network but one in which emergent properties arise from the interplay among interconnected components, that cannot be deduced from analyzing individual elements in isolation [17].

To represent this complexity, in [3] authors adopt a multiplex network perspective, in which each layer corresponds to a different type of interaction or functional domain (e.g., control, data, user context). The approach allows to capture interdependencies and feedback loops between layers, uncovering hidden structural patterns and facilitating multi-layer optimization [42, 43]. It also enables the study of topological adaptation in response to strategic decisions taken by nodes modeled as game players. The framework proposed in [3], self-organization and collective adaptation represent necessary mechanisms to manage the complexity of 6G systems. Strategic interactions among nodes–for example, in deciding where to offload tasks or how to allocate bandwidth–can be modeled using replicator dynamics and evolutionary equilibria [11, 44]. The result is a system capable of converging toward stable, cooperative configurations, even in the presence of local conflicts or noise. It is crucial to rethink on design approach in order to integrate tools and methods proper of complex systems theory, multiplex network representations, machine learning, and evolutionary game theory to support the next generation of network and communication paradigms.

The next sections delve into technological directions for 6G. These include the transformation of communication systems into distributed, cognitive environments through concepts like the network of networks, the rise of digital twins as virtual counterparts of physical infrastructures, and the integration of edge intelligence, federated learning, and adaptive machine learning algorithms. Together, these perspectives demonstrate how a complex-based approach can enable robust, scalable, and intelligent architectures for the future of communication.

3.3 The Networks of Networks

Recent research underscores that Networks of Networks (NoNs) represent the next critical frontier in network science, offering a more realistic and comprehensive framework for understanding the complexity of interconnected systems [45–47]. Traditional single-layer models fall short when attempting to capture the diversity of interdependencies and cascading dynamics that emerge in multi-network systems. From communication infrastructures and energy grids to financial markets and ecological systems, it is evident that disruptions in one domain can propagate unpredictably to others. Such interconnections call for advanced modeling paradigms capable of representing heterogeneous link types, cross-layer dependencies, and the co-evolution of dynamics [48]. NoNs enable this complexity to be structurally encoded, providing a foundation to study real-world systems more accurately [48–51].

NoNs, also referred to as multilayer or interdependent networks, capture the interplay between distinct networks where each network represents a subsystem (e.g., transportation, communication, energy). Nodes in one layer may depend on nodes in other layers for functionality. This leads to asymmetric, directed dependencies, and creates feedback mechanisms capable of initiating systemic cascades [51]. In this framework, dependency links such as $i \rightarrow j$ imply that node j cannot function without i. The removal or failure of node i could thus compromise multiple systems simultaneously. Empirical evidence supports this: communication networks rely on power systems, just as economic markets depend on both logistics and communication infrastructures [47, 48]. Percolation theory adapted for NoNs has revealed that traditional second-order percolation transitions in single-layer networks are replaced by first-order, discontinuous transitions in interdependent systems [46, 51]. A failure in one layer can recursively propagate across dependent layers, leading to systemic collapse [52, 53]. The robustness of NoNs is heavily shaped by the degree distribution and topology of each layer. While scale-free networks are robust in isolation, their interdependencies dramatically reduce their resilience [54]. The concept of a Mutually Connected Giant Component (MCGC) becomes central: the system collapses if the MCGC vanishes [55]. The coupling between network layers introduces significant changes in diffusion and spreading processes. Epidemics that remain isolated in one layer under SIR models can synchronize across layers in SIS models [56]. Random walkers, diffusion, and information flow also behave differently depending on inter-layer dependencies, revealing new bottlenecks and congestion phenomena. These altered thresholds and global behaviors demonstrate the inadequacy of isolated models in systems where multiple modalities (e.g., physical contact, online interaction) operate concurrently [51].

Synchronization in multiplex and temporal multiplexed NoNs exhibits novel properties. In temporal multiplexing, network connections evolve over time, and specific switching patterns can enhance synchronization [57]. In simultaneous multiplexing, synchronization can emerge globally even if local coherence is absent, emphasizing the amplifying role of inter-layer connections [41]. This distinction is vital for power grids, robotics, and sensor networks, where partial coherence is often sufficient for robust performance. Future studies are expected to further explore the effects of topological symmetry and inter-layer alignment on synchronization stability. The study of evolutionary games has expanded into NoNs, where strategic interactions are influenced by dependencies across layers [58, 59]. Interdependent network reciprocity allows cooperative clusters in one layer to support cooperation in another, enhancing resilience [60]. Mechanisms such as probabilistic interconnections [61], cross-layer information sharing [62], and reward-based interdependence [63] have shown that cooperation can be sustained more effectively in multilayer environments. These models have also been applied to social dilemmas, climate governance [64], and crime prevention [65].

The NoNs paradigm is central to the architectural vision of 6G networks. Future infrastructures are expected to integrate heterogeneous sub-networks–terrestrial, satellite, aerial–into a unified and adaptive NoN [66, 67]. This shift enables the management of diverse services, from vehicular mobility and edge intelligence to smart agriculture [68, 69]. Advanced models such as the A2I2 framework offer metaphors for understanding modular sub-networks (islands), their aggregation (atolls), and their federation (archipelagos) [70]. This abstraction supports distributed orchestration and AI-native decision-making in large-scale communication systems. The Networks of Networks paradigm provides a rigorous and flexible foundation for modeling, analyzing, and designing resilient, adaptive infrastructures. Its applicability spans from theoretical studies in synchronization and cascading failures to real-world scenarios in 6G, smart cities, and decentralized intelligence. As communication technologies evolve, embracing this multilayered approach will be essential for understanding and engineering the next generation of complex systems.

3.3.1 Future Directions, Perspectives and Research Challenges in Networks of Networks

The study of Networks of Networks (NoNs) and multilayer structures has unveiled critical dimensions of complexity and interdependence that are reshaping our understanding of resilience and vulnerability in communication infrastructures. The research agenda in this domain unfolds across three major trajectories: theoretical advancements, the development of adaptive and resilient mechanisms, and the integration of cognitive profiling to enable intelligent behavior in interconnected systems. Continued efforts are needed to expand analytical frameworks capable of capturing the emergent behavior of interdependent systems. Key challenges include modeling failure propagation, cascading dynamics, and synchronization in dynamic environments. For instance, percolation theory has been generalized to interdependent networks, where recursive failures reveal abrupt, discontinuous transitions not observed in single-layer settings [51, 53, 71]. Likewise, studies on temporal networks show how evolving connectivity affects synchronization and stability [52]. Another important avenue involves the extension of evolutionary game theory to the

NoN framework. While cooperative strategies have long been explored in monoplex networks, recent findings show that interdependencies enable the persistence of cooperation through mechanisms such as interdependent reciprocity [60, 63]. These insights have broad implications for modeling behavioral dynamics in social, biological, and technological systems.

To move beyond diagnosis toward mitigation, attention must turn to adaptive and recovery strategies in NoNs. Cascading failures, while now better understood, require equally sophisticated recovery protocols. Research has started to explore spontaneous recovery [72], functional reorganization [73], and coevolutionary adaptation [74], where networks dynamically reconfigure to maintain operability. The interplay between robustness and adaptability becomes central: rather than resisting change, systems must evolve roles and connections to survive under perturbations. Evolutionary dynamics offer a conceptual basis to explore how decentralized mechanisms can support role reassignment, load redistribution, or connectivity adjustments in response to failures.

To address the limitations of static NoN models, cognitive profiling introduces a dynamic, context-aware layer that enhances self-organization and decision-making across networked infrastructures. In this paradigm, each node is embedded with a cognitive profile reflecting its current role, resources, and interaction states. The state evolves based on local sensing and global stimuli:

$$C(v_i, t+1) = f(C(v_i, t), I(t), S(v_i, t)) \tag{3.1}$$

Here, $C(v_i)$ is the cognitive state vector, $I(t)$ represents global conditions, and $S(v_i, t)$ denotes local sensory input. Role-based adaptation allows nodes to reassign their functions dynamically within layers:

$$R(v_i, \ell_k, t) = g(\eta_i, \rho_i, \gamma_i, \ell_k) \tag{3.2}$$

Synchronization and resilience emerge from distributed coordination:

$$\frac{dC(v_i, t)}{dt} = \alpha \sum_{v_j \in N(v_i)} w_{ij} \left[C(v_j, t) - C(v_i, t) \right] + \beta F(t) \tag{3.3}$$

This framework supports robust performance in dynamic environments such as smart cities, IoT ecosystems, and critical infrastructures [75, 76].

Applications of cognitive-driven NoNs span diverse domains. In urban infrastructures, profiling can facilitate priority routing during emergencies and congestion management [64]. In IoT, it enables context-aware clustering and self-healing under resource constraints [76]. In edge computing, profiling supports adaptive caching and load balancing [72]. These implementations pave the way for future research focused on:

- **ML Integration**: Reinforcement and federated learning to guide profile evolution.
- **Cross-Layer Optimization**: Protocols for dynamic coordination across layers.

- **Large-Scale Simulations**: Evaluating cognitive models in realistic infrastructures.
- **Interoperability with NextG**: Adapting the framework for 6G, edge intelligence, and AI-native environments [70, 77].

In conclusion, these directions highlight the potential of cognitive profiling and role-based adaptation to transform static NoN representations into intelligent, resilient, and scalable infrastructures. As the digital landscape evolves, such adaptive paradigms will be essential for building future-ready communication systems.

3.4 Edge Intelligence in 6G Networks

The global standardization and deployment of 5G networks have set the stage for envisioning the next evolution in wireless communication: 6G. This vision is driven by key projects initiated by major global economies, including the United States [78] and the European Union [79]. These initiatives collectively aim to achieve the transformative concept of *connected intelligence*, where seamless integration of Artificial Intelligence (AI), the Internet of Everything (IoE), and next-generation network infrastructures redefine global connectivity. In alignment with this vision, the International Telecommunication Union (ITU) has outlined the system requirements and technical drivers for Network 2030 [80], underscoring the importance of distributed intelligence and autonomous operations at the network edge.

The transition from 5G to 6G is characterized by new usage scenarios that extend beyond enhanced mobile broadband (eMBB), ultra-reliable low-latency communications (URLLC), and massive machine-type communications (mMTC) [81, 82]. Huawei, for instance, has introduced the concept of 5.5G with three novel application scenarios: Uplink Centric Broadband Communication (UCBC), Real-Time Broadband Communication (RTBC), and Harmonized Communication and Sensing (HCS) [83]. These advancements indicate a shift towards ubiquitous AI services and IoE applications [84, 85], enabling transformative applications such as sustainable smart cities, connected autonomous systems, brain-computer interfaces, high-fidelity holographic society, and extended reality (XR) [18, 23, 24].

3.4.1 A Complex Systems Perspective for 6G Evolution

The development of 6G introduces novel service paradigms, including Computation-Oriented Communications (COC), Contextually Agile eMBB Communications (CAeC), and Event-Defined uRLLC (EDuRLLC) [18]. These paradigms aim to integrate intelligence, coordination, sensing, and computing within a hyper-connected cyber-physical world. From the perspective of complex systems theory, these paradigms reflect fundamental principles such as *emergence* and *self-organization*. In multi-layered 6G architectures, edge nodes autonomously adjust their communication, computation, and sensing tasks based on local interactions, mirroring decentralized decision-making observed in biological systems and social networks. This adaptive capability allows the network to dynamically reconfigure itself, enhancing resilience and optimizing resource utilization, much like biological ecosystems that self-regulate in response to environmental changes.

The concept of *emergence* is particularly evident in COC and CAeC, where individual nodes optimize their operations through local information exchange, leading to globally coherent behaviors without centralized control. This mirrors the behavior of ant colonies or neural networks, where simple local interactions yield complex global patterns. For example, in a 6G-enabled smart city, edge nodes distributed across traffic lights, surveillance cameras, and environmental sensors can collaboratively optimize traffic flow and energy consumption through localized data processing and decision-making. This self-organization mechanism, driven by edge AI, is fundamental for achieving latency-free, context-aware communication and computing.

To meet the stringent requirements of 6G, interdisciplinary research and disruptive technologies are crucial. These include spectrum exploration, advanced device technologies, and novel networking architectures that support learning, sensing, and computing capabilities. Among these technologies, Artificial Intelligence (AI), particularly Deep Learning (DL), plays a pivotal role in optimizing 6G networks across various layers [86]. DL-driven solutions enable the reconfiguration of the radio environment [87], the enhancement of communication algorithms [88], and the optimization of hardware and applications [89, 90]. In the context of complex systems, these innovations act as catalysts for *emergent behaviors* in communication patterns, such as real-time adaptation to fluctuating network loads or dynamic service reconfigurations. These behaviors are akin to self-organizing processes in complex adaptive systems, where distributed agents, represented by edge nodes and devices, interact and adjust based on local information flows, resulting in scalable and resilient network structures. Moreover, the interplay between sensing, communication, and computing in 6G reflects the principles of *synergetics*, where the cooperative action of components results in higher-order capabilities. In multi-access edge computing (MEC) scenarios, for instance, edge nodes can collectively process and analyze massive datasets, dynamically adjusting resource allocation and service provisioning without the need for centralized orchestration. This aligns with complex systems theory, where collective intelligence and synchronization emerge from local interactions among distributed components.

In a broader perspective, the evolution of 6G can be seen as an embodiment of complex system principles, where decentralized intelligence, self-organization, and emergent communication protocols drive network adaptability and scalability. These features are not merely enhancements but necessary evolutionary steps to manage the hyper-connected, data-driven environments expected in 6G smart cities, industrial automation, and autonomous vehicular networks. By adopting a complex systems perspective, the design of 6G architectures can leverage decentralized learning, edge-driven computation, and adaptive reconfiguration to meet the unprecedented demands of latency, reliability, and massive connectivity. Thus, the advancement of 6G is not only a technological leap but also a paradigm shift towards biologically inspired networking principles, where the self-organizing properties of complex systems provide the foundation for resilient, efficient, and intelligent communication infrastructures. This alignment with complex system dynamics signifies a transformative approach in the engineering of next-generation networks, where adaptive capacity and local decision-making are core design principles rather than mere optimizations.

3.4.2 Edge AI as an Enabler of Distributed Intelligence

Edge AI emerges as a key enabler of connected intelligence in 6G, addressing challenges related to latency, privacy, and network congestion. By performing data processing, learning, and inference at the network edge, Edge AI reduces traffic load, enhances data privacy, and optimizes computational resources [91]. This shift towards edge-based processing alleviates the dependency on centralized cloud infrastructures, thereby reducing latency and enhancing real-time responsiveness. The edge-centric architecture not only optimizes bandwidth utilization but also aligns with the principles of *self-organization* and *decentralization* found in complex systems, where intelligence and decision-making are distributed across interconnected nodes rather than centralized in a single location.

One of the key technologies facilitating this distributed intelligence is *over-the-air federated learning* (FL), which supports collaborative model training without requiring direct access to raw data, thereby preserving confidentiality and reducing latency [92]. FL enables multiple edge devices to independently train local models on decentralized datasets, while only transmitting model updates to a central server for aggregation. This approach significantly mitigates privacy concerns, as raw data remains local, and optimizes communication efficiency by only exchanging minimal gradient information. The collaborative nature of FL mirrors the emergent behavior seen in distributed complex networks, where local interactions collectively contribute to global learning optimization.

Moreover, the concept of *edge inference*, wherein AI models are executed directly at edge nodes, further empowers low-latency, high-reliability services while minimizing computation and communication overhead [92]. In traditional cloud-based AI, the inference process involves sending raw data to centralized servers for processing, which introduces latency and potential security vulnerabilities. Edge inference overcomes these limitations by executing models locally, near the data source, enhancing both security and response time. This decentralized inference mechanism reflects the principles of *locality* and *adaptation* found in complex systems, where decisions are made based on real-time local states rather than waiting for centralized instructions.

To enhance scalability and efficiency, *co-inference* strategies are employed, where the execution of large neural models is distributed across both edge devices and edge servers. This hierarchical processing framework allows critical layers of neural networks to be processed at the edge, while more complex computations are offloaded to more resource-rich edge servers. This multi-layered orchestration reduces bottlenecks and optimizes latency, facilitating real-time applications such as autonomous driving, industrial automation, and augmented reality. From a complex systems perspective, this layered approach mirrors the multi-scale interactions observed in natural ecosystems, where different subsystems operate semi-independently while contributing to global system coherence. As Edge AI becomes more prevalent, it is expected to generate novel data flows that demand efficient resource allocation and advanced wireless techniques. Consequently, next-generation wireless systems must be co-designed with edge intelligence to seamlessly integrate communication, computation, and learning, propelling the vision of *connected intelligence* into reality. This co-design principle is deeply rooted in the foundations of complexity science, where communication, processing, and adaptation are intertwined in an evolving, self-regulating ecosystem. Much like biological networks that adapt and optimize based on local stimuli, Edge AI in 6G networks facilitates self-regulation and emergent intelligence, enhancing robustness and adaptability in dynamically changing environments.

Furthermore, the distributed architecture of Edge AI supports fault tolerance and resilience, crucial for mission-critical applications in 6G such as telemedicine, smart grids, and autonomous transportation. By decentralizing processing tasks across multiple edge nodes, the system is inherently less vulnerable to single points of failure. This resilience is analogous to the concept of *redundancy and adaptation* in complex networks, where local failures do not necessarily compromise the global functionality due to interconnected paths and distributed control.

In summary, Edge AI not only enhances the efficiency of distributed intelligence in 6G networks but also embodies the principles of complex systems, such as self-organization, emergence, and decentralized control. These principles enable the creation of resilient, adaptive, and intelligent network infrastructures, laying the groundwork for the hyper-connected, AI-driven ecosystems envisioned for next-generation wireless communication.

3.4.3 Complex Systems-Inspired Resilience and Adaptability

The principles of complex systems, such as robustness, adaptability, and scalability, are central to the deployment of Edge AI in 6G networks. Edge computing nodes, acting as localized intelligent agents, possess the ability to dynamically adjust to network variations, process failures, or demand surges. This adaptability is critical for enabling applications such as autonomous vehicles, smart cities, and real-time healthcare, where latency and reliability are non-negotiable. In the context of complex systems, this behavior mirrors the principles of *emergence* and *self-organization*, where local interactions among distributed agents lead to the formation of global patterns without the need for centralized control.

Furthermore, the distributed nature of Edge AI aligns with concepts of *resilience* found in complex systems. Edge nodes can reconfigure their processing tasks, redistribute loads, and optimize pathways autonomously, ensuring continuous operation even in the face of disruptions. This mirrors the self-healing properties observed in biological networks and ecological systems, where redundancy and distributed control prevent catastrophic failures. Such resilience is achieved through decentralized decision-making, where each edge node autonomously assesses its local state and interacts with neighboring nodes to reach a consensus on optimal pathways, much like swarm intelligence in nature [91, 92].

In addition, Edge AI's ability to adapt to real-time changes in network topology and user demand reflects the *adaptive capacity* seen in complex adaptive systems. For example, in a smart city scenario, edge nodes can prioritize emergency response communications during critical events, reroute network traffic, and optimize resource allocation dynamically. This form of distributed intelligence is fundamental to the design of next-generation networks, enabling the system to recover and re-optimize in real time, similar to how biological systems react to environmental changes [18]. The decentralized and distributed structure of Edge AI also promotes *scalability*. In traditional cloud-based architectures, scaling up resources often introduces bottlenecks and latency due to centralized processing. In contrast, Edge AI allows for horizontal scaling by adding more edge nodes without affecting overall performance, embodying principles of *distributed growth* typical of complex systems. This approach ensures that as more IoT devices and intelligent applications are integrated into the network, the system scales organically, maintaining efficiency and performance [87, 88].From a complex systems perspective, the emergent behaviors of Edge AI are not simply pre-programmed responses but result from the interactions among distributed nodes, local learning processes, and real-time environmental feedback. This fosters a form of collective intelligence, enabling 6G networks to perform self-optimization, fault tolerance, and rapid adaptation without central oversight. Thus, the edge layer in 6G can be seen as a manifestation of complex adaptive networks, where robustness and adaptability are achieved through decentralized interactions and local intelligence.

The evolution of 6G through Edge AI is not merely an incremental advancement over its predecessors but a paradigm shift towards decentralized intelligence and emergent network behaviors. By harnessing the principles of complex systems, 6G networks can achieve unprecedented levels of efficiency, robustness, and adaptability, paving the way for the fully interconnected, intelligent environments envisioned for the future. This shift towards distributed intelligence mirrors the dynamics observed in natural ecosystems and biological networks, where local interactions give rise to global order and stability. Consequently, the development of Edge AI within 6G architectures can be seen as a crucial step towards realizing self-organizing, resilient, and scalable communication networks [18, 86].

3.4.4 Communication-Efficient Edge Training

Communication efficiency is a critical factor in deploying edge learning models, particularly within the resource-constrained environments typical of edge networks. To address this, various distributed optimization algorithms have been designed to minimize the communication overhead during the training process. These algorithms facilitate efficient synchronization of model updates across distributed edge nodes while reducing latency and conserving bandwidth. One prominent example is Federated Learning (FL), which allows decentralized edge devices to collaboratively train a global model without sharing raw data, preserving user privacy and reducing data transmission costs [91]. Over-the-air FL further enhances this by performing updates directly over wireless channels, reducing communication rounds and optimizing spectrum utilization [92]. This paradigm reflects the *self-organizing* capabilities of complex systems, where local agents interact with their immediate environment, contributing to global learning without centralized coordination.

In addition to FL, model compression techniques, such as quantization and sparsification, have been introduced to further reduce communication overhead. These techniques minimize the size of gradient updates during synchronization, enabling faster model convergence with lower bandwidth consumption [87]. Decentralized learning approaches also align with the principles of complex networks, where learning tasks are distributed across multiple nodes, enabling robustness and adaptability in non-ideal network conditions [88].

Finally, hierarchical edge learning structures, which distribute tasks between edge nodes and cloud servers, embody the principles of *multiscale adaptation* seen in complex systems. In this configuration, localized edge nodes handle real-time data processing and initial model updates, while cloud servers manage global optimization and large-scale analytics. This layered architecture enhances scalability and resilience, allowing 6G networks to adapt dynamically to changes in traffic loads and network topology, further solidifying the connection between Edge AI and complex systems theory [18].

3.4.5 Edge Learning Models and Algorithms

The training process for edge AI models generally involves minimizing a loss or empirical risk function to fit a global model from decentralized data generated by a massive number of intelligent devices. The goal of distributed optimization in edge training is to minimize the global loss function L, defined as:

$$\min_{\theta \in \mathbb{R}^d} L(\theta) := \sum_{k \in S} w_k L_k(\theta; D_k), \tag{3.4}$$

where θ represents the model parameters, L_k is the local loss function for device k over its local dataset D_k, S denotes the set of participating edge nodes, and $w_k \geq 0$ with $\sum w_k = 1$ is the weight for each local loss function. Distributed optimization algorithms must address the challenges of heterogeneous datasets, varying device participation, dynamic communication, and computation environments, along with privacy concerns and adversarial attacks. The structure of this distributed training process reflects core principles of *self-organization* and *adaptability* found in complex systems, where local interactions among edge nodes contribute to the optimization of a global objective without centralized coordination.

Federated Learning

Federated Learning (FL) is a decentralized machine learning framework that allows the training of a global statistical model without accessing the raw data from edge devices. Instead, a dedicated edge server aggregates local learning model updates and disseminates global learning model updates [93]. FL has been widely adopted in real-world applications, such as Google's Gboard for next-word prediction, Apple's QuickType for vocal classifiers, and NVIDIA for predicting COVID-19 patients' oxygen needs [94]. The distributed nature of FL mimics the *emergent properties* of complex systems, where global learning emerges from the collaboration of decentralized entities. One major challenge in FL is the high communication cost of frequent model exchanges over wireless networks. Federated Averaging (FedAvg) [94] addresses this by performing multiple local updates before aggregating at the central server, significantly reducing the number of communication rounds. Techniques like quantization and sparsification further optimize communication by reducing the message size during each update [95–97]. Advanced methods such as vector quantization have further enhanced these approaches [98, 99]. These optimizations allow FL to maintain high scalability, a core feature of complex adaptive systems where nodes can efficiently manage local tasks while contributing to global learning.

Decentralized Learning

Decentralized learning eliminates the need for a central server by leveraging peer-to-peer communications, allowing edge devices to collaboratively learn a global model [100]. This architecture is particularly promising for autonomous industrial applications, such as cooperative automated driving and collaborative robotics [101]. Key methods include consensus-based techniques [102] and diffusion strategies [103] for efficient model aggregation.

In the context of complex systems, decentralized learning aligns with the concept of *distributed control*, where each node independently optimizes its local parameters while periodically exchanging updates with its neighbors. This mirrors the behavior of biological and social networks, where collective behaviors emerge from local interactions without centralized oversight. Communication efficiency is enhanced through methods like variance reduction with gradient tracking [104], periodic averaging [105, 106], and model compression. These methods reduce the number of communication rounds required for model synchronization, reflecting the *self-organizing* nature of decentralized systems.

Model Split Learning

Model split learning partitions model parameters across edge nodes, enabling collaborative training between edge devices and servers [107]. Each edge node is responsible for updating only a specific subset of parameters, optimizing both privacy and communication efficiency. Split Deep Learning (Split DL) further divides deep neural networks into segments processed at the edge and server sides, reducing communication bottlenecks and improving computation scalability [108].

The concept of model partitioning in Split DL echoes the *modularity* found in complex networks, where different segments of a task are handled autonomously by separate agents. This distributed execution enables localized processing and minimizes communication overhead, similar to how subsystems in nature operate independently yet contribute to the global function of the system.

Distributed Reinforcement Learning

Distributed Reinforcement Learning (DRL) facilitates decision-making in dynamic environments, leveraging a multi-agent setup where edge devices interact with their surroundings to optimize a global objective [109]. Actor-critic methods [110] and decentralized multi-agent RL algorithms [27] enable efficient learning in autonomous systems like self-driving cars.

From a complex systems viewpoint, DRL represents a form of *collective intelligence*, where agents operate semi-independently while contributing to a shared goal. Techniques like lazily aggregated policy gradients [109] improve communication efficiency by only transmitting critical updates, much like how biological systems communicate only essential information to conserve energy and optimize responses.

Trustworthy Learning

To ensure the reliability of edge learning in high-stakes applications like autonomous driving, robust and privacy-preserving methods are essential. Differential privacy mechanisms [111] enhance security by adding noise to local updates, preventing sensitive information leakage. Techniques such as geometric median and trimmed mean provide robustness against Byzantine attacks [112, 113]. These mechanisms ensure that the system remains resilient even in the presence of malicious nodes or data corruption, mirroring the *fault tolerance* observed in natural ecosystems and engineered complex systems.

Moreover, blockchain-based architectures have been proposed to secure the integrity of global model updates by validating each learning iteration through consensus mechanisms [114]. These methods prevent tampering and enhance trust in decentralized learning processes, reinforcing the principles of *distributed trust* and *emergent security*, which are key attributes of resilient complex networks.

The edge learning models and algorithms presented demonstrate the need for novel wireless communication techniques and strategies to support massive and flexible edge device participation. They also highlight the importance of efficient function computation for model aggregation, such as weighted sum aggregation in FL, consensus models in decentralized learning, and robust model aggregation in secure learning. Future wireless network designs must address these requirements to unlock the full potential of Edge AI in next-generation communication networks.

The distributed nature of these learning architectures exemplifies the influence of complex systems principles, where local interactions lead to global learning and resilience. As Edge AI continues to evolve, the integration of these paradigms with complex systems thinking will be fundamental for achieving self-organizing, adaptive, and resilient 6G networks. The next generation of edge learning algorithms will not only enhance communication efficiency but will also transform network intelligence, aligning with the emergent properties of complex adaptive systems.

3.5 Digital Twins

3.5.1 Introduction

The concept of Digital Twin (DT) technology has rapidly emerged as a transformative innovation in modern network design, evolving from its roots in industrial manufacturing to become a cornerstone of next-generation wireless communication networks. DTs create real-time, virtual representations of physical assets, enabling enhanced monitoring, simulation, and optimization across diverse domains, including smart cities, industrial automation, and transportation systems [32, 115–117].

The synergy between Edge Intelligence and Digital Twins marks a pivotal advancement in the 6G landscape. As outlined in the previous section, Edge Intelligence enables distributed processing, real-time decision-making, and adaptive control directly at the network edge. This decentralized architecture not only optimizes latency and bandwidth but also provides the computational foundation for real-time synchronization and simulation, which are critical for effective Digital Twin Networks (DTNs). In essence, Edge AI serves as the computational layer that supports the dynamic mirroring and real-time updates of DTs, reinforcing their self-adaptive and resilient nature. This mirrors the concept of *emergence* in complex systems, where local interactions among edge nodes aggregate into coherent, system-wide behaviors. As we move towards the deployment of sixth-generation (6G) wireless networks, the need for more sophisticated management tools has become apparent. The anticipated scale and complexity of 6G networks, characterized by multi-vendor components, ultra-low latency, and massive connectivity requirements, demand innovative solutions for design, analysis, and operation [32, 115]. DT technology is set to address these challenges by providing digital replicas of network infrastructure, user devices, and service mechanisms. These replicas are not static models but are dynamic and real-time synchronized with their physical counterparts, enabling real-time decision-making and predictive maintenance [32].

In the context of 6G, Digital Twin Networks (DTNs) are envisioned as foundational elements to realize the concept of a cyber-physical continuum. A DTN is a digital replica of the full lifecycle of a physical network, using real-time data and computational models to simulate network behaviors and predict future states [115]. Unlike conventional network simulators, DTNs enable two-way communication between the physical and virtual layers, creating a seamless interaction that facilitates closed-loop control and optimized resource allocation. This real-time bi-directional exchange allows for proactive network management, anomaly detection, and intelligent traffic optimization, crucial for maintaining the stringent Quality of Service (QoS) expected from 6G services [116].

The development of DTNs is not purely theoretical; major standardization bodies are already establishing guidelines and frameworks to support its implementation. The International Telecommunication Union (ITU) and the Institute of Electrical and Electronics Engineers (IEEE) have initiated projects to define architecture, sensor interfaces, and interoperability standards for DTNs, reflecting the strategic importance of DT technology in achieving the goals of 6G [115]. The realization of DT in 6G networks is deeply intertwined with various technological advancements. Innovations in cloud/edge computing, IoT, and Internet-of-Vehicles (IoV) serve as the foundational infrastructure to support massive real-time data streams, crucial for synchronizing physical and digital representations [32, 116]. Moreover, breakthroughs in AI/ML enhance the DT's capacity for real-time decision-making, predictive analytics, and autonomous operations, enabling it to dynamically adapt to changing network conditions and user demands [117]. One of the most promising aspects of DT technology in 6G is its application to real-time optimization of network performance. By leveraging big data analytics and AI-driven insights, DTs are capable of simulating various network scenarios to optimize throughput, latency, and resource allocation, preemptively identifying bottlenecks and preventing service degradation [32, 116]. This capability is particularly relevant for handling the exponentially increasing mobile traffic projected for 2030 and beyond, where current 5G networks are expected to struggle under the load [32].

The concept of emergent behavior becomes evident when considering how DTs operate within a 6G infrastructure. Much like in complex systems, where macro-level patterns arise from micro-level interactions, DTNs leverage localized edge processing to build global awareness. This distributed intelligence is inherently resilient, capable of reconfiguring itself in response to network anomalies, cyber threats, or traffic surges without centralized coordination. In this sense, DTNs are not just passive reflections of the physical network–they are active participants in real-time decision-making and optimization.

The strategic integration of DT into 6G is not merely an enhancement but a necessary evolution to meet the ambitious Key Performance Indicators (KPIs) set for next-generation networks, including ultra-low latency, massive machine-type communications (mMTC), and enhanced mobile broadband (eMBB). Thus, Digital Twins will serve as both a driver and an enabler of the hyper-connected, intelligent infrastructure envisioned for 2030 and beyond. This marks a clear step towards achieving the self-organization and adaptability typical of complex systems, positioning DTNs as critical infrastructures for resilient and intelligent 6G networks.

3.5.2 Related Works

The concept of the Digital Twin (DT) originates from early space exploration efforts, notably NASA's Apollo 13 mission, where a mirrored system was used to simulate the spacecraft's conditions and devise recovery strategies in real time [118]. The first formal introduction of the term occurred in 2002, when Michael Grieves presented a three-dimensional model comprising a physical object, its virtual representation, and the data links between them. This was further refined by Framling, who emphasized the role of virtual agents within Product Life Management (PLM) systems. A decade later, NASA defined DTs as integrated, multiphysics and multiscale simulations of real systems, leveraging physical models, sensor data, and historical information to replicate system lifecycles [118].

Digital Twin Networks (DTNs) have recently emerged as key enablers of 6G systems. Sheraz et al. [116] identified the integration of AI with DTNs as a transformative advancement for optimizing communication networks across domains such as healthcare and urban intelligence. Khan et al. [119] proposed a DTN architecture incorporating edge-cloud twinning strategies, highlighting the importance of virtualized components for scalability. Shen et al. [120] expanded this notion by integrating DTs with AI-enhanced network slicing to enable intelligent, data-driven 6G infrastructures. Tang et al. [121] focused on DTNs within edge computing environments, particularly in smart cities and Internet of Vehicles (IoVs) scenarios. Security issues and interoperability remain critical barriers to DTN deployment. Alcaraz and Lopez [122] categorized data security threats in DT environments and proposed mechanisms to ensure data trust and integrity. Barricelli et al. [123] and Minerva et al. [124] explored interoperability and virtualization requirements, emphasizing the need for regulatory frameworks and standards. Wu et al. [29] and Khan et al. [125] called for further investigation into real-time scalability and cross-domain data integration.

DTs are increasingly adopted in manufacturing, healthcare, and autonomous systems. Masaracchia et al. [32] discussed DTs as virtual replicas enhancing lifecycle management, predictive maintenance, and resilience in industrial applications. Bellavista et al. [126] introduced the Application-Driven Digital Twin Networking (ADTN) middleware for dynamic resource allocation in heterogeneous industrial environments. Girletti et al. [127] proposed an edge-based DT framework for robotics, offloading tasks to edge servers to support automation and control. In vehicular applications, Hu et al. [128] introduced a real-time traffic prediction system using DTs and locality-sensitive hashing to overcome data sparsity. Xie et al. [129] developed a vehicular edge computing system with DTs and multi-agent reinforcement learning to optimize task offloading. Similarly, Zhang et al. [130] designed a content caching framework for vehicular networks using LSTM and DDPG algorithms to improve caching efficiency. The synergy between DTs and AI significantly advances DT capabilities. Tao et al. [131] reviewed AI-enhanced DT systems in Industry 4.0, highlighting predictive maintenance and process optimization. Dai and Zhang [132] proposed a DRL-powered computation offloading framework for DTNs, optimizing latency and resource distribution. Van Huynh et al. [133] introduced a DT-empowered edge architecture for ultra-reliable low-latency communications (URLLC), using an alternating optimization algorithm to minimize total latency.

The reviewed literature collectively illustrates that DTNs are central to the vision of 6G, offering solutions for system virtualization, intelligent decision-making, and resilience. However, critical gaps remain, particularly regarding standardization, secure data sharing, real-time adaptability, and scalable deployment. Further research is needed to establish robust frameworks that ensure interoperability, facilitate AI-driven DT orchestration, and address latency and privacy concerns across increasingly heterogeneous communication environments.

3.5.3 AI-Enabled Digital Twin Networks IN 6G

This section provides an in-depth analysis of the foundational elements and critical concepts necessary for the effective integration of Digital Twin (DT) technology within 6G communication networks. Initially, it is discussed the concept of digital twinning, highlighting its significance and practical applications. Following this, it is discussed the transformative impact of DTs on the progression of wireless networks. Lastly, the exploration continues with the complexities involved in merging AI with DTs, revealing the synergistic potential of these technologies in the realm of 6G.

Recently, there has been a surge in research focusing on the implementation of Digital Twin Networks (DTNs) to enhance wireless network optimization. Acknowledging the promising applications of DT technology across various sectors–including healthcare, aviation, and manufacturing–the telecommunications industry is increasingly adopting DT solutions [134]. The advent of 6G is expected to leverage advanced technologies to bolster network service capacity, minimize data access latency, and ensure massive connectivity and reliability [135]. Nonetheless, 6G networks face several challenges, such as spectrum allocation, privacy concerns, and cost implications. DTNs can play a pivotal role in addressing these issues by providing accurate modeling and real-time control of network processes. Deploying DTNs in wireless networks necessitates consideration of the complexity and scale of both physical and logical network assets, including base stations, user devices, network services, and transport infrastructures [136]. The move towards DT adoption is driven by the understanding that advanced digital modeling and the replication of physical objects and transport links within telecom systems offer real-time control over wireless networks. This not only enhances network performance but also ensures a superior Quality of Experience (QoE) for users [137]. Moreover, integrating DTs into wireless networks facilitates predictive network planning, efficient maintenance, and customization across various network layers and processes. This comprehensive approach leads to optimal network performance [138], significantly improving operational efficiency while reducing costs, time, and resource requirements.

In wireless networks, interactions between network infrastructure and end-users are dynamic and complex. These networks must adapt to diverse and evolving user behaviors, device preferences, and varying network conditions. To manage this complexity, the development of two distinct DTs is essential, providing a holistic view of the network and enabling operators to optimize both technical aspects (infrastructure) and user-centric aspects (end-user experiences) [136]. The first DT focuses on network infrastructure, serving as a digital replica of the wireless network's technical components, such as base stations, antennas, spectrum allocation, and communication protocols. This allows network operators to optimize these technical elements for enhanced performance and efficiency. The second DT centers on end-users, capturing and analyzing user behaviors, preferences, and Quality of Service (QoS) expectations. By adopting a user-centric perspective, operators can fine-tune services to align precisely with user expectations, ensuring that the network not only performs efficiently but also meets user demands. These two DTs operate independently, and data associated with one should not be exposed to the other, preserving data privacy and security. This separation enables network operators to conduct virtual network testing, facilitating optimal decision-making in physical 6G networks. This iterative, data-driven approach is crucial for achieving optimal network performance and enhancing the overall wireless communication experience. DT technology enables the replication of physical wireless network components, from the Core Network (CN) to end-users, allowing for autonomous control and management of network elements [136]. Achieving end-to-end twinning of network infrastructure involves several sub-twinning phases. For the 6G infrastructure layer, an end-to-end DT can be developed by integrating DTs for the core network, transport network, and Radio Access Network (RAN) [139]. For example, DT technology can optimize 6G RAN administration by replicating base stations, including elements like antennas, cables, sensors, and power systems. The RAN management function collects extensive data on network performance, functionality, and operational characteristics. AI can then analyze this data before feeding it into the RAN's DT, enabling network operators to optimize predictive maintenance, network operations, and performance [140].

Furthermore, DTs can assess the 6G network's status and alleviate congestion by simulating network coverage areas in real-time, predicting QoS and QoE for users [125]. For instance, DTs can visualize coverage area demographics, installed base stations, and end-user devices, allowing network administrators to model and predict operational environments, wireless interfaces, data traffic, and frequency allocations precisely. Similarly, DTs for Non-Public Networks (NPNs), such as vehicular networks, can visualize roads, traffic congestion, and signals [141]. These applications extend beyond network optimization into user-centric domains, including personalized content delivery based on user preferences, streamlining operational processes, and enhancing workflow efficiencies. DT technology holds the promise of not only optimizing data delivery but also improving personalized content and value-added services, such as data privacy and entertainment, for end-users. This approach concurrently optimizes operational workflows and processes [125].

Developing DTNs for 6G networks involves several key steps. First, accurately virtualizing physical objects while capturing 6G network configurations, architectures, data traffic, and diversity is essential [125]. Second, models defined by Standards Development Organizations (SDOs) can be utilized to enable software and tools for DTs to be developed and managed efficiently, minimizing time and resource expenditure [138]. Third, merging DTs with live software models ensures that network operators have real-time, accurate information for optimal decision-making to enhance network intelligence and operations. Fourth, deploying DTs promptly and ensuring rapid execution is imperative. To achieve execution rates faster than real-time, DTs in a 6G network must employ advanced simulation techniques, such as quantum computing and parallel and distributed computing [140]. DTs must also seamlessly synchronize live and simulated components in real-time, establishing a dynamic, integrated test environment that incorporates DT equipment to improve environment monitoring, testing, and adaptability. Finally, DTs must include tools for accurately and effectively visualizing network traffic, underlying infrastructure, and end-users, enabling analysts and planners to gain a detailed understanding of 6G network functionalities [138]. These advantages allow network operators to maximize operational effectiveness, including network planning, control, management, and termination, while minimizing resource and time expenditure.

3.5.4 Role of AI-Enabled DTN in 6G Fundamental Components

The integration of Artificial Intelligence (AI) with Digital Twin Networks (DTNs) is crucial for optimizing the efficiency and functionality of 6G communication networks. AI-driven techniques enhance critical aspects such as resource allocation, caching, data offloading, and security, which are essential for the advancement of 6G.

- Resource Allocation: The dynamic and heterogeneous nature of 6G networks necessitates intelligent resource allocation strategies. AI-enabled DTNs facilitate real-time monitoring and predictive analysis, enabling efficient distribution of network resources. For instance, Liu et al. [142] proposed a Bidirectional Gated Recurrent Unit (Bi-GRU) algorithm based on Federated Learning (FL) to predict traffic patterns and optimize Virtual Network Function (VNF) migrations. Their approach effectively reduces energy consumption and maintains network stability through load balancing and predictive modeling. Moreover, a Distributed Proximal Policy Optimization (DPPO)-based Deep Reinforcement Learning (DRL) model was employed to optimize VNF migration decisions, demonstrating significant improvements in energy efficiency and resource usage.
- Caching: Efficient caching mechanisms are essential to reduce latency and improve Quality of Service (QoS) in 6G networks. AI-driven DTNs enable smart caching by analyzing user behavior and content popularity to optimize placement strategies.

Sheraz et al. [143] demonstrated that integrating AI in wireless caching reduces access delays and conserves energy by bringing content closer to end users. In this direction, Hui et al. [144] proposed an on-demand content delivery scheme for DT-enabled Heterogeneous Vehicular Networks (DT-HetVNets). The solution leverages DT to facilitate interactions between Vehicular Users (VUs) and Roadside Units (RUs), modeling these interactions as a double auction game to optimize content delivery. Similarly, Zhang et al. [145] introduced a social-aware vehicular edge caching mechanism that uses Deep Deterministic Policy Gradient (DDPG) learning to dynamically manage cache resources based on user preferences and social awareness, significantly enhancing caching efficiency and content distribution.

- Data Offloading: It is a critical strategy for alleviating the load on network infrastructure and improving data rates. In AI-enabled DTNs, proactive resource management facilitates efficient offloading, optimizing both latency and energy consumption. Zheng et al. [146] proposed a DT-enhanced framework using Deep Reinforcement Learning (DRL) with an Asynchronous Advantage Actor-Critic (A3C) algorithm to optimize offloading in the Internet of Vehicles (IoV). The solution employs a cloud-based DT as a software agent to access global network insights and historical data for predictive offloading. In a similar effort, Chen et al. [147] introduced a DT-empowered MEC architecture that manages computation offloading and resource allocation using an A3C-based optimization scheme. This architecture minimizes energy consumption and adapts dynamically to network conditions. Furthermore, Zhang et al. [145] and Dai et al. [148] extended these concepts to vehicular edge networks, proposing adaptive DT models and multi-agent DRL techniques for efficient task offloading and resource distribution.
- Data Security: It is a crucial concern in 6G networks due to their pervasive connectivity and interlinked components. AI-enabled DTNs offer robust security measures by creating virtualized environments to simulate security scenarios. Lu et al. [149] proposed a blockchain-empowered Federated Learning (FL) scheme for Digital Twin Edge Networks (DITEN), enhancing data security and privacy. Their approach employs asynchronous aggregation and reinforcement learning to optimize user scheduling and spectrum allocation securely. Similarly, Liu et al. [150] introduced a DT-assisted task offloading scheme (DTTOS) that integrates blockchain and Channel State Information (CSI) to secure Mobile Edge Server (MES) selection. The scheme utilizes Decision Tree Algorithms (DTA) and Double Deep-Q-learning Networks (DDQN) to optimize offloading decisions while maintaining data integrity. In the healthcare domain, Qu et al. [151] proposed a Digital Twin-assisted Quantum Federated Learning (DTQFL) algorithm for intelligent diagnosis in 5G mobile networks. The algorithm leverages Internet of Medical Things (IoMT) devices to create real-time DTs of patients, enhancing data privacy and computational efficiency.

3.5.5 I-DTN Applications: Smart Cities and Urban Planning

The smart city market is projected to exceed USD 2.376 trillion by 2025 [152]. Achieving the status of a "smart city" requires the implementation of people-centric technological innovations that enhance the quality of life for all inhabitants, extending beyond mere efficiency improvements. Traditional urban planning tools often fall short in addressing the intricate dynamics of urban systems and the diverse needs of citizens. Digital technologies present a substantial opportunity to bridge this gap, fostering intelligent and sustainable governance. By actively involving citizens and supporting democratic urban planning, these technologies offer a holistic perspective on city development. In particular, Digital Twins (DTs) play a pivotal role in tackling the multifaceted challenges of smart cities, especially through applications in virtual reality (VR). DTs promote widespread citizen participation, encourage collaboration, and facilitate consensus-building among various stakeholders, leading to innovative approaches in urban planning [153, 154].

In the realm of 6G Digital Twin Networks (DTNs), data serves as the cornerstone, seamlessly flowing from the physical environment of smart cities to their virtual counterparts represented by DTs. Utilizing sensor technologies like the Internet of Things (IoT), cities acquire real-time, location-specific data to address emerging urban planning challenges, including urban morphology, weather events, climate conditions, accessibility, and proximity. Advanced techniques such as Artificial Intelligence (AI) and Edge computing process this data swiftly, providing valuable insights, patterns, and predictions. The integration of 6G DTNs with immersive technologies like VR and Augmented Reality (AR) fosters hyper-connectivity, creating a highly interactive environment for citizens and stakeholders to collaborate and optimize urban systems effectively. Furthermore, the advent of geo-spatial "big data" has opened new possibilities for enhanced connectivity between communities, the environment, and high-tech devices [155, 156]. Overall, 6G DTNs empower cities to leverage data and advanced technologies to construct accurate virtual representations of their urban environments. These DTs serve as invaluable tools for smart city planners, administrators, and citizens to collaboratively address complex challenges, devise smart solutions, and enhance the overall sustainability and livability of urban areas. Dembski et al. [157] introduced a prototype of an urban DT designed for the town of Herrenberg, Germany. This prototype encompasses several key elements: a 3D-modeled urban DT, street design through graph theory, simulations of urban mobility and wind flow, and a collection of empirical quantitative and qualitative data obtained through volunteered geographic information (VGI). To ensure accessibility and engagement, the urban DT is integrated into a VR visualization platform. The authors conducted public demonstrations of their prototype during various participatory events, such as the Morgenstadt Werkstatt (Tomorrow's Cities Workshop). The results highlight the significant role of the urban DT in facilitating participatory and collaborative processes for smart city planning, empowering citizens and stakeholders in decision-making, and enhancing the overall urban planning process.

Furthermore, combining DT technology with AI algorithms can develop dynamic and real-time virtual representations that accurately reflect urban systems and interact with the physical world. This integration enables data-driven decision-making, efficient resource management, and optimized traffic flow, leading to reduced congestion, lower fuel consumption, and improved air quality. Salunke et al. [158] explored the potential of Reinforcement Learning (RL)-enabled DTs for advancing smart cities towards optimal urban dynamics. Their approach demonstrated enhanced traffic flow, reduced congestion, shortened commute times, and improved overall transportation efficiency. Additionally, projected improvements in air quality, green spaces, and renewable energy adoption underscore the positive environmental implications of this approach. Adreani et al. [159] proposed an automated method for generating 3D city models from diverse datasets and integrating them into the open-source Smart City framework, Snap4City. Their solution facilitates comprehensive visualizations of 3D city elements combined with various Smart City data, including time-series data from IoT devices, heatmaps, traffic flows, bus routes, and cycling paths. Utilizing a deep learning method based on a U-Net architecture, they achieved accurate rooftop detection and alignment. The implementation has been successfully incorporated into the Snap4City platform, enabling intelligent navigation of the city's global DT and its associated data.

Ensuring the security and trustworthiness of digital assets used in a DT is paramount. Pervez et al. [160] introduced the Smart cIty diGital twiN vErifiable Data (SIGNED) framework, built on principles of data ownership, selective disclosure, and verifiability. By employing Verifiable Credentials, SIGNED ensures the authenticity of DT data, encompassing aspects of provenance, transparency, and reliability through verifiable presentation. The evaluation of SIGNED demonstrates its success in facilitating trusted and verifiable data exchange with minimal performance cost, thereby enhancing security and addressing privacy concerns when sharing data among various functional units in a smart city.

3.5.6 AI-Enabled Digital Twin Networks in 6G: Adaptability, Data Exchange, and Energy Efficiency

The rapid evolution of communication services necessitates the development of flexible Digital Twin Networks (DTNs) capable of adapting to emerging services. Integrating Artificial Intelligence (AI) into DTNs enhances their adaptability across various scenarios, significantly reducing design time and costs [29, 125]. Looking forward, 6G is anticipated to be integral to services and applications reliant on wireless resources and user interactions with DTN virtual entities [68]. Efficient deployment of DTNs within communication systems, whether on cloud or edge servers, requires a co-design approach encompassing both software and hardware, with careful consideration of cost implications and precise mathematical modeling of the physical environment [120]. AI, as a data-driven model, facilitates real-time adjustments and precision, ensuring effective communication between DTs. In the context of 6G DTN standardization, AI will focus on optimization and compression to enhance performance in resource-constrained environments [29]. With a multitude of users and devices, including mobile ones, interacting within the network, accurately modeling user mobility becomes crucial for optimal DTN performance. Therefore, standardizing AI algorithms for adaptive learning and decision-making is essential to enable dynamic adjustments based on evolving network conditions.

In 6G, the integration of DTNs and AI promises to enhance real-time and dynamic representations of the physical environment. This integration relies on the seamless availability and exchange of data among diverse components, including sensors, IoT devices, and cloud platforms. However, challenges arise due to the heterogeneous nature of data formats, structures, and protocols, necessitating solutions for uniform data exchange throughout the network [124]. Interoperability across various platforms within DTNs further increases these challenges, demanding the establishment of shared data semantics and consistent data interpretation. For instance, scenarios where sensor data from different manufacturers employ distinct data formats and communication protocols, such as JSON and XML [161], require the development of middleware or translation layers capable of harmonizing data formats and protocols for coherent interpretation across the network.

Real-time synchronization of data is critical for DTNs to ensure a continuous, accurate, and updated depiction of physical objects. Delays in data synchronization can compromise the effectiveness of representations, leading to inefficient decision-making. For example, an industrial DTN aiming to synchronize real-time data from various IoT devices and machines to monitor and optimize production processes may face sub-optimal decision-making due to delays in synchronizing critical parameters like temperature or pressure [162]. Implementing efficient edge computing solutions and low-latency communication protocols becomes essential to ensure timely synchronization and maintain the accuracy of the DTs [91, 92].

The exchange of data between different components introduces challenges related to data privacy, reliability, and security, especially concerning sensitive information. Robust data encryption, access controls, and data anonymization approaches are imperative to address these concerns and mitigate unauthorized access or data breaches [122]. For instance, in a healthcare application of a DTN where patient monitoring devices continuously exchange sensitive health data, robust encryption methods, access controls, and data anonymization techniques are imperative to ensure data privacy and security. Compliance with stringent healthcare regulations, such as HIPAA, further emphasizes the need for robust security measures to protect patient identities and mitigate unauthorized access. In 6G DTNs, precise DT modeling is crucial to provide efficient services such as network management, task offloading, and resource allocation. However, communication and computation constraints hinder precise modeling of physical environments. Therefore, real-time processing of physical object information necessitates the utilization of advanced technologies such as THz communications, Reconfigurable Intelligent Surfaces (RIS) [163], and AI integration into wireless systems to ensure ultra-high capacity and reliable communications.

The integration of AI into DTNs plays a pivotal role in addressing these constraints by leveraging AI and predictive analytics to enhance the precision and efficiency of DT modeling [68]. Additionally, there is a pressing need to develop lightweight and efficient modeling approaches to overcome computational resource limitations [161]. AI techniques, including lightweight machine learning models and edge AI, significantly contribute to addressing this challenge by reducing computational overhead and enabling real-time modeling and decision-making with minimal resource consumption [92].

Moreover, 6G DTNs heavily depend on large and high-quality data, necessitating energy-efficient mechanisms for model training and decision-making. Therefore, data-driven models need to prioritize minimal data collection while ensuring the accuracy and reliability of DT tasks. Periodic and event-triggered data collection approaches can be employed for optimal resource utilization. Periodic data collection enables DTNs to request the network for the desired data at regular intervals, with periods adjusted based on the cost of data collection. Event-triggered approaches transmit data during specific events. Developing adaptive data collection methods, further enhanced with AI-driven techniques, is essential to control the volume and frequency without compromising performance [132].

Deploying AI approaches with lower power consumption is necessary to optimize predictions for maximizing energy conservation. Recently, Federated Learning (FL)-based DT modeling has been explored in 6G DTNs [91, 92]. FL enables local model training and transmits local model parameters to the server for aggregation, providing high-level performance gains in data transmission volume. However, incentivizing physical objects and servers to actively contribute their resources to DT modeling remains a challenge. AI plays a crucial role in optimizing FL-based modeling and resource allocation, ensuring that AI models are energy-efficient while achieving the desired precision. Achieving a balance between modeling precision and cost requires considering the selection of participating devices with varying communication and computation times. AI-driven optimization algorithms assist in achieving this balance by dynamically adjusting the participation of devices based on their capabilities and resource availability. Given the potential of AI in widespread applications across various fields such as medical, space technology, and manufacturing, extensive exploration of AI in 6G DTNs is imperative to foster trust-based joint learning and avoid redundant learning tasks across domains within the DTN [29]. AI's ability to facilitate cross-domain learning and knowledge transfer is essential for achieving efficiency and synergy in 6G DTNs.

The integration of Digital Twin Networks (DTNs) within the 6G ecosystem represents a paradigm shift towards hyper-connected, intelligent communication systems. By providing real-time, synchronized digital replicas of physical components, DTNs enable unprecedented capabilities for predictive maintenance, resource optimization, and seamless orchestration across heterogeneous network environments. As 6G continues to evolve, the convergence of DT technology with Edge Intelligence will play a pivotal role in extending computational capabilities closer to the network edge, further minimizing latency and enhancing system reliability [68]. This strategic synergy will mark a transformative step toward achieving the ambitious goals of 6G, unlocking new possibilities for smart cities, autonomous vehicles, and next-generation industrial automation.

3.6 Machine Learning and AI in 6G Networks

3.6.1 Introduction

The rapid evolution of communication technologies has paved the way for the integration of Machine Learning (ML) and Artificial Intelligence (AI) as foundational enablers of 6G networks. Unlike previous generations, 6G aims to support a fully intelligent and interconnected ecosystem, characterized by massive machine-type communications (mMTC), enhanced mobile broadband (eMBB), and ultra-reliable low-latency communications (URLLC) [23]. The complexity and heterogeneity of 6G require innovative approaches for network management, resource allocation, and real-time optimization, where AI and ML emerge as key solutions [18].

ML-driven algorithms enable dynamic adaptation and optimization across network layers, enhancing spectral efficiency, latency reduction, and energy optimization. In particular, techniques such as Federated Learning (FL) [164], Multi-Agent Reinforcement Learning (MARL) [27], and Deep Learning (DL) are employed to manage decentralized network slicing, resource orchestration, and predictive maintenance. These methodologies align with principles of *self-organization* and *emergence* in complex systems, where local decisions drive global optimizations, mirroring adaptive processes found in natural systems [165].

The integration of AI in 6G networks mirrors the behavior of complex adaptive systems, where multiple autonomous entities–such as edge devices, base stations, and user equipment–continuously interact and adapt to changing network conditions. This decentralized intelligence enables the network to self-configure, self-optimize, and self-heal, enhancing resilience and adaptability [23]. The collective intelligence of these distributed agents emerges from local interactions, optimizing communication paths, resource allocation, and load balancing in real-time without centralized control [18].

3.6.2 Techniques of Machine Learning for 6G Networks

Supervised Learning

Supervised learning is extensively used in 6G for tasks such as channel estimation, traffic prediction, and interference management. Predictive models are trained using historical data to optimize handovers, anticipate traffic congestion, and improve

Quality of Service (QoS). For instance, regression models are utilized for real-time estimation of link quality, while classification models enhance spectrum sensing capabilities [18].

One prominent example is the application of Convolutional Neural Networks (CNNs) for detecting spectrum holes in cognitive radio networks, optimizing bandwidth usage, and minimizing interference. These learning mechanisms operate similarly to biological learning processes in neural systems, where repeated exposure and feedback enhance predictive accuracy and decision-making capabilities [23]. The use of supervised learning in communication networks also reflects the concept of *self-optimization* found in complex systems, where adaptive learning strategies enhance network efficiency through experience and feedback loops.

Unsupervised Learning

Unsupervised learning techniques are fundamental for network anomaly detection and clustering in network slicing. In scenarios where labeled data is scarce, clustering methods enable the identification of hidden patterns, facilitating the dynamic allocation of resources based on usage patterns and fault detection [23]. These methods resonate with the concept of *self-organization*, where distributed nodes adapt to network changes autonomously.

For example, clustering algorithms like K-means and DBSCAN are employed to detect traffic anomalies in dense urban environments, optimizing load distribution across edge nodes. The emergent properties of these models are analogous to collective behaviors in biological ecosystems, where individual agents, without centralized control, dynamically adjust their roles and interactions based on environmental feedback [18]. In 6G, these clustering mechanisms enable intelligent network slicing, where isolated segments of the network independently optimize resources while maintaining global service quality.

Reinforcement Learning (RL)

Reinforcement Learning (RL) introduces decision-making capabilities within network management. Through policy optimization and value-based methods, RL agents learn optimal strategies for load balancing, handover management, and interference mitigation [27]. In 6G, RL is particularly effective in multi-agent environments such as cooperative vehicular networks and decentralized edge computing, enabling distributed intelligence and *emergent behaviors* in network optimization.

A concrete example is the use of Deep Q-Learning for optimizing routing in Software-Defined Networks (SDNs). Agents explore different routing paths and receive feedback based on latency and packet loss, gradually learning the optimal routes for data transmission. This reflects the concept of *stigmergy* observed in ant colonies, where individual agents leave markers to guide collective optimization without centralized coordination [165]. This decentralized approach enhances the network's ability to adapt to congestion, optimize resource usage, and prevent bottlenecks dynamically.

Federated Learning (FL)

Federated Learning (FL) offers a decentralized approach to training machine learning models across distributed edge devices without requiring raw data exchange [164]. This paradigm significantly enhances data privacy and reduces bandwidth consumption, crucial for large-scale 6G deployments. By leveraging edge computing, FL allows for localized model updates, which are then aggregated at a central server, optimizing global models while preserving user privacy [93].

In 6G networks, FL enables edge devices to collaboratively learn models for traffic prediction, network optimization, and anomaly detection without central data aggregation. This decentralized intelligence fosters a form of *self-regulation*, where edge nodes optimize communication based on local observations, mirroring the principles of distributed decision-making seen in complex adaptive systems [18].

Hybrid Learning Approaches

To address the diverse requirements of 6G, hybrid learning approaches that combine supervised, unsupervised, and reinforcement learning are emerging. These methods integrate real-time analytics with predictive modeling, enabling the seamless orchestration of network slices, service differentiation, and resource allocation [18]. Hybrid models enhance the network's adaptability, mirroring *complex adaptive systems* where multiple learning mechanisms operate in tandem.

A practical example of hybrid learning in 6G is the combined use of RL and FL for predictive maintenance and edge resource optimization. By integrating the decision-making capabilities of RL with the decentralized learning of FL, 6G networks can achieve both global optimization and local adaptability. This synergy reflects the principle of *emergent intelligence*, where global performance emerges from locally optimized components, ensuring robust and resilient communication infrastructures.

3.6.3 Network Slicing Through Machine Learning and AI

Network slicing is a pivotal concept in 6G that enables the logical segmentation of physical network infrastructure into multiple virtual networks, each tailored to meet specific application requirements, such as latency, bandwidth, and reliability [18]. This paradigm allows a single physical network to operate as multiple isolated virtual networks, optimized for different use cases like enhanced Mobile Broadband (eMBB), massive Machine Type Communication (mMTC), and Ultra-Reliable Low Latency Communication (URLLC) [23].

From the perspective of complex systems, network slicing mirrors the concept of *modularity*, where subsystems operate semi-independently while contributing to the overall functionality of the larger system. This modular design enhances robustness and scalability, much like how biological systems compartmentalize functions to increase adaptability and resilience to perturbations [164].

Machine Learning for Dynamic Slice Optimization

Machine Learning (ML) algorithms are instrumental in optimizing the allocation of network resources across slices dynamically. Techniques such as Reinforcement Learning (RL) and Deep Reinforcement Learning (DRL) allow the network to learn optimal slicing policies through interaction with the environment, improving resource utilization and reducing latency [27]. For instance, Q-learning and Deep Q-Learning models are applied to adjust bandwidth and computing power allocation in real-time, adapting to fluctuating traffic loads and service demands.

A notable example is the use of DRL for optimizing radio resource management (RRM) in network slicing. Agents are trained to maximize resource efficiency by learning optimal policies for spectrum allocation, interference mitigation, and user handover across slices [165]. This decentralized decision-making process reflects the principles of *stigmergy* in complex adaptive systems, where local interactions lead to emergent global optimization.

Moreover, federated learning (FL) is increasingly employed to enhance slice optimization by enabling distributed edge nodes to collaboratively train machine learning models without sharing raw data [164]. This decentralized learning paradigm ensures privacy preservation and reduces communication overhead, crucial for latency-sensitive applications in 6G networks [93].

Adaptive Network Slicing with Multi-Agent Systems

Multi-Agent Reinforcement Learning (MARL) extends the capabilities of RL by allowing multiple agents to learn and cooperate in a shared environment, optimizing network slicing through distributed intelligence [27]. In 6G networks, MARL-based approaches enable distributed edge nodes to negotiate and adapt resource allocation autonomously, improving overall network efficiency and reducing bottlenecks.

For example, in vehicular networks, agents representing roadside units (RSUs) and autonomous vehicles collaborate to manage spectrum resources dynamically, optimizing latency and throughput for high-mobility scenarios [18]. This cooperative learning process is analogous to *self-organization* in biological ecosystems, where local agents adjust behaviors based on environmental feedback, leading to optimal global outcomes. MARL also enhances the resilience of network slices by introducing mechanisms for fault detection and recovery. Agents trained with reinforcement signals can detect performance degradation and autonomously reconfigure resource allocation to maintain service continuity. This capability mirrors the *self-healing* properties observed in complex systems, where local adaptations contribute to the restoration of global equilibrium [23].

Context-Aware Slicing and Predictive Analytics

The dynamic nature of 6G applications demands context-aware slicing, where network parameters are adjusted based on real-time context, such as user mobility, device density, and traffic patterns. AI-driven predictive analytics enable the network to anticipate load fluctuations, optimize resource allocation, and prevent congestion [18].

Contextual data is analyzed through machine learning models, providing insights for adaptive slice reconfiguration. For instance, mobility prediction models leverage historical data to forecast handovers in high-mobility scenarios, reducing latency and enhancing service reliability [165]. This predictive capability resembles the anticipatory behaviors observed in complex adaptive systems, where components preemptively adjust to maintain stability.

Furthermore, AI-based anomaly detection models continuously monitor slice performance to identify and mitigate potential bottlenecks or malicious attacks in real-time. Techniques like Autoencoders and Recurrent Neural Networks (RNNs) are employed to detect deviations from normal traffic patterns, enabling proactive slice adjustments [164]. This self-regulatory mechanism exemplifies the *homeostasis* principle of complex systems, where local adaptations stabilize global network performance.

Future Perspectives and Challenges

The integration of machine learning and AI into network slicing for 6G networks presents several promising avenues for research and development. Future studies are expected to explore more sophisticated models for distributed optimization, particularly in multi-domain slicing scenarios involving heterogeneous networks and cross-layer orchestration [23]. Additionally, advancements in Federated Learning (FL) will further enhance privacy-preserving network slicing, minimizing data transmission while optimizing global models collaboratively [164]. However, several challenges remain. These include achieving reliable slice isolation to prevent cross-interference, managing the scalability of distributed learning models in ultra-dense networks, and optimizing the trade-off between local adaptation and global optimization. Addressing these challenges will require innovative approaches rooted in complex systems theory, where decentralized control, emergent behavior, and adaptive learning are leveraged to enhance robustness and resilience in 6G network slicing [27].

3.6.4 Network Management Through AI and Machine Learning

Effective network management is a cornerstone of 6G wireless communication systems, ensuring optimal resource allocation, fault tolerance, and service reliability. As network complexity continues to escalate with the proliferation of IoT devices, smart infrastructure, and ultra-dense connectivity, traditional centralized management approaches prove insufficient. Machine Learning (ML) and Artificial Intelligence (AI) introduce a paradigm shift, enabling autonomous, distributed, and adaptive network management that aligns with the principles of complex systems [18, 23].

Self-Organizing Networks (SON)

Self-Organizing Networks (SON) are a critical innovation in AI-driven network management, designed to automate configuration, optimization, and fault recovery tasks without human intervention [164]. In 6G networks, SON leverages AI models to dynamically adapt to changes in traffic load, user mobility, and spectrum availability. Key functionalities of SON include:

- **Self-Configuration**: Upon deployment, base stations and network nodes autonomously configure their operational parameters based on local environmental data and pre-trained machine learning models [165]. This process reduces the time and complexity associated with manual configuration, mirroring the principles of *self-assembly* in biological systems, where components organize themselves based on local information.
- **Self-Optimization**: ML algorithms continuously analyze network performance metrics, such as throughput, latency, and packet loss, to optimize parameters in real-time. Techniques like Reinforcement Learning (RL) and Deep Reinforcement Learning (DRL) allow the system to learn optimal policies for spectrum allocation, load balancing, and interference management [27]. This adaptive behavior resembles the *stigmergy* observed in ant colonies, where local decisions collectively enhance global efficiency.
- **Self-Healing**: AI-driven fault detection and recovery mechanisms allow the network to identify and respond to disruptions autonomously. For example, when a base station fails, the network dynamically reroutes traffic through alternative paths, ensuring continuity of service [164]. This is analogous to *resilience* in ecological systems, where redundancy and distributed control prevent cascading failures.

These self-organizing capabilities are underpinned by distributed learning models, where each network node acts as an intelligent agent, capable of adjusting its configuration based on local and global state observations. This decentralized approach enhances scalability and fault tolerance, crucial for the ultra-dense and heterogeneous environments envisioned for 6G [93].

Autonomous Resource Orchestration

Resource orchestration in 6G networks involves the dynamic allocation of computational, storage, and bandwidth resources to meet the quality of service (QoS) requirements of diverse applications. AI-driven resource orchestration optimizes these processes by predicting network demand and adapting resource distribution accordingly [18].

Machine Learning (ML) techniques such as Supervised Learning and Reinforcement Learning (RL) are employed to predict traffic patterns and optimize resource scheduling. For instance, Long Short-Term Memory (LSTM) networks are utilized for time-series forecasting of bandwidth demands, enabling proactive resource allocation [165]. Moreover, Multi-Agent Reinforcement Learning (MARL) allows distributed network nodes to collaboratively manage resources, ensuring load balancing and minimizing congestion [27].

This decentralized control mechanism mirrors the principles of *emergence* in complex systems, where local optimizations contribute to global efficiency without centralized oversight. Such emergent behavior enables 6G networks to self-adapt to fluctuating traffic loads and varying application demands, enhancing overall network reliability and scalability [23].

Predictive Maintenance and Fault Management

One of the most transformative applications of AI in 6G network management is predictive maintenance, where machine learning models are employed to anticipate equipment failures before they occur. Techniques such as anomaly detection, time-series analysis, and supervised learning are used to identify deviations from normal operation, triggering preemptive maintenance actions [164].

In this context, Digital Twin (DT) technology, as introduced in the previous section, plays a crucial role. Digital replicas of network components continuously synchronize with their physical counterparts, providing real-time monitoring and predictive insights. This capability allows network operators to anticipate failures and optimize maintenance schedules, significantly reducing downtime and operational costs [18]. The underlying approach parallels the concept of *homeostasis* in complex systems, where feedback mechanisms maintain stability despite external perturbations. By monitoring key performance indicators and adjusting configurations in real-time, the network sustains its operational integrity even under adverse conditions [27].

Cognitive Radio and Spectrum Management

Efficient spectrum management is paramount in 6G, where the demand for wireless bandwidth is exponentially increasing. Cognitive Radio (CR) systems, empowered by AI, enable dynamic spectrum sharing by identifying underutilized frequency bands and reallocating them to active users [23].

Reinforcement Learning (RL) algorithms are applied to optimize spectrum sensing and access, allowing CR systems to dynamically adjust transmission parameters based on environmental feedback [165]. Furthermore, Federated Learning (FL) enables distributed edge devices to collaboratively improve spectrum allocation strategies without transmitting raw data, preserving privacy and reducing communication overhead [164].

This adaptive spectrum management mimics the *resource distribution* mechanisms in natural ecosystems, where organisms adapt resource consumption based on local availability. Such self-optimization enhances spectrum efficiency and minimizes interference, critical for maintaining high data rates in 6G networks [93].

Future Directions and Challenges

While AI-driven network management introduces unprecedented levels of automation and efficiency, several challenges persist. These include ensuring robust slice isolation to prevent cross-traffic interference, scaling distributed learning models across ultra-dense deployments, and addressing latency constraints in real-time decision-making [27].

Future research is expected to explore the integration of Distributed Ledger Technologies (DLTs) for enhanced trust and transparency in decentralized learning, as well as Quantum Machine Learning (QML) for faster optimization of network policies [18]. Furthermore, advancing explainability and interpretability of ML models will be critical for regulatory compliance and operational reliability in mission-critical 6G applications [23].

3.6.5 Security and Trust Management with AI and Machine Learning

As 6G networks evolve towards hyper-connected, intelligent infrastructures, ensuring robust security and trust becomes paramount. The exponential increase in connected devices, coupled with decentralized architectures such as Edge Computing and Digital Twin Networks (DTNs), introduces new vulnerabilities that traditional security mechanisms are not equipped to handle [18, 23]. Artificial Intelligence (AI) and Machine Learning (ML) have emerged as key enablers for addressing these challenges, providing autonomous, adaptive, and resilient security solutions that align with the principles of complex systems.

Decentralized Trust Management and Blockchain

The distributed nature of 6G communication networks necessitates decentralized trust mechanisms to ensure data integrity and security across nodes without relying on centralized authorities. Blockchain technology has emerged as a promising solution for trust management in decentralized environments [93]. By leveraging distributed ledger technology (DLT), blockchain enables secure, transparent, and tamper-proof transactions, which is critical for applications like Federated Learning (FL) and multi-agent collaborative systems [27].

In the context of FL, blockchain facilitates secure aggregation of local model updates from edge devices without requiring raw data to be transmitted [164]. This decentralized learning framework mitigates risks associated with data breaches and enhances privacy, enabling collaborative learning across multiple stakeholders. Furthermore, smart contracts–self-executing agreements on the blockchain–automate trust verification processes, reducing latency and enhancing scalability for large-scale 6G networks [165].

The application of blockchain in network slicing further exemplifies its capacity for enhancing security. Each slice can be registered as a separate blockchain instance, enabling real-time monitoring of resource allocation and usage. This distributed ledger approach prevents malicious tampering and provides transparent auditing of slice performance and security [18].

Zero Trust Architecture (ZTA)

The transition towards ultra-dense, multi-tenant 6G networks has prompted a paradigm shift from perimeter-based security models to Zero Trust Architecture (ZTA) [23]. Unlike traditional models, ZTA operates under the assumption that no entity, internal or external, is inherently trustworthy. Every access request is verified, authenticated, and authorized, irrespective of its origin.

Machine Learning (ML) plays a pivotal role in enhancing ZTA in 6G networks. By continuously monitoring access patterns, anomaly detection algorithms can identify deviations from normal behavior, signaling potential threats in real time [27]. Furthermore, reinforcement learning techniques enable adaptive access control mechanisms that dynamically adjust permissions based on real-time threat analysis [93].

The principles of *self-organization* and *adaptability* in complex systems are mirrored in ZTA. Each node or micro-segment in the network independently monitors and defends itself, collaborating with neighboring nodes to form a resilient security perimeter. This decentralized strategy enhances fault tolerance and mitigates the risk of single points of failure [165].

Federated Learning for Privacy-Preserving AI

The massive data exchange required for AI-driven applications in 6G networks introduces significant privacy concerns. Federated Learning (FL) addresses these concerns by enabling collaborative model training without exposing raw data [164]. Instead, only model updates are shared with a central aggregator, ensuring that sensitive information remains localized on edge devices [18].

In addition to preserving privacy, FL is resilient to data breaches and adversarial attacks. Local training at edge nodes reduces the attack surface, while secure aggregation protocols protect model updates from tampering during transmission [93]. Recent advancements in Secure Multi-Party Computation (SMPC) and Differential Privacy (DP) have further strengthened FL, ensuring that collaborative learning can proceed securely even in untrusted environments [27].

The architecture of FL mirrors the *distributed coordination* observed in natural complex systems, where decentralized agents (nodes) independently adapt to local conditions while contributing to a global objective. This structure not only enhances privacy and scalability but also improves network robustness against cyber threats [165].

Anomaly Detection and Threat Mitigation

Detecting anomalies in real time is crucial for maintaining the security and stability of 6G networks. Traditional rule-based intrusion detection systems struggle to adapt to the dynamic, high-dimensional nature of 6G data flows. Machine Learning models, particularly Deep Learning (DL) techniques, have proven effective in identifying complex attack patterns and zero-day vulnerabilities [23].

Autoencoders, Recurrent Neural Networks (RNNs), and Convolutional Neural Networks (CNNs) are employed to detect anomalies by learning normal traffic patterns and flagging deviations indicative of cyber threats [164]. Moreover, Graph Neural Networks (GNNs) have been applied to model the interdependencies between network nodes, identifying anomalous behaviors that spread across network topologies [27].

This real-time anomaly detection capability reflects the *emergent behavior* found in biological immune systems, where local anomaly signals trigger global responses to neutralize threats [18]. By continuously learning and adapting to evolving threats, AI-driven security models contribute to a self-healing network infrastructure.

Future Directions and Challenges

While AI-driven security solutions offer promising enhancements for 6G networks, challenges remain. Ensuring the scalability of distributed learning models, maintaining robust privacy in decentralized environments, and addressing the energy costs of blockchain implementations are key areas of ongoing research [93].

Future research directions include the integration of Quantum Key Distribution (QKD) for enhanced cryptographic security, advancements in Zero-Knowledge Proofs (ZKPs) for confidential transaction verification, and the exploration of bio-inspired algorithms for adaptive threat response [23]. Furthermore, the convergence of blockchain with Federated Learning opens new pathways for transparent, verifiable learning processes in highly distributed environments [27].

3.6.6 Resource Allocation and Optimization in 6G Networks

The unprecedented expansion of connected devices and ultra-dense network deployments envisioned for 6G networks demands advanced resource allocation and optimization strategies [18, 23]. Resource allocation in this context encompasses the efficient distribution of spectrum, computational power, storage, and energy across a highly dynamic and heterogeneous network environment. Traditional centralized resource management approaches are no longer viable due to latency constraints, scalability issues, and the increasing complexity of distributed network topologies [164]. To address these challenges, AI and Machine Learning (ML) techniques have emerged as key enablers for optimizing resource distribution in a decentralized and adaptive manner.

Network Slicing and Virtualized Resource Management

Network slicing is a fundamental enabler of 6G, allowing multiple virtualized and independent networks to coexist on a shared physical infrastructure [18]. Each slice is tailored to specific application requirements–such as Ultra-Reliable Low-Latency Communication (URLLC), Enhanced Mobile Broadband (eMBB), and Massive Machine-Type Communication (mMTC)–by dynamically allocating network resources in a self-organizing manner.

Machine Learning (ML) models are crucial for managing these slices efficiently. Reinforcement Learning (RL) algorithms, for example, optimize the allocation of radio resources by learning from network state feedback to maximize throughput and minimize latency [27]. Multi-Agent Reinforcement Learning (MARL) further extends this capability by enabling distributed decision-making across multiple network slices, ensuring optimal resource sharing and minimizing inter-slice interference [93].

The concept of *emergence*, typical of complex systems, is evident in network slicing: individual slices operate independently, yet collectively contribute to the global optimization of network performance. This decentralized coordination mirrors the behavior of distributed biological systems, where local decisions influence global behavior without centralized control [23].

Edge Intelligence for Real-Time Optimization

The integration of Edge Computing with 6G architectures introduces new paradigms for real-time resource allocation. Edge nodes, equipped with processing capabilities, can handle local computations, reducing the burden on central cloud servers and minimizing latency [164]. Federated Learning (FL) enables edge devices to collaboratively train machine learning models without sharing raw data, enhancing privacy and reducing bandwidth consumption [93].

In terms of optimization, Deep Reinforcement Learning (DRL) models have been applied to edge-based resource allocation. These models learn optimal offloading policies, balancing computational tasks between edge nodes and cloud servers based on network congestion, device capabilities, and latency constraints [165]. This distributed learning mechanism reflects the principles of *self-organization* and *resilience* found in natural ecosystems, where local optimization contributes to global stability and efficiency.

Dynamic Spectrum Allocation

Efficient spectrum utilization is crucial for the success of 6G networks, particularly given the expansion into the millimeter-wave (mmWave) and terahertz (THz) frequency bands [18]. Traditional static spectrum allocation methods are inefficient in highly dynamic environments. Instead, AI-driven spectrum sharing mechanisms allow for real-time adaptation to traffic demand and environmental changes [23].

Cognitive Radio Networks (CRNs), empowered by Reinforcement Learning, enable dynamic spectrum access by sensing unused spectrum bands and reallocating them to devices in real-time [27]. This approach not only maximizes spectrum utilization but also reduces congestion and interference, optimizing Quality of Service (QoS) across heterogeneous network segments.

Furthermore, multi-agent approaches have been employed to enhance spectrum sharing in dense urban environments, where spectrum scarcity is most critical. These agents autonomously negotiate spectrum access, dynamically adapting to usage patterns and network conditions [93]. This distributed intelligence fosters *self-regulation* and *emergent order*, akin to adaptive behaviors observed in biological swarms.

Energy Efficiency and Green Networking

The ambitious goals of 6G also extend to energy efficiency, with the objective of minimizing the environmental impact of massively interconnected networks [18]. AI-driven optimization plays a central role in achieving energy-efficient communication by intelligently managing power consumption at base stations, edge nodes, and mobile devices [164].

Deep Learning (DL) models have been applied to predict energy consumption patterns and optimize load balancing across distributed network components [165].

Reinforcement Learning (RL) further enhances energy efficiency through adaptive power control, dynamically adjusting transmission power based on traffic load and channel conditions [27].

Green Networking principles, aligned with complex systems theory, advocate for distributed energy optimization, where each node autonomously adjusts its power consumption based on local interactions and network-wide feedback [93]. This self-organizing mechanism reduces energy wastage, promotes sustainability, and enhances the resilience of the network infrastructure.

Future Directions and Challenges

Despite significant advancements, several challenges remain for AI-driven resource allocation in 6G. These include:

- **Scalability**: The massive scale of 6G networks demands highly scalable learning models capable of real-time adaptation.
- **Privacy Concerns**: Distributed learning models like FL need robust mechanisms to protect user data during collaborative training.
- **Security Vulnerabilities**: Decentralized resource allocation models are susceptible to adversarial attacks and data poisoning.
- **Energy Constraints**: The high computational demands of AI-driven optimization require energy-efficient algorithms to minimize carbon footprints.
- **Interference Management**: Coordinating multiple slices and spectrum allocations without degrading performance remains an open challenge.

Future research directions include the integration of Quantum Machine Learning (QML) for enhanced optimization, the exploration of bio-inspired algorithms for resilient resource distribution, and the development of robust federated mechanisms for privacy-preserving resource allocation [18, 23]. Additionally, the convergence of Digital Twin Networks (DTNs) with AI-driven resource management is expected to revolutionize real-time monitoring and predictive optimization, further aligning with the principles of *self-adaptation* and *emergence* in complex systems [164].

3.7 Floating Content Architectures

3.7.1 Introduction

Floating Content (FC) is an innovative communication paradigm designed to enable the distributed storage and retrieval of data in geographically localized regions without the need for centralized infrastructure [166, 167]. Unlike traditional caching or cloud-based storage solutions, FC allows information to "float" within a specified area through opportunistic communication among mobile nodes, persisting only within the boundaries of its geographic relevance. This mechanism eliminates the dependency on backhaul networks, making it highly suitable for scenarios where connectivity is intermittent or unreliable, such as disaster recovery, rural communications, and densely populated urban environments [168, 169]. The concept of Floating Content emerged as a solution to address the limitations of distributed communication in highly dynamic and resource-constrained environments. Its primary objective is to maintain the availability of information within a defined geographical area, where mobile devices collectively act as both storage and communication relays. This localized replication process is inspired by epidemic spreading models in complex systems, where information propagates through proximity-based interactions without centralized control [170]. In this context, nodes intermittently exchange data when they come into contact, ensuring that the content "floats" within the target area as long as a sufficient density of nodes is maintained [166].

One of the defining features of FC is its reliance on opportunistic networking principles. In opportunistic communication, nodes do not require a continuous connection to the network backbone. Instead, they dynamically establish peer-to-peer links based on proximity, enabling data transfer even in disconnected environments [171]. This mode of communication aligns with the principles of self-organization observed in complex systems, where local interactions collectively influence global behavior without centralized coordination [172, 173].

From a network architecture perspective, Floating Content introduces a paradigm shift by decoupling content persistence from static storage points, leveraging mobile nodes as transient storage units [168]. This design is particularly advantageous for edge-centric scenarios where rapid data dissemination and low latency are critical. For instance, in vehicular networks, FC can support real-time traffic updates by distributing information locally among vehicles without relying on centralized servers. Similarly, in urban IoT deployments, sensor data can be shared across localized networks, reducing backhaul traffic and enhancing real-time data availability. The relevance of Floating Content in the context of 6G networks is increasingly evident. As communication infrastructures move towards decentralized and self-organizing models, FC represents a viable solution for distributed content dissemination in smart cities, autonomous vehicular systems, and edge-based computing environments. Furthermore, the integration of Floating Content with Digital Twins and Edge Intelligence, as discussed in previous sections, can enhance the robustness and scalability of next-generation communication architectures [32, 115]. The subsequent sections of this chapter will delve into the theoretical foundations, communication models, applications, and complex systems perspectives of Floating Content, highlighting its transformative potential in resilient and scalable 6G networks.

3.7.2 Theoretical Foundations of Floating Content

Floating Content (FC) is a distributed communication paradigm designed to facilitate the opportunistic exchange and localized storage of data in specific geographical areas without the dependency on a centralized infrastructure [166]. Unlike traditional caching or cloud-based storage solutions, FC leverages the mobility and density of mobile devices to maintain data availability in a predefined geographic region, known as the anchor zone. This mechanism relies on direct device-to-device (D2D) communication, where information "floats" among nodes that come within communication range, persisting only as long as the required node density is maintained [167].

The conceptual framework of Floating Content is heavily inspired by epidemic spreading models observed in complex systems [172]. In these models, information propagates similarly to how diseases spread through populations–through random, localized contacts that allow information to hop from one node to another. This distributed form of communication follows the principles of self-organization and emergent behavior, where the collective actions of individual nodes create a global communication fabric without centralized coordination [173]. The stochastic nature of these interactions mirrors the dynamics of complex networks, where the availability of content is probabilistically ensured by the local interactions of mobile nodes. From a theoretical standpoint, FC can be represented as a form of geocasting, where content is bound to a geographic location rather than to specific users or devices [166]. This binding is maintained through opportunistic forwarding, a communication method where data is exchanged only upon physical proximity between devices [171]. The effectiveness of this mechanism is dependent on three critical parameters:

- Node Density: A minimum number of nodes is required within the anchor zone to maintain content availability [170].
- Mobility Patterns: The movement of nodes impacts the probability of contact and thus the spread of content [167].
- TTL (Time-to-Live): The duration for which the content is allowed to persist within the network is predefined, aligning with the transient nature of floating data [168].

The resilience of Floating Content is an emergent property stemming from its decentralized nature. Since there is no reliance on centralized infrastructure, FC is inherently robust against single points of failure. This makes it particularly well-suited for scenarios where infrastructure may be compromised, such as in disaster recovery or highly congested urban environments [166]. Furthermore, the opportunistic exchange among mobile nodes supports self-healing capabilities. If nodes carrying the content leave the area, the information is dynamically redistributed among remaining or newly entering nodes to sustain availability [169]. Complex network topologies, such as small-world and scale-free networks, provide a natural representation of Floating Content's communication patterns [172, 173]. In small-world networks, the average path length between nodes is minimized, which facilitates rapid content dissemination even with sparse connectivity. Meanwhile, scale-free properties ensure robustness to random node failures, as highly connected nodes (hubs) maintain the overall structure of the network.

The application of these topological principles enhances the efficiency and reliability of Floating Content. For example, in urban deployments, where mobile devices frequently converge in high-density zones (e.g., public transport, marketplaces), Floating Content leverages these hubs for efficient data replication and persistence [166]. This emergent behavior mirrors natural complex systems, where decentralized interactions foster resilience and adaptability in dynamic environments.

The propagation of Floating Content can be mathematically modeled using differential equations that account for node mobility, density, and interaction rates. One common approach is the SIR model (Susceptible-Infected-Recovered), adapted to represent the spreading and eventual disappearance of floating data [172]. Extensions of this model consider factors such as:

- Buffer Management: Controlling the data replication process to prevent overflow.
- Data Prioritization: Introducing mechanisms for prioritizing critical information in disaster scenarios.
- Spatial Constraints: Optimizing content spread based on urban geography and movement patterns [171].

These theoretical aspects provide the foundation for understanding how Floating Content achieves localized persistence and resilience, drawing directly from principles in complex systems and network science.

3.7.3 Communication Models for Floating Content in 6G Networks

The deployment of Floating Content (FC) in 6G networks necessitates a robust understanding of communication models that optimize data dissemination and persistence within geographically constrained regions. FC's communication paradigm is fundamentally decentralized, relying on opportunistic networking and distributed storage across mobile devices, rather than traditional infrastructure-based architectures [166, 171]. This design is inherently aligned with the principles of self-organization and emergent behavior observed in complex systems, where local interactions between nodes collectively contribute to global communication efficiency [172, 173]. In Floating Content, data dissemination occurs through opportunistic communication, where nodes exchange information during transient contacts. This form of communication is facilitated by Device-to-Device (D2D) interactions, which eliminate the need for centralized coordination and instead leverage the physical mobility of devices to propagate content [168]. When two nodes come within transmission range, they opportunistically share their locally stored data, ensuring that content "floats" in the anchor zone even as individual nodes move in and out of range [166].

The efficiency of this mechanism is influenced by:

- Node Mobility Patterns: The movement trajectories of nodes significantly impact the spread and persistence of floating content. Studies have shown that certain

mobility models, such as the Random Waypoint Model and Levy Walks, optimize contact rates and prolong data availability [167].
- Contact Duration and Frequency: The amount of time nodes spend within communication range and the frequency of such encounters affect the likelihood of successful data exchange [171].
- Density and Distribution of Nodes: Higher node densities increase the probability of contact, enabling more robust and continuous data propagation within the anchor zone [170].

These characteristics are closely tied to epidemic spreading models in complex systems, where local interactions drive the global propagation of information. Floating Content adopts this epidemic-based mechanism to ensure that information persists within the anchor zone as long as the node density remains above a critical threshold [172]. The effectiveness of Floating Content is contingent upon its data replication and consistency strategies. Unlike traditional distributed systems, FC does not guarantee strong consistency; instead, it prioritizes eventual consistency, where nodes converge to a consistent state over time through continuous interactions [169]. This approach reduces the overhead associated with strict synchronization and aligns with the principles of complex systems, where global stability emerges from local, asynchronous updates.

Three main replication strategies are typically employed:

- Epidemic Replication: Inspired by biological contagion, this method allows information to spread by "infecting" neighboring nodes upon contact [170]. The redundancy introduced by this method enhances data availability but may lead to resource consumption if not managed efficiently.
- Probabilistic Replication: To mitigate redundancy, probabilistic models are applied, where nodes decide to replicate content based on predefined probability metrics [171]. This mechanism reduces the replication burden while still maintaining coverage.
- Context-Aware Replication: In scenarios such as urban centers or high-density events, context-aware mechanisms adjust replication rates based on environmental variables like node density, mobility, and connectivity patterns [168].

These strategies are analogous to self-organizing principles in complex adaptive systems, where local decision-making drives global outcomes without centralized control. A key challenge for Floating Content is ensuring spatial and temporal persistence within the anchor zone. Spatial persistence refers to the ability of content to remain accessible throughout the defined geographic region, while temporal persistence denotes the duration for which content is retrievable as nodes move dynamically [166].

To model this, several approaches are used:

- Anchor Zone Definition: The geographic boundary where content is intended to float is predefined. Nodes within this area participate in content propagation, while those outside are excluded [166].

- Time-to-Live (TTL) Mechanisms: To prevent indefinite propagation, data is tagged with a TTL value, ensuring it is deleted after a specified duration or upon leaving the anchor zone [167].
- Density-Based Persistence: Content remains available as long as node density within the anchor zone is sufficient to support data transfer. When density drops below a critical threshold, content decay begins, mirroring phase transitions in complex systems [172].

These persistence strategies are vital for urban applications, emergency scenarios, and vehicular networks where reliable data availability is crucial [168].

The deployment of Floating Content in 6G networks introduces new possibilities for ultra-reliable low-latency communication (URLLC) and massive machine-type communication (mMTC). Floating Content's decentralized nature reduces the dependency on backhaul communication, optimizing edge-based data propagation and enhancing network resilience [169]. Furthermore, its integration with Reconfigurable Intelligent Surfaces (RIS) and Edge Intelligence enables dynamic adaptation to real-time network conditions, reinforcing the self-organizing capabilities of next-generation communication architectures.

For instance, in vehicular networks, FC can facilitate real-time traffic updates and hazard alerts through local D2D interactions, bypassing centralized delays [170]. In smart city infrastructures, sensor data can be opportunistically shared among connected devices, supporting seamless communication even during infrastructure failures [171].

3.7.4 Complex Systems Perspective in Floating Content for 6G Networks

The emergence of Floating Content (FC) as a decentralized communication paradigm aligns closely with the theoretical principles of complex systems. At its core, FC leverages self-organization, emergent behavior, and robustness–hallmarks of complex adaptive systems–to enable scalable and resilient data dissemination in 6G networks [172, 173]. This decentralized model is inspired by biological networks and epidemic spreading, where local interactions collectively drive global outcomes without centralized control [16].

In complex systems, self-organization describes the ability of a system to spontaneously arrange its components into structured patterns through local interactions, often in the absence of centralized coordination. Floating Content adopts this principle by enabling data to "float" within a specified anchor zone through device-to-device (D2D) communication, dynamically adjusting to node density and mobility patterns [166]. This self-organizing behavior ensures that information remains accessible within the target area as long as a sufficient density of nodes is present, even as individual devices join or leave the network [167].

Emergent properties in Floating Content are observed in the way data spreads across mobile nodes. Similar to epidemic models, content dissemination in FC follows probabilistic replication paths that depend on opportunistic contacts among devices [170]. The local interactions between mobile nodes contribute to the global persistence of data within the defined geographic boundary, mimicking the contagion processes in social and biological networks [172]. A distinctive feature of Floating Content is its reliance on decentralized control, where the responsibility for data replication and storage is distributed across all participating nodes. Unlike traditional communication paradigms that depend on centralized servers for synchronization and updates, FC operates autonomously, using peer-to-peer (P2P) mechanisms to maintain content availability [171]. This distributed consensus is analogous to swarm intelligence in biological systems, where individual agents operate based on local rules to achieve global coordination [174].

In 6G networks, this decentralized model reduces backhaul traffic and enhances scalability by minimizing the reliance on core network infrastructure. Nodes dynamically adjust their storage and forwarding strategies based on local observations, contributing to a self-healing and adaptive communication architecture [168]. This capability is particularly advantageous in environments with intermittent connectivity, such as rural areas, disaster recovery zones, and vehicular networks [167].

The scalability of Floating Content is inherently linked to its distributed architecture. As node density increases, the probability of opportunistic contacts rises, enhancing the persistence and reachability of floating data [169]. This scalability is crucial for urban IoT applications, where thousands of devices communicate within dense environments. Furthermore, the robustness of FC is derived from its redundancy mechanisms: multiple copies of data are replicated across diverse nodes, ensuring that content remains accessible even if individual nodes fail or move out of range [171].

Additionally, the stochastic nature of Floating Content propagation–rooted in epidemic spreading models–provides resilience against unpredictable disruptions. When nodes move unpredictably or experience connectivity loss, the distributed copies of the content continue to float within the anchor zone, maintaining data availability [170]. This behavior mirrors concepts of network robustness in complex systems, where decentralized control enhances resilience against localized failures [16].

The transition to 6G networks introduces new opportunities for enhancing Floating Content with advanced technologies:

- Reconfigurable Intelligent Surfaces (RIS): RIS can dynamically manipulate wireless propagation environments to improve signal coverage and reliability for FC, enabling more stable opportunistic contacts [39].
- Edge Intelligence: Integrating FC with edge-based AI allows for real-time optimization of content replication and storage strategies, reducing latency and improving data retrieval rates [18].

- Digital Twin Networks (DTN): FC can be synchronized with DTNs to simulate real-time floating content propagation across virtual replicas of urban environments, enabling predictive analytics and optimized resource allocation [32].

In 6G smart city scenarios, Floating Content can facilitate real-time sharing of environmental data (e.g., pollution levels, traffic congestion), public safety alerts, and hyper-local news through decentralized P2P dissemination. Vehicles in vehicular ad-hoc networks (VANETs) can exchange hazard warnings and traffic information directly with nearby cars, enhancing safety and traffic efficiency without the need for centralized coordination.

3.7.5 Security and Privacy in Floating Content for 6G Networks

As Floating Content (FC) becomes a critical enabler for decentralized communication in 6G networks, ensuring its security and privacy remains a paramount concern. Unlike traditional communication models that rely on centralized control and infrastructure-based security measures, FC operates through opportunistic interactions and distributed storage among mobile devices [171]. This decentralized nature introduces unique security challenges, such as data integrity, confidentiality, node trustworthiness, and resistance to malicious attacks.

Security Challenges in Floating Content

The decentralized architecture of FC exposes it to several vulnerabilities, primarily due to the lack of a centralized authority to monitor and verify data exchanges [168]. Key challenges include.

Data Integrity and Authenticity

In FC, data is replicated and disseminated across multiple mobile nodes. This replication process, while enhancing availability, also increases the risk of data corruption and unauthorized modification. Malicious nodes can inject false information or tamper with legitimate data, leading to widespread misinformation in applications like vehicular networks or emergency communication systems [170]. Techniques such as cryptographic hashing and digital signatures are essential to validate the authenticity of floating content during opportunistic exchanges. Recent work has suggested the integration of blockchain technology to establish immutable records of data exchanges, ensuring integrity without relying on centralized verification [175].

Confidentiality and Privacy

Since FC operates on peer-to-peer communication without centralized encryption, sensitive information can be intercepted by unauthorized nodes. This is particularly critical in urban IoT settings or vehicular networks where personal and location-based data may be exchanged opportunistically [171]. Mao et al. [77, 176] highlighted the need for lightweight cryptographic protocols and privacy-preserving data sharing mechanisms tailored for edge-based architectures [77]. Techniques like Attribute-Based Encryption (ABE) [177] and Homomorphic Encryption [178] are emerging as viable solutions to protect data confidentiality in distributed environments.

Trust Management

In the absence of centralized verification, assessing the trustworthiness of participating nodes is challenging. Opportunistic interactions rely on local decisions for data dissemination, which can be exploited by malicious actors to spread corrupted or misleading information [169]. Distributed trust models, inspired by reputation-based systems in complex networks, are proposed as solutions to evaluate node reliability based on historical interactions. Blockchain-based trust mechanisms have been suggested to maintain a distributed ledger of node behavior, enhancing accountability in Floating Content environments [175].

Sybil and Replay Attacks

Due to its distributed nature, FC is particularly susceptible to Sybil attacks, where a single malicious entity fabricates multiple fake identities to manipulate data flows or disrupt network stability [171]. Similarly, replay attacks can occur when malicious nodes capture data packets and retransmit them at later times, causing outdated or misleading information to propagate in the network. Advanced Sybil-resistant mechanisms, such as decentralized identity verification and multi-factor authentication, are critical to mitigating these risks [179].

Node Impersonation and Spoofing

The lack of strong authentication mechanisms in floating content networks exposes them to node impersonation, where malicious entities present themselves as legitimate nodes to intercept or alter communications [168]. To address this, identity-based encryption and decentralized authentication protocols are necessary to validate node legitimacy during opportunistic exchanges [175].

Privacy-Preserving Mechanisms for Floating Content

To address these vulnerabilities, several privacy-preserving mechanisms are proposed.

Distributed Ledger Technologies (DLT)

Blockchain and Directed Acyclic Graph (DAG) architectures have been suggested as methods for enhancing the security of decentralized communication [175]. By leveraging DLT, each data exchange within the FC environment is recorded on an immutable ledger, ensuring traceability and integrity without relying on centralized control. This approach also facilitates decentralized trust, mitigating the risks associated with data manipulation and unauthorized modifications.

Attribute-Based Encryption (ABE)

This cryptographic approach allows data to be encrypted based on node attributes rather than individual identities, enhancing privacy during data dissemination [177]. For instance, in a smart city application, only nodes with specific geographic attributes or access privileges can retrieve and decrypt sensitive information.

Homomorphic Encryption

Homomorphic encryption enables computation on encrypted data without the need for decryption, preserving privacy even during processing [178]. This approach is particularly useful in FC scenarios where data needs to be aggregated or processed locally without exposing sensitive information.

Trust-Based Routing Protocols

To mitigate the risk of malicious nodes in opportunistic communication, trust-based routing protocols evaluate the reliability of nodes based on past interactions and network behavior. These protocols prioritize data transmission through trusted pathways, minimizing the influence of compromised devices [169].

Smart Contracts for Autonomous Security

Smart contracts can automate security policies and data validation without the need for central coordination. They allow for predefined rules to execute security measures autonomously when certain conditions are met, enhancing both reliability and trustworthiness of data exchanges within localized FC regions [175].

Future Directions and Research Challenges

Security and privacy in Floating Content for 6G are still nascent fields with numerous open research challenges:

- Decentralized Trust Management: Developing scalable, trust-based mechanisms that do not rely on centralized verification is crucial for FC's expansion into smart cities and vehicular networks.
- Secure Opportunistic Data Exchange: Enhancing security in peer-to-peer data sharing without compromising latency or energy efficiency remains an open problem.
- Integration with Digital Twin Networks (DTNs): As discussed previously, DTNs in 6G will require seamless and secure synchronization of real-time data. Research on how FC's decentralized dissemination can be securely integrated with DTNs is ongoing.
- Quantum-Resistant Security Protocols: The advent of quantum computing poses a significant threat to classical encryption schemes. Future FC architectures must explore quantum-resistant cryptographic techniques to safeguard opportunistic communications [179].

3.7.6 Resource Management and Scalability in Floating Content for 6G Networks

The effective management of network resources and scalability is critical challenge in the deployment of Floating Content (FC) architectures for 6G networks. The inherently decentralized nature of FC requires efficient mechanisms to optimize resource allocation, data persistence, and replication within highly dynamic and geographically localized communication environments [166, 167]. Unlike traditional caching or cloud-based storage solutions, FC relies on mobile nodes to act as temporary storage units, maintaining content availability through opportunistic communication [168]. This section explores the techniques for resource optimization and the scalability of FC in the context of next-generation communication networks, highlighting the role of complex systems theory in enhancing resilience and adaptability.

Dynamic Resource Allocation in Floating Content

One of the primary challenges in FC architectures is the dynamic allocation of network resources, particularly in environments characterized by fluctuating node density and mobility patterns. Efficient resource allocation in FC is closely tied to the principles of complex systems, where local interactions between nodes influence the global persistence and dissemination of content [172, 173]. In this distributed model, nodes autonomously manage data replication based on their local view of network density and connectivity patterns, mimicking the self-organizing behaviors observed in biological and ecological systems [170].

To address the limitations of traditional allocation strategies, recent research has proposed decentralized optimization algorithms inspired by swarm intelligence and game theory. These approaches enable nodes to collaboratively decide on caching strategies, replication factors, and transmission priorities based on real-time environmental conditions [171]. For instance, local congestion awareness mechanisms can be employed to prevent network overloads by dynamically adjusting the replication rate and prioritizing content that is spatially and temporally relevant. This adaptability reduces latency and optimizes bandwidth usage, aligning with the principles of self-organization and emergent behavior characteristic of complex networks [172].

Scalability Challenges in 6G Floating Content Networks

Scalability in FC is primarily influenced by the ability of the architecture to maintain data persistence and availability as the number of mobile nodes and the volume of transmitted data increase. In 6G scenarios, characterized by massive device connectivity and ultra-dense networks, FC must efficiently scale to accommodate rapid changes in node mobility and geographical density [167, 168]. This requirement is compounded by the introduction of IoT devices, autonomous vehicles, and smart city infrastructures, which significantly increase the demand for localized data exchange and low-latency communication [169].

A promising approach to enhance scalability in FC systems is the integration of blockchain-based mechanisms for decentralized trust and data integrity [180]. Blockchain technology enables secure, transparent, and verifiable record-keeping of data exchanges between nodes, reducing the risks associated with malicious data manipulation and unauthorized access. Furthermore, the use of attribute-based encryption (ABE) [177] and fully homomorphic encryption (FHE) [178] allows for fine-grained access control and secure data sharing in opportunistic networks, ensuring that only authorized nodes can access or modify floating content.

Additionally, complex systems theory suggests that scalability in distributed networks can be achieved through the principles of modularity and redundancy. By organizing FC nodes into overlapping clusters or communities, it is possible to enhance fault tolerance and load balancing, mimicking the resilient behaviors observed in natural ecosystems [173]. This approach not only improves the persistence of content within defined geographical regions but also reduces the overall communication overhead, thereby optimizing energy consumption and extending the lifespan of mobile nodes.

Future Directions and Open Challenges

As 6G networks evolve, the scalability of FC architectures will depend increasingly on advanced resource optimization techniques and intelligent data management strategies. Future research must focus on the development of predictive algorithms that leverage machine learning and artificial intelligence to anticipate node mobility, network congestion, and content demand [172]. Moreover, integrating Digital Twins with FC could enable real-time simulation of network dynamics, offering enhanced capabilities for predictive maintenance and optimization in urban environments and autonomous vehicular networks [115].

In addition, the convergence of Edge Intelligence and FC represents a promising avenue for achieving ultra-low latency and high data availability in distributed communication scenarios. By offloading computational tasks to the network edge, FC can enhance its capacity to handle large volumes of data while reducing latency and backhaul dependency [168]. This synergy is particularly relevant for mission-critical applications such as disaster recovery, vehicular ad hoc networks (VANETs), and real-time urban monitoring.

Complex systems provide a robust theoretical framework for addressing the challenges of resource management and scalability in Floating Content architectures. By leveraging decentralized optimization, modularity, and redundancy, FC can achieve high levels of resilience and adaptability, paving the way for its integration into next-generation 6G networks. Floating Content (FC) emerges as a transformative paradigm for localized, resilient, and infrastructure-independent communication in the era of 6G networks. By leveraging the mobility and density of mobile nodes, FC enables distributed data dissemination in geographically constrained areas without the dependency on traditional backhaul infrastructures. This innovative approach not only enhances communication efficiency in urban IoT deployments, vehicular networks, and disaster recovery scenarios but also reduces backhaul load and latency by maintaining information persistence through opportunistic peer-to-peer interactions [166–168].

The theoretical underpinnings of FC are deeply rooted in complex systems science, particularly in concepts such as epidemic spreading, self-organization, and emergent behavior. Drawing parallels with natural systems, FC exhibits robustness, scalability, and decentralized control that are critical for next-generation communication networks. Epidemic spreading models illustrate how localized interactions can propagate information efficiently, mimicking the natural diffusion of data across mobile nodes without the need for centralized coordination [172, 173]. This opportunistic exchange mechanism allows FC to self-heal and adapt to changes in node density and mobility patterns, ensuring content persistence even in highly dynamic environments [169].

One of the most compelling advantages of Floating Content is its alignment with emerging 6G paradigms such as Digital Twin Networks (DTNs) and Edge Intelligence. FC seamlessly integrates with these technologies to enhance real-time monitoring, predictive maintenance, and distributed decision-making in smart cities and autonomous vehicular systems [32, 115]. The ability to localize content within specific geographic regions not only optimizes bandwidth consumption but also enables low-latency communication critical for ultra-reliable and massive machine-type communications (mMTC) envisioned for 6G.

Despite its advantages, FC also introduces critical challenges in terms of security, privacy, and resource management. Its decentralized architecture, while resilient, is inherently vulnerable to data tampering, unauthorized access, and malicious attacks. To address these issues, blockchain-based distributed ledgers, attribute-based encryption (ABE), and homomorphic encryption are proposed as effective mechanisms for enhancing data integrity and confidentiality [77, 175, 179]. Furthermore, decentralized trust models and trust-based routing protocols are emerging to ensure node reliability and mitigate risks associated with Sybil and replay attacks in opportunistic networks [169].

From a resource management perspective, FC demonstrates scalability through self-organizing principles. By optimizing resource allocation dynamically and leveraging swarm intelligence and game-theoretic models, FC can effectively manage replication and persistence in densely populated or highly mobile environments [171]. These mechanisms, inspired by complex systems theory, ensure that the network remains robust and adaptive to node failures and congestion, mimicking the resilience observed in biological and ecological systems [172].

Looking forward, the full potential of Floating Content in 6G networks hinges on continued advancements in opportunistic communication models, distributed ledger technologies, and privacy-preserving mechanisms. Future research directions should explore the integration of quantum-resistant cryptographic techniques to secure decentralized data exchanges and the adoption of Digital Twin models to enhance predictive analytics and resource optimization. Moreover, machine learning-driven predictive algorithms can further enhance Floating Content's capacity to anticipate network congestion, optimize replication strategies, and improve latency in real-time applications [115].

In conclusion, Floating Content represents a paradigm shift towards resilient, self-organizing, and infrastructure-independent communication in the 6G era. Its integration with complex systems theory and edge-based intelligence paves the way for scalable, secure, and low-latency data dissemination in next-generation networks. As 6G technologies mature, Floating Content is poised to become a cornerstone of decentralized, adaptive, and hyper-connected communication infrastructures.

3.8 Complexity and Quantum Technologies

3.8.1 Introduction

The convergence of complexity science and quantum technologies is driving a paradigm shift in the design and optimization of next-generation telecommunications networks. Modern infrastructures, such as those envisioned for 6G and beyond, are not merely expansions of previous generations; they represent a transformative leap towards self-organizing, decentralized, and resilient architectures capable of handling unprecedented scales of connectivity and data throughput [181]. At the same time, quantum technologies are emerging as critical enablers for enhanced security, synchronization, and computational efficiency, addressing key challenges in network reliability and data privacy [182]. This section explores how the principles of complex systems, such as emergence, self-organization, and non-linearity, integrate with quantum communication technologies to support ultra-reliable, low-latency, and secure communications. The analysis of how Quantum Key Distribution (QKD) and Quantum Random Number Generators (QRNGs) can be leveraged to enhance the robustness and security of highly interconnected network infrastructures follows.

3.8.2 Complex Systems Perspective in Telecommunications

The telecommunications industry exemplifies the properties of complex adaptive systems. It is characterized by multi-layered interactions, feedback loops, and emergent behavior that arise from the non-linear dynamics of its components [181]. Unlike traditional centralized architectures, next-generation networks are expected to be distributed, edge-centric, and capable of self-organization. These properties mirror the behavior of natural complex systems, where local interactions yield global stability without centralized oversight. Self-organization in telecommunications networks enables decentralized nodes to autonomously manage routing, resource allocation, and fault recovery. This mirrors the principles observed in biological systems, where local interactions among agents (e.g., ants, neurons) lead to globally coherent behaviors [172]. For example, in Device-to-Device (D2D) communication, nodes dynamically establish links based on proximity and data demand, optimizing data dissemination in real-time without the need for a central controller [166]. This decentralized architecture enhances resilience by reducing single points of failure and enabling adaptive recovery mechanisms in response to network disruptions. The intrinsic resilience of complex systems is derived from their distributed nature and redundancy. In telecommunications, the adoption of Network-of-Networks (NoNs) architectures allows for multiple overlapping communication paths, reducing vulnerability to localized failures. These multi-layered interconnections enable traffic rerouting and load balancing, akin to biological ecosystems that maintain stability through species diversity and adaptive interactions [181].

3.8.3 Quantum Technologies in Telecommunications

Quantum technologies are emerging as game-changers for telecommunications, promising enhancements in security, synchronization, and computational capabilities. Unlike classical networks, which rely on encryption protocols that are vulnerable to algorithmic breakthroughs, quantum-based communications leverage the principles of quantum mechanics to achieve unprecedented levels of security and efficiency [182].

Quantum Key Distribution (QKD) is a cryptographic protocol that leverages quantum entanglement and Heisenberg's uncertainty principle to establish secure communication channels. In QKD, any attempt to eavesdrop on the communication introduces detectable anomalies, ensuring data integrity [182]. This method is particularly relevant for edge-based architectures in 6G, where decentralized communication needs to be both secure and resilient against cyber threats. The deployment of QKD in Network-of-Networks (NoNs) enables secure data exchange across distributed network segments. This integration ensures that each sub-network can autonomously verify the authenticity of data packets without the need for a central authority. Moreover, the inherent properties of quantum entanglement allow for near-instantaneous synchronization between nodes, drastically reducing latency in cross-network communication. QRNGs are pivotal for enhancing the security of encryption mechanisms in telecommunications. Traditional pseudo-random number generators (PRNGs) rely on deterministic algorithms, making them susceptible to prediction and cyber attacks. Conversely, QRNGs exploit quantum indeterminacy to produce truly random numbers, ensuring the unpredictability of cryptographic keys [182].

In 6G networks, QRNGs are integrated into distributed edge nodes, facilitating secure, on-the-fly key generation for encrypted communication. This decentralization aligns with the principles of complex systems, where local node behavior contributes to global security without centralized control. Furthermore, QRNGs support time-sensitive applications, such as real-time vehicular networks and smart grid communications, by ensuring that security is maintained across dynamically changing network topologies.

3.8.4 Complexity-Inspired Quantum Networks

The fusion of complexity theory and quantum technologies is poised to redefine the architecture of telecommunications networks. This convergence enables the development of Quantum-Enhanced Networks (QENs), which exhibit:

- **Self-Healing Capabilities**: Complexity-based models allow for adaptive recovery from node failures, while quantum entanglement facilitates instantaneous state synchronization across distributed nodes.
- **Decentralized Coordination**: Quantum teleportation techniques enable instant state synchronization, reducing the need for intermediary routing and minimizing latency in distributed ledger technologies.
- **Scalable Security Layers**: QRNGs integrated with complex network overlays provide entropy-rich security layers that adapt to real-time cyber threats.

The architecture of QENs reflects the principles of complex adaptive systems. Through decentralized quantum key exchanges and self-organizing data replication, these networks can maintain global coherence and fault tolerance, even under adversarial conditions [182].

3.8.5 Future Directions: Quantum Network-of-Networks (Q-NoNs)

The logical evolution of complexity and quantum integration is the development of Quantum Network-of-Networks (Q-NoNs). In this architecture:

- **Ultra-Low Latency Communication**: Entanglement-based state propagation ensures real-time synchronization across geographically distributed nodes.
- **Autonomous Resource Allocation**: Quantum-based algorithms optimize bandwidth and node utilization based on real-time network conditions.
- **Enhanced Fault Tolerance**: Quantum error correction, combined with the modular redundancy of complex networks, reduces susceptibility to localized failures.

Q-NoNs represent the pinnacle of decentralized, secure, and adaptive telecommunications infrastructures, aligning with the principles of self-organization and emergent resilience found in complex systems. These networks are anticipated to form the backbone of smart cities, autonomous vehicles, and global IoT ecosystems in the 6G era and beyond. The integration of complexity science and quantum technologies represents a transformative shift in the architecture of next-generation telecommunications networks. By adopting self-organizing principles and leveraging quantum mechanics for enhanced security and synchronization, telecommunications infrastructures can achieve unprecedented levels of resilience, scalability, and efficiency. As 6G and subsequent generations continue to evolve, the synergy between complexity and quantum principles will be crucial for enabling decentralized, robust, and ultra-secure global communication networks.

3.8.6 Quantum Internet and Its Role in Next-Generation Telecommunications

The advent of the *Quantum Internet* represents a fundamental shift in global communication networks, enabling the seamless exchange of quantum information across distributed systems. Unlike classical networks, which transmit bits, the Quantum Internet allows the transmission of quantum bits (qubits), leveraging phenomena such as superposition and entanglement to achieve unparalleled levels of security and computational efficiency [182]. This new paradigm promises to unlock transformative applications in cryptography, distributed computing, and ultra-secure communications, all of which are pivotal for the future of 6G and beyond.

At the core of the Quantum Internet lies the principle of *quantum entanglement*, wherein the state of one qubit instantaneously influences the state of another, irrespective of physical distance [183]. This property enables the creation of entangled networks where information can be shared with absolute synchronization and minimal latency. In telecommunications, this could revolutionize synchronization protocols, enabling real-time data alignment across global networks without the need for intermediary relay stations.

Quantum entanglement also facilitates *Quantum Teleportation*, a process that allows the state of a qubit to be transmitted from one location to another without physical movement of the qubit itself [183]. This is particularly significant for ultra-low latency applications, such as autonomous vehicular networks and smart city infrastructure, where instantaneous synchronization is crucial for safety and efficiency.

The Quantum Internet not only enhances communication but also paves the way for *Distributed Quantum Computing*. In this framework, quantum processors located at the network edge can collaborate to solve complex optimization problems that are computationally infeasible for classical systems [184]. This distributed model aligns with the principles of complex adaptive systems, where local processing contributes to global optimization in a decentralized manner [172].

For telecommunications, distributed quantum computing enables:

- **Real-Time Network Optimization**: Quantum algorithms optimize routing, load balancing, and spectrum allocation with near-instantaneous feedback.
- **Dynamic Spectrum Sharing**: Quantum-enhanced edge nodes dynamically adjust spectrum usage based on traffic demand, reducing latency and optimizing bandwidth.
- **Fault-Tolerant Network Slicing**: Quantum error correction allows for resilient network slicing, ensuring continuous service even under fault conditions.

A critical challenge for the Quantum Internet is the seamless integration with existing classical telecommunications infrastructures. Hybrid quantum-classical protocols are being developed to enable compatibility, allowing quantum data streams to coexist with traditional IP-based traffic [183]. Techniques such as *entanglement swapping* and *quantum repeaters* are employed to extend quantum links over vast distances, making global quantum communication a viable reality.

The deployment of quantum repeaters, which regenerate quantum states without measuring them, extends the range of quantum communication far beyond current optical fiber limitations [182]. This allows for the creation of continent-spanning quantum networks capable of secure, real-time data exchange.

3.8.7 *Quantum-Safe Networking for 6G*

The transition to 6G networks introduces unprecedented levels of connectivity, data throughput, and distributed intelligence. However, it also exposes vulnerabilities to emerging threats, particularly from quantum computing capabilities that can undermine classical cryptographic protocols [182]. To address these challenges, the concept of *Quantum-Safe Networking* has emerged, leveraging quantum-resistant cryptography and decentralized trust mechanisms to secure next-generation communication infrastructures.

As quantum computing advances, traditional cryptographic methods such as RSA and ECC (Elliptic Curve Cryptography) become increasingly vulnerable to quantum attacks. Quantum algorithms, such as Shor's algorithm, can efficiently factor large integers and solve discrete logarithm problems, compromising the security of widely-used cryptographic protocols [185].

To mitigate this, Post-Quantum Cryptography (PQC) aims to develop cryptographic algorithms resilient to quantum-based attacks. Techniques such as *Lattice-based Cryptography*, *Multivariate Polynomials*, and *Hash-based Signatures* are at the forefront of PQC research [186]. These methods leverage mathematical problems that are computationally infeasible for both classical and quantum computers, ensuring long-term data confidentiality.

Quantum Key Distribution (QKD) enables the secure generation and exchange of cryptographic keys by exploiting the principles of quantum mechanics. The most prominent QKD protocol, BB84, uses photon polarization states to transmit key information securely [187]. Any eavesdropping attempt disturbs the quantum states, alerting the communicating parties to potential intrusions.

In the context of 6G networks, QKD can be deployed at edge nodes to establish secure communication channels across decentralized infrastructures. This ensures that data integrity and confidentiality are maintained even in highly dynamic environments, such as mobile ad hoc networks (MANETs) and vehicular networks [183].

To further enhance security in 6G quantum networks, decentralized trust models based on *blockchain technology* are being explored. Blockchain enables immutable record-keeping of quantum key exchanges and routing updates, reducing the risk of tampering and unauthorized modifications [180]. The integration of blockchain with quantum communication facilitates:

- **Distributed Trust Management**: Secure verification of node behavior without centralized control.
- **Smart Contract-Based Security**: Automated execution of security policies based on predefined conditions.
- **Fault-Tolerant Consensus Mechanisms**: Ensuring network integrity even in the presence of Byzantine nodes or malicious attacks.

The principles of complexity science offer a robust framework for designing resilient and adaptive quantum-safe topologies in 6G networks. These networks can leverage:

- **Self-Organizing Security Mechanisms**: Localized decision-making for autonomous threat detection and response.
- **Emergent Defense Strategies**: Distributed nodes collaborate to detect anomalies and mitigate attacks without centralized intervention.
- **Adaptive Resource Management**: Quantum algorithms dynamically allocate resources to optimize bandwidth and energy efficiency under real-time conditions.

The integration of complexity science with quantum technologies fosters the development of communication networks that are not only secure but also scalable and self-healing, mirroring the robustness observed in biological ecosystems [181]. The confluence of complexity science and quantum technologies marks a significant leap in the evolution of telecommunications networks. Through the implementation of Quantum Internet architectures and Quantum-Safe Networking principles, next-generation networks can achieve unprecedented levels of security, resilience, and decentralized control. These advances are not merely incremental but represent a paradigm shift towards intelligent, self-organizing infrastructures capable of withstanding emerging cyber threats and adapting to dynamic communication landscapes. As 6G and beyond become a reality, the principles of complexity and quantum mechanics will be foundational to achieving the vision of a hyper-connected, secure, and resilient global communication network.

3.8.8 Quantum Network Slicing in 6G Architectures

The concept of *Network Slicing* has been a pivotal innovation in 5G, enabling the partitioning of a single physical network into multiple virtual networks, each optimized for specific services such as enhanced Mobile Broadband (eMBB), massive Machine-Type Communications (mMTC), and Ultra-Reliable Low-Latency Communications (URLLC) [188]. The integration of *Quantum Network Slicing* emerges as a transformative approach that not only extends this concept but fundamentally enhances its capabilities through quantum principles.

Quantum Network Slicing leverages the unique properties of quantum mechanics, such as superposition and entanglement, to enable more dynamic and efficient partitioning of network resources. Unlike classical network slicing, which relies on static allocation of bandwidth and computational resources, quantum slicing dynamically adjusts based on instantaneous network demand and quantum state measurements [184]. This allows for:

- **Real-Time Adaptation**: Quantum state monitoring enables real-time adjustments to bandwidth and processing power across slices, optimizing latency and throughput [183].
- **Enhanced Resource Efficiency**: Quantum entanglement allows for the simultaneous use of network paths, improving resource utilization by enabling parallel data streams over entangled links.

- **Quantum State Handoff**: In mobile networks, quantum states can be handed off between nodes without physical relocation, maintaining service continuity during high-mobility scenarios such as vehicular networks [182].

A significant advantage of Quantum Network Slicing is its capacity for *multi-domain slicing* across different geographical and infrastructural boundaries. Quantum entanglement creates direct, real-time links between network segments that may be geographically distant, allowing for seamless cross-domain communication without traditional bottlenecks [183]. This is particularly critical for global 6G services, such as:

- **Global IoT Deployments**: Enabling synchronized communication between IoT devices across continents without latency degradation.
- **Interconnected Smart Cities**: Facilitating instantaneous data exchange between urban infrastructures for real-time monitoring and emergency response.
- **Autonomous Transportation Networks**: Ensuring low-latency coordination between self-driving vehicles in different cities.

In classical network slicing, security is managed through encryption and centralized control. Quantum Network Slicing enhances this paradigm with *Quantum Key Distribution (QKD)* and *Post-Quantum Cryptography (PQC)* to secure inter-slice communication [187]. Since quantum states cannot be measured or duplicated without detection, any interception attempt disrupts the entangled state, alerting the network to potential eavesdropping [183]. Furthermore, PQC mechanisms are integrated to safeguard the classical control plane, ensuring both quantum and classical layers are secure from quantum-based attacks.

Quantum Network Slicing aligns with the principles of *self-organization* and *emergence* inherent to complex systems [172]. Unlike rigidly structured classical slices, quantum slices are adaptive, dynamically forming and dissolving based on real-time demand and quantum measurements. This decentralized control mechanism allows for:

- **Dynamic Resource Scaling**: Quantum slices can expand or contract in response to user demand without centralized orchestration.
- **Fault Tolerance and Self-Healing**: Distributed quantum entanglement ensures continuity even when individual nodes or paths fail, mirroring the robustness seen in biological networks.
- **Multi-Modal Communication**: Quantum states can facilitate hybrid communication models that blend classical and quantum data streams seamlessly.

These capabilities make Quantum Network Slicing a foundational component of resilient, adaptive, and intelligent 6G infrastructures, extending beyond the static partitioning of network resources into a dynamic, quantum-enhanced paradigm.

3.8.9 Quantum Edge Computing for 6G Networks

As 6G networks continue to push the boundaries of decentralized intelligence, *Quantum Edge Computing* emerges as a transformative enabler for distributed processing and machine learning at the network edge [184]. This paradigm extends the computational capabilities of traditional edge nodes by integrating quantum processing units (QPUs), allowing for near-instantaneous data processing and advanced optimization tasks. Quantum Edge Computing leverages quantum algorithms for edge-based data analysis and decision-making, significantly improving computational efficiency and reducing latency [183]. Key innovations include:

- **Quantum Machine Learning (QML)**: Edge nodes equipped with QPUs can perform rapid data classification, anomaly detection, and predictive analytics, enhancing real-time decision-making in IoT and smart city applications [184].
- **Quantum-Assisted Federated Learning**: Quantum nodes at the edge can collaborate to train machine learning models without centralized data pooling, preserving privacy while accelerating convergence [182].
- **Distributed Quantum Inference**: Edge devices execute quantum inference on locally gathered data, reducing the need for backhaul communication and enhancing latency-sensitive applications like augmented reality (AR) and autonomous driving.

The application of quantum principles at the edge introduces robust fault-tolerant mechanisms. Quantum entanglement allows for redundant data paths and error correction, ensuring data integrity and service continuity even under adverse conditions [183]. This is particularly critical for mission-critical applications where low-latency and high reliability are non-negotiable.

The decentralized and self-organizing nature of Quantum Edge Computing reflects principles of complex systems [172]. Local quantum nodes operate independently yet maintain global coherence through entangled states. This mirrors complex adaptive systems where global behavior emerges from localized interactions without centralized coordination. In the context of 6G:

- **Self-Optimizing Edge Nodes**: Quantum algorithms enable edge nodes to autonomously adjust processing tasks based on real-time demand and network conditions.
- **Decentralized Control and Self-Healing**: Quantum entanglement facilitates error correction and autonomous recovery, mirroring resilience strategies observed in biological networks.
- **Dynamic Load Balancing**: Quantum-enhanced routing algorithms optimize traffic distribution and minimize latency without centralized oversight.

The fusion of Quantum Edge Computing with 6G infrastructures opens new research frontiers, particularly in:

- **Quantum-AI Convergence**: Integrating quantum processing with artificial intelligence to enable real-time learning and decision-making at the edge.

- **Edge-Enabled Quantum Teleportation**: Applying quantum teleportation principles to facilitate instantaneous data synchronization between geographically distributed edge nodes.
- **Quantum-Secure Edge Networking**: Implementing quantum key distribution (QKD) at the edge to ensure end-to-end encryption and data integrity.

As 6G evolves, Quantum Edge Computing is poised to redefine latency-sensitive applications, providing a scalable, resilient, and secure framework for decentralized intelligence.

The exploration of *Quantum Network Slicing* and *Quantum Edge Computing* represents a paradigm shift in the design and operation of 6G communication networks. By leveraging quantum principles such as superposition, entanglement, and quantum state measurement, these technologies extend the traditional capabilities of network slicing and edge intelligence, paving the way for unprecedented levels of adaptability, scalability, and security [183, 184].

Quantum Network Slicing enhances the classical concept of virtualized network partitions by introducing real-time, state-dependent adjustments that transcend the limitations of static bandwidth allocation. Through quantum entanglement, it enables multi-domain slicing, facilitating seamless communication across geographically distant infrastructures while ensuring robust data security via quantum key distribution (QKD) [187]. This capability is crucial for global-scale applications such as interconnected smart cities, autonomous vehicular networks, and next-generation IoT ecosystems, where latency, reliability, and scalability are paramount [182].

Simultaneously, Quantum Edge Computing brings computation closer to data sources through quantum-enhanced processing at the network edge. This model optimizes latency-sensitive applications by reducing the dependency on centralized cloud processing, allowing edge nodes to execute complex tasks in real-time [184]. The integration of Quantum Machine Learning (QML) and quantum-assisted federated learning introduces a new dimension of intelligence at the edge, where local decision-making is accelerated, and privacy concerns are mitigated through decentralized training [183].

The theoretical underpinnings of these technologies resonate deeply with the principles of *complex systems*. Both Quantum Network Slicing and Quantum Edge Computing exhibit self-organization, decentralized control, and resilience–key features that mirror biological networks and ecological systems [172, 173]. In this perspective, the dynamic partitioning of resources, the adaptive optimization of computational tasks, and the fault-tolerant nature of quantum-based communication reflect the emergent behaviors observed in complex adaptive systems. These properties are instrumental in addressing the scalability and reliability challenges that 6G networks are expected to encounter.

Looking forward, the convergence of quantum communication paradigms with edge intelligence and network slicing sets the foundation for a fully distributed, self-optimizing 6G infrastructure. Future research should prioritize the seamless integration of quantum principles with digital twins, reconfigurable intelligent surfaces (RIS), and machine learning-based optimization to further enhance network resilience and adaptability. Furthermore, the development of quantum-resilient cryptographic protocols remains essential to secure the expanding landscape of decentralized edge computing and virtualized network slices [188].

Ultimately, Quantum Network Slicing and Quantum Edge Computing represent more than technological advancements; they signify a conceptual shift towards embracing complexity and emergence in communication networks. This shift is not merely an optimization of current paradigms but a rethinking of how networks are designed, managed, and evolved–marking a crucial step towards realizing the ambitious vision of 6G as a globally interconnected, resilient, and intelligent communication ecosystem.

3.9 6G Integrated Architectures and Self-Organizing Systems

3.9.1 Introduction

The advent of sixth-generation (6G) networks heralds a transformative era in telecommunications, characterized by unprecedented demands for ultra-low latency, massive connectivity, and pervasive intelligence. Traditional network architectures, predominantly centralized and static, are increasingly inadequate to meet these evolving requirements. Consequently, there is a paradigm shift towards decentralized, adaptive, and intelligent systems that can self-organize and evolve in response to dynamic environments.

3.9.2 Hybrid Self-Organizing Networks (H-SONs)

Self-Organizing Networks (SONs) have been instrumental in automating network configuration, optimization, and healing processes. However, the complexity and heterogeneity of 6G networks necessitate more advanced self-management capabilities. Hybrid Self-Organizing Networks (H-SONs) emerge as a compelling solution, integrating centralized and distributed control mechanisms to enhance scalability and resilience [189].

H-SONs leverage loosely coupled architectures, wherein different network layers and components interact with minimal dependencies. This design mitigates the risk of cascading failures and enhances system stability. By adopting a hybrid approach, H-SONs can dynamically balance global oversight with local autonomy, facilitating efficient resource allocation and fault management in complex network environments.

3.9.3 Artificial Intelligence of Things (AIoT)

The convergence of Artificial Intelligence (AI) and the Internet of Things (IoT) gives rise to the Artificial Intelligence of Things (AIoT), a paradigm that imbues IoT devices with cognitive capabilities. AIoT enables devices to perceive, learn, and make decisions, thereby fostering intelligent and context-aware applications [190].

In the context of 6G, AIoT plays a pivotal role in achieving self-organizing and adaptive network behaviors. By harnessing machine learning algorithms, AIoT devices can collaboratively optimize network performance, predict and mitigate faults, and adapt to user behaviors and environmental changes. This distributed intelligence is essential for managing the complexity and dynamism inherent in 6G networks.

3.9.4 Self-Evolving and Transformative Protocol Architectures

To accommodate the diverse and stringent requirements of 6G applications, there is a need for protocol architectures that are not only flexible but also capable of self-evolution. The Self-Evolving and Transformative (SET) protocol architecture addresses this need by incorporating intelligent agents that can autonomously reconfigure protocol stacks in response to changing network conditions and service demands [191].

SET architectures facilitate in-network intelligence, allowing for real-time adaptation and optimization of communication protocols. This adaptability is crucial for supporting emerging 6G use cases, such as immersive virtual reality, autonomous vehicles, and remote healthcare, which demand high reliability and low latency.

3.9.5 Complex Systems and Evolutionary Game Theory in 6G

The integration of complex systems theory and Evolutionary Game Theory (EGT) provides a robust framework for modeling and analyzing the dynamic interactions within 6G networks. Complex systems theory offers insights into emergent behaviors, self-organization, and adaptability, which are essential characteristics of next-generation networks.

EGT, in particular, is instrumental in understanding the strategic interactions among autonomous network entities. By modeling these interactions as evolutionary games, network designers can predict and influence the emergence of cooperative behaviors, resource sharing strategies, and stability in decentralized environments. This theoretical foundation is vital for designing 6G networks that are resilient, efficient, and capable of self-optimization.

3.9.6 Self-Organizing Systems in 6G

Self-organizing systems (SOS) represent a paradigm shift in communication networks, wherein network elements autonomously manage their configuration, optimization, and recovery processes without centralized control [192]. This capability is particularly significant for 6G architectures, which are envisioned to support massive device connectivity, ultra-low latency, and highly dynamic environments. In this context, self-organization emerges as a critical enabler of scalability, resilience, and adaptability in next-generation networks.

Principles of Self-Organization

The concept of self-organization is rooted in the capacity of a system to structure itself, adapt to environmental changes, and recover from disruptions autonomously [192]. This principle draws inspiration from natural systems, such as biological networks and ecological systems, where local interactions among individual components give rise to global order and stability. Key attributes of self-organizing systems include:

- **Decentralized Control**: Unlike traditional networks that rely on centralized control planes, self-organizing networks (SONs) operate through distributed decision-making. Nodes independently assess their environment and optimize parameters such as power levels, spectrum usage, and routing paths [192].
- **Emergence**: Local interactions among network elements lead to emergent behaviors that enhance global performance. For instance, autonomous load balancing in mobile networks emerges from individual nodes optimizing their connectivity based on local traffic conditions [193].
- **Adaptability and Robustness**: Self-organizing systems dynamically adapt to network changes, such as node mobility, link failures, and traffic surges. This adaptability improves robustness, allowing the network to maintain service continuity despite disruptions [192].
- **Self-Healing**: In distributed architectures, self-healing mechanisms allow nodes to detect faults and reconfigure their operations autonomously, minimizing service interruptions [194].

The realization of these principles in 6G networks extends beyond conventional optimization strategies, incorporating machine learning (ML) and artificial intelligence (AI) to enhance decision-making processes. ML models enable nodes to predict network states and optimize configurations proactively, reducing latency and improving throughput in dynamically shifting environments.

Self-Organization in 6G Architectures

6G networks, characterized by hyper-connectivity and multi-tier architectures, demand sophisticated self-organizing capabilities to meet performance expectations. Three primary domains benefit from self-organization in 6G:

- Self-Organizing Radio Access Networks (RANs): The increasing density of small cells in 6G networks necessitates efficient management of radio resources. Self-organizing RANs autonomously configure handovers, adjust power levels, and optimize spectrum utilization based on real-time traffic patterns [192]. This decentralized approach mitigates interference and enhances spectral efficiency, crucial for achieving the low latency and high throughput targets of 6G.
- Self-Organizing Edge Computing: The edge layer in 6G architectures is expected to process vast amounts of data generated by IoT devices and smart sensors. Self-organizing edge nodes dynamically allocate computational resources and storage based on local demand, reducing the reliance on centralized cloud infrastructure [193]. This self-adaptive behavior minimizes latency and optimizes bandwidth usage, supporting real-time applications such as augmented reality (AR) and autonomous driving.
- Self-Healing and Fault Management: Network reliability is a critical concern in 6G, where service disruptions can significantly impact critical applications. Self-healing mechanisms, inspired by distributed fault-tolerant systems, enable 6G networks to detect and resolve failures autonomously [194]. Techniques such as decentralized anomaly detection and cooperative recovery protocols enhance network resilience, ensuring continuous service availability even in highly dynamic environments.

Emergence and Complexity in Self-Organizing Systems

A distinctive feature of self-organizing systems is the emergence of complex behaviors from simple local rules. In 6G networks, this emergence manifests in adaptive load balancing, efficient resource allocation, and resilience to node failures. Emergent phenomena are particularly evident in scenarios involving large-scale IoT deployments and vehicular networks, where thousands of autonomous devices coordinate seamlessly through local interactions [193].

The study of emergence in self-organizing systems is closely linked to complexity science, which explores how distributed components collectively solve problems without centralized oversight. In 6G, these principles are leveraged to design networks that self-configure, optimize, and heal, mimicking the distributed intelligence found in natural ecosystems [192]. This alignment with complex systems theory allows 6G networks to scale efficiently and adapt to real-time fluctuations in network demand.

Future Directions and Challenges

Despite its promise, the implementation of self-organizing systems in 6G faces several challenges:

- **Scalability**: As the number of connected devices scales into the billions, maintaining efficient self-organization without overwhelming communication overhead remains a challenge.
- **Security and Trust**: Decentralized decision-making introduces vulnerabilities to malicious nodes and untrusted interactions. Future research must address secure self-organization mechanisms that can authenticate and validate node behavior.
- **Interoperability**: Self-organizing mechanisms must seamlessly integrate with legacy systems and heterogeneous network infrastructures, ensuring backward compatibility.
- **Energy Efficiency**: The constant reconfiguration and optimization of network elements in self-organizing architectures can increase energy consumption. Strategies for sustainable energy use are crucial for 6G's environmental goals.

Emerging technologies such as Federated Learning (FL), Blockchain, and Reconfigurable Intelligent Surfaces (RIS) are expected to enhance the security, scalability, and adaptability of self-organizing systems in 6G. These technologies, combined with advanced machine learning models, will drive the evolution of communication networks towards fully autonomous, resilient infrastructures capable of meeting the demands of next-generation services.

3.9.7 Cross-Layer Design in 6G Architectures

The emergence of 6G networks introduces unprecedented levels of complexity and heterogeneity across communication layers, from the Physical (PHY) and Medium Access Control (MAC) layers to the Network layer. Traditional siloed approaches to network optimization, where each layer is independently managed, fall short in addressing the dynamic and interconnected nature of next-generation wireless environments. In response, cross-layer design principles are gaining traction, advocating for the integration and cooperative optimization of multiple network layers to enhance throughput, minimize latency, and improve resilience [195].

Cross-layer design is particularly critical in 6G due to the convergence of emerging technologies such as Edge AI, Digital Twins, and Quantum Communication, which necessitate tight synchronization and real-time adaptation across the communication stack [125]. Unlike traditional layered models, cross-layer architectures allow for real-time information exchange and joint decision-making between layers, enabling dynamic resource allocation, interference management, and fault tolerance in highly volatile environments [196].

Integration of PHY, MAC, and Network Layers

The Physical (PHY), Medium Access Control (MAC), and Network layers form the backbone of wireless communication. In traditional architectures, these layers operate independently with minimal inter-layer communication, often leading to suboptimal resource usage and increased latency. However, in 6G, where ultra-reliable low-latency communication (URLLC) and massive machine-type communication (mMTC) are prerequisites, such isolation is infeasible [195].

To address this, cross-layer design principles are employed to enable seamless information exchange and joint optimization:

- **PHY-MAC Integration**: At the physical layer, resource blocks are dynamically allocated based on real-time channel state information (CSI) gathered at the MAC layer. This enables adaptive modulation, coding schemes, and dynamic spectrum access, reducing interference and optimizing throughput [195].
- **MAC-Network Coordination**: At the MAC layer, decisions related to scheduling, resource allocation, and congestion control are informed by network-layer routing metrics and traffic patterns. This cooperative approach minimizes packet collisions and reduces end-to-end latency, particularly in dense urban environments [125].
- **Network Layer Adaptation**: Cross-layer feedback from the PHY and MAC layers enables the Network layer to optimize routing paths dynamically, adapting to node mobility and varying traffic loads. In vehicular ad-hoc networks (VANETs) and drone-based communication, this adaptability is crucial for maintaining consistent quality of service (QoS) [196].

These multi-layer interactions are crucial for supporting advanced 6G use cases such as autonomous vehicles, smart cities, and industrial IoT, where real-time data exchange and ultra-low latency are non-negotiable.

Interactions Between Edge AI, Digital Twins, and Quantum Technologies

The deployment of Edge AI, Digital Twins, and Quantum Technologies in 6G networks further emphasizes the need for cross-layer optimization:

- **Edge AI Integration**: Edge AI enables distributed intelligence at the network's edge, reducing latency and optimizing resource allocation. Cross-layer feedback mechanisms facilitate real-time updates on network conditions, enhancing the accuracy of machine learning models deployed at the edge [195].
- **Digital Twins Synchronization**: Digital Twins create real-time virtual replicas of network components, enabling predictive maintenance and fault detection. For Digital Twins to be effective, data synchronization across PHY, MAC, and Network layers must be maintained, allowing the virtual model to reflect the physical state accurately [125].

- **Quantum Technologies for Secure Communication**: Quantum-enhanced communication introduces mechanisms for ultra-secure key distribution and low-latency data exchange. These capabilities necessitate fine-grained synchronization between communication layers to manage quantum key exchanges and dynamic path adjustments in real-time [196].

The interplay between these technologies is enhanced through cross-layer coordination, where data from the PHY and MAC layers are utilized to dynamically adjust the operation of Digital Twins and Edge AI models. This integration ensures optimal performance in scenarios requiring real-time decision-making and enhanced security.

Multi-layer Optimization for Latency and Resilience

One of the critical challenges in 6G is achieving ultra-low latency and resilience against network disruptions. Cross-layer design principles address these challenges through:

- **Latency Minimization**: By enabling real-time feedback loops between the PHY, MAC, and Network layers, latency is significantly reduced. Adaptive scheduling and resource allocation minimize transmission delays, while multi-hop routing optimization reduces path length [195].
- **Resilience through Multi-layer Redundancy**: Cross-layer design introduces redundancy mechanisms across communication layers. For example, the MAC layer can switch transmission frequencies upon detecting interference, while the Network layer reroutes traffic to avoid congestion. This multi-layer redundancy is inspired by self-healing properties observed in complex systems, where local adaptations collectively maintain global stability [125].
- **Quantum-enhanced Security**: The integration of quantum technologies into the communication stack enhances resilience against cyber threats. Quantum key distribution (QKD) protocols secure data exchanges at the PHY layer, while cross-layer synchronization ensures real-time path optimization to mitigate interception risks [196].

The evolution towards 6G demands a paradigm shift in network architecture, moving from isolated, layer-specific optimization to holistic cross-layer design. By integrating the PHY, MAC, and Network layers, alongside advanced technologies such as Edge AI, Digital Twins, and Quantum Communication, 6G networks can achieve unprecedented levels of latency reduction, resilience, and security. Cross-layer design emerges as a foundational principle, enabling seamless communication, enhanced adaptability, and robust defense mechanisms in the face of dynamic and distributed network topologies.

3.9.8 Multi-layer Optimization for Latency and Resilience

The demand for ultra-low latency and high resilience in 6G networks necessitates a rethinking of traditional network optimization strategies. In highly dynamic communication environments–characterized by massive device connectivity, real-time edge computing, and autonomous systems–latency and reliability become critical parameters for maintaining Quality of Service (QoS) and Quality of Experience (QoE) [195]. To meet these stringent requirements, 6G introduces the concept of *Multi-Layer Optimization*, where the Physical (PHY), Medium Access Control (MAC), Network, and Application layers are jointly optimized to enhance communication efficiency and robustness [125].

Multi-layer optimization draws inspiration from complex systems theory, where interactions at multiple scales and layers influence the global behavior of the system [196]. In the context of 6G, these interactions are not limited to single-layer adaptations but involve coordinated adjustments across all network layers, leveraging real-time feedback loops and distributed intelligence. This cross-layer coordination ensures that latency is minimized and resilience is enhanced, even in highly volatile network conditions.

Optimization Techniques Across Layers

In a multi-layer architecture, each communication layer plays a critical role in optimizing latency and resilience:

At the physical layer, optimization techniques focus on enhancing signal propagation, reducing interference, and improving spectral efficiency:

- **Adaptive Modulation and Coding (AMC)**: To dynamically adjust modulation schemes based on real-time channel state information (CSI), thereby optimizing throughput and reducing latency [195].
- **Massive MIMO and Beamforming**: These technologies enhance spatial diversity and signal strength, ensuring robust communication paths in dense urban environments and vehicular networks [125].
- **Reconfigurable Intelligent Surfaces (RIS)**: RIS technologies optimize the wireless environment by dynamically altering signal reflection properties, reducing path loss and mitigating interference [196].

The MAC layer coordinates access to the communication medium, playing a vital role in reducing collision rates and optimizing bandwidth utilization:

- **Dynamic Spectrum Access (DSA)**: Real-time allocation of spectrum resources based on network demand and channel conditions, enhancing spectrum efficiency [195].
- **Time-Sensitive Networking (TSN)**: Ensuring deterministic latency for critical applications through time-synchronized transmissions and traffic prioritization [125].

- **Multi-Channel Coordination**: Efficiently managing multiple frequency channels to avoid interference and balance traffic loads across densely populated network areas [196].

At the network layer, optimization focuses on routing, congestion control, and energy efficiency:

- **Software-Defined Networking (SDN)**: SDN decouples the control and data planes, enabling centralized management of traffic flows and dynamic path reconfiguration to minimize latency [195].
- **Network Slicing**: By creating virtualized network segments, specific QoS requirements for URLLC, eMBB, and mMTC applications can be met with high reliability and low latency [125].
- **Multi-Path Routing**: Employing multi-path strategies to distribute traffic loads and ensure redundancy in case of link failures [196].

At the application layer, optimization strategies aim to enhance user experience and service reliability:

- **Edge Caching and Computing**: By offloading computations to edge nodes, latency is significantly reduced, enabling real-time processing for critical applications [195].
- **Service-Oriented Architectures (SOA)**: Modular service designs enable distributed deployment and real-time adaptation based on changing network conditions [125].
- **Predictive Analytics and Machine Learning**: Leveraging AI models to predict network congestion, optimize routing, and allocate resources dynamically [196].

Quantum-Enhanced Communication for Resilience

The integration of quantum technologies in 6G architectures introduces new paradigms for achieving resilience and low-latency communication:

- **Quantum Key Distribution (QKD)**: Provides unbreakable encryption for data transmission, ensuring security and resilience against cyber threats [196].
- **Quantum Repeater Networks**: Enhance long-distance communication by enabling quantum entanglement over extended ranges, reducing latency and signal loss [196].
- **Quantum Teleportation**: Allows instantaneous transmission of information between nodes, promising breakthroughs in real-time communication [196].

The convergence of quantum technologies with traditional multi-layer optimization paves the way for communication systems that are not only faster but also inherently secure and resilient against eavesdropping and data corruption.

Future research in multi-layer optimization should explore:

- **AI-Driven Cross-Layer Coordination**: Implementing machine learning models that autonomously manage resource allocation and path optimization across layers.
- **Integrated Edge Intelligence**: Developing edge-based algorithms that predict network congestion and optimize data flows in real time.
- **Quantum-Enhanced Routing Protocols**: Designing quantum-aware routing strategies that leverage QKD and teleportation for ultra-secure and low-latency communication.

These research directions aim to enhance the robustness, adaptability, and security of 6G networks, setting the foundation for hyper-connected, self-organizing infrastructures.

3.9.9 Self-Organizing and Self-Healing Networks in 6G

The evolution towards 6G communication networks introduces unprecedented levels of complexity and dynamicity in network architecture. To manage this complexity, the concepts of *self-organizing networks* (SON) and *self-healing mechanisms* have emerged as fundamental paradigms. These mechanisms enable 6G networks to autonomously optimize, recover, and adapt to changing environmental conditions and unexpected failures without human intervention [197–199].

Self-Organizing Networks (SON) in 6G

Self-Organizing Networks (SON) represent a key architectural innovation designed to enhance the autonomy and resilience of wireless networks. Initially conceived for 4G LTE, SON capabilities have progressively evolved, positioning them as essential components in 5G and upcoming 6G infrastructures. The primary functions of SON include:

- **Self-Configuration**: Automatic configuration of new nodes and integration into the existing network topology.
- **Self-Optimization**: Continuous optimization of network parameters to improve performance metrics such as latency, throughput, and energy efficiency.
- **Self-Healing**: Automated detection and recovery from failures, minimizing service disruptions and maintaining Quality of Service (QoS) [199].

In 6G, SON technologies are expected to leverage advanced machine learning algorithms and distributed intelligence to achieve real-time optimization and fault recovery [199]. Unlike previous generations, 6G networks will incorporate edge computing and AI-driven orchestration to enhance real-time decision-making capabilities. This distributed intelligence is inspired by the principles of complex systems, where local interactions among nodes lead to global stabilization and enhanced resilience [198].

Self-Healing Mechanisms for 6G Networks

Self-healing mechanisms are designed to detect, diagnose, and recover from network anomalies autonomously. These mechanisms are crucial in 6G networks, where the density and heterogeneity of devices significantly increase the probability of faults and interruptions [197]. Self-healing in 6G encompasses:

- **Fault Detection and Isolation**: Identifying network anomalies, such as node failures or communication bottlenecks, using AI-based predictive analytics.
- **Automated Recovery**: Enabling the network to reconfigure itself by rerouting traffic, reallocating resources, and initiating redundant paths to maintain service continuity.
- **Graph-Based Learning for Self-Healing**: Advanced methods like Graph Neural Networks (GNNs) allow the modeling of 6G networks as graph-based structures, where anomalies can be identified and localized based on network topology and node interactions [197].

In the context of 6G, Digital Twin (DT) technology plays a pivotal role in enhancing self-healing capabilities. Through real-time mirroring of the physical network in a virtual environment, DTs enable continuous monitoring and predictive maintenance. Graph Neural Networks (GNNs) integrated into DTs facilitate real-time fault detection and adaptive recovery by analyzing connectivity patterns and identifying structural vulnerabilities [197].

Complex Systems Perspective

The design and operation of self-organizing and self-healing mechanisms in 6G align with the principles of complex systems. These networks exhibit properties such as:

- **Emergence**: Global stability and optimized performance emerge from local, decentralized interactions among nodes.
- **Resilience**: The ability to adapt to faults and reconfigure pathways without centralized control mirrors biological ecosystems' capacity for self-repair.
- **Self-Organization**: Nodes dynamically adjust configurations and optimize resource allocation based on real-time environmental feedback [198, 199].

These properties ensure that 6G networks are not only robust to failures but also capable of self-improvement over time. Future research is expected to deepen the integration of artificial intelligence and distributed ledger technologies to enhance trust, security, and autonomy in self-organizing and self-healing 6G networks.

Future Research Directions

Future research on self-organizing and self-healing 6G networks is expected to explore:

- **Quantum-Enhanced Self-Healing**: Leveraging quantum computing for faster anomaly detection and optimization in complex network environments.
- **Federated Learning for Fault Detection**: Employing federated learning across distributed edge nodes to improve fault detection accuracy without compromising data privacy.
- **Adaptive Resource Allocation**: Developing AI-driven models for real-time optimization of resources based on predicted network behavior.
- **Integration with Digital Twins**: Enhancing self-healing capabilities through real-time digital replicas that predict and prevent faults proactively [197].

These directions highlight the transformative potential of self-organizing and self-healing mechanisms as foundational elements of 6G's resilient, autonomous communication infrastructures. Self-organizing and self-healing mechanisms are crucial to the operation of 6G networks, enabling autonomous adaptation, fault recovery, and optimization in highly dynamic environments. By drawing inspiration from complex systems theory, these technologies allow networks to operate resiliently without centralized oversight, supporting the robust, self-optimizing communication fabrics envisioned for the next generation of wireless networks.

3.9.10 Towards AI-Native Architectures in 6G: From Network Softwarization to Intelligentization

One of the key enablers of future 6G communication systems is the envisioned shift from softwarization–pioneered in 5G through Software-Defined Networking (SDN) and Network Functions Virtualization (NFV)–towards a paradigm of full network intelligentization [200]. While softwarization enables virtualization and dynamic resource slicing, it is no longer sufficient to meet the ultra-heterogeneous, low-latency, and high-capacity demands expected in 6G environments. The next-generation networks must become structurally adaptive, data-intelligent, and self-organizing. As illustrated in Fig. 3.1, the architectural vision for 6G converges into a modular, AI-native, and topology-aware system composed of four interrelated functional domains. These domains–centered around AI–hardware abstraction, intelligent service delivery, distributed edge learning, and cloud-based policy optimization–work in synergy to ensure robust, context-aware, and structure-informed operation in dynamic environments.

In the Top-Left quadrant of Fig. 3.1 *Topology-Aware AI–Hardware Abstraction Architecture* is the e core of the 6G design and it is a refined abstraction architecture that separates software intelligence from physical infrastructure while embedding

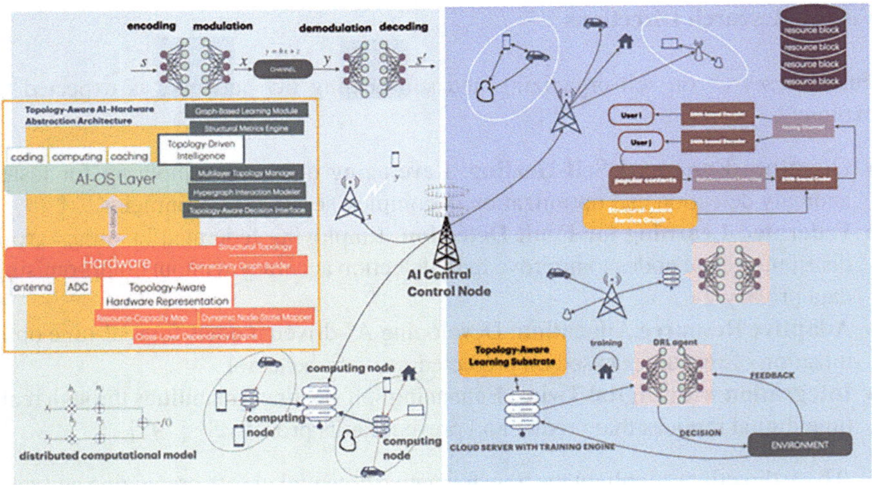

Fig. 3.1 Flow chart illustrating a topology-aware AI-hardware architecture. The left section shows an AI-OS layer with components like coding, computing, and caching, connected to hardware elements such as an antenna and ADC. The central part features an AI central control node linked to users and a structural-aware service graph, highlighting DNN-based decoders and reinforcement learning. The right section depicts a topology-aware learning substrate with a cloud server, DRL agent, and environment feedback loop. Key processes include encoding, modulation, demodulation, and decoding, with neural network representations and mathematical expressions

structural awareness across both. This architectural layer is composed of *AI–OS Layer* that is traditionally responsible for core AI operations like coding, computing, and caching, now extended with the block *Topology-Driven Intelligence*. It includes the *Graph-Based Learning Module*, which infers network state from structural data, and the *Multilayer Topology Manager*, which captures the interdependence of services, users, and protocols across multiple layers. Furthermore, the *AI-OS layer* includes the *Hypergraph Interaction Modeler*, which models higher-order interactions among groups of nodes or resources; the *Topology-Aware Decision Interface*, which integrates topological insights into AI decision pipelines. The *Hardware Layer*, which originally is composed of hardware primitives such as antennas, ADCs, and phase shifters, now augmented with the *Topology-Aware Hardware Representation*, including: the *Structural Topology Mapper*, which visualizes node interconnections; the *Connectivity Graph Builder* that dynamically builds network maps based on current conditions; the *Resource-Capacity Map* that links physical limitations with service demand; the *Cross-Layer Dependency Engine* that models vertical interdependencies (e.g., between link-layer states and application-layer needs) and the *Dynamic Node-State Mapper* that monitors status and roles of devices in real time. These additions enable fine-grained co-design between algorithms and hardware, allowing for context-sensitive adaptation and reconfiguration based on topological complexity.

In the Top-Right Quadrant, the *Intelligent Service Delivery Plane* is described by capturing how services are delivered to users through adaptive mechanisms. Users

i and j represent mobile or stationary agents consuming personalized content. A resource pool composed of *Resource Blocks (RBs)* is allocated dynamically, while the *Popular Contents* (e.g., trending media or frequently requested data) are proactively cached. The *DNN-based Coders and Decoders* apply deep learning to optimize encoding and decoding in real-time based on user context and channel condition (e.g., fading channels). The block of *Reinforcement Learning* governs the allocation of content and channels, balancing throughput and latency. The added module *Structural-Aware Service Graph* provides structural context (e.g., user clusters, path diversity, group centrality) to the learning process, enabling smarter decisions on resource allocation, content delivery paths, and transmission strategies, following the complex-based approach. This part of the architecture exemplifies the synergy between user behavior, network topology, and machine learning, demonstrating how structure-aware intelligence can improve quality of service. In the Bottom-Right Quadrant, the *Cloud-Based Policy Optimization and Training* is highlighted. This section focuses on the edge of the network, where computational capacity is distributed across user devices, vehicles, and small-scale servers. The *Distributed Computational Model* provides a simplified mathematical abstraction in which data inputs x_i and local noise or interference n_i are processed through a set of functional transformations f_1, f_2, and f_3, and aggregated via logic gates or neural processing elements. This formulation enables localized processing prior to central coordination. *Computing nodes* interact to collaboratively execute distributed neural networks, particularly in environments characterized by non-IID (non-independent and non-identically distributed) data–conditions typical of heterogeneous user and device behavior in real-world deployments. Local training processes contribute to model updates that are then communicated to centralized or hierarchical orchestration mechanisms, such as in *federated learning*, allowing the global model to evolve without raw data exchange. This architectural block thus enables real-time inference and on-device intelligence, ensuring privacy preservation, reducing latency, and meeting the stringent responsiveness requirements of mission-critical 6G applications.

At the back-end of the intelligent loop, in the Bottom-Right Quadrant, the *Cloud-Based Policy Optimization and Training*, a *Cloud Server with Training Engine* executes centralized model training, parameter aggregation, and policy evolution. The *Training Data and feedback* from the environment are collected from distributed agents and users. A *DRL Agent (Deep Reinforcement Learning)* receives this data to optimize decision policies, such as dynamic resource allocation or energy-efficient routing. The newly introduced *Topology-Aware Learning Substrate* injects structural awareness into the training process. It provides the DRL agent with topological features–such as modularity, percolation threshold, node centrality, or community overlap–which enrich the state space representation and guide learning toward globally optimal strategies. This quadrant closes the learning loop by integrating environmental feedback, structural features, and model updates, which are then propagated across the AI–OS and service layers. The integration of complex network modeling into this 6G architecture significantly enhances its ability to adapt, self-organize, and reason under uncertainty. By incorporating topology-aware modules across the full stack–from physical layer representation to decision-making agents and service-level learning–this framework shifts the paradigm from flat, reactive networks to hierarchical, structurally adaptive intelligent systems. Complex systems theory thus becomes a foundational enabler for 6G design, allowing architectures to learn from, adapt to, and even anticipate the evolution of network behavior over time and scale.

3.10 Conclusion

The third chapter of this book, "A Complex-based Rethinking of Communications", has provided an extensive exploration of the emerging paradigms and technologies that will define the communication networks of the future. By embracing the principles of complex systems, 6G architectures are set to move beyond the constraints of traditional design models, shifting towards self-organizing, resilient, and adaptive infrastructures. Through concepts such as Networks of Networks (NoNs), Edge Intelligence, Digital Twins, Machine Learning, AI-driven Network Slicing, Floating Content Architectures, and the convergence of Quantum Technologies, this chapter has outlined a multi-faceted vision for next-generation networks.

The rise of Edge Intelligence reflects the growing need for decentralized processing and real-time data analysis, enabling edge nodes to operate autonomously with minimal latency. By integrating machine learning models directly at the edge, 6G architectures achieve not only rapid decision-making but also enhanced data privacy and reduced bandwidth consumption, crucial for latency-sensitive applications like autonomous driving and augmented reality. Similarly, the introduction of Digital Twins into telecommunications marks a paradigm shift towards hyper-accurate virtual representations of physical networks. These digital replicas serve as real-time mirrors, facilitating predictive maintenance, anomaly detection, and optimization of network performance. When combined with AI and edge computing, Digital Twins enable a dynamic orchestration of network resources that optimizes performance while reducing overhead costs. The exploration of Machine Learning and AI in 6G Networks highlights the transformative impact of artificial intelligence on resource management, security, and optimization. The application of deep learning models for network slicing, anomaly detection, and predictive maintenance illustrates how AI can autonomously govern complex communication environments, enhancing both efficiency and resilience.

Furthermore, the integration of Floating Content Architectures introduces new possibilities for distributed and resilient data sharing. By decoupling content persistence from centralized servers, Floating Content leverages opportunistic communication between mobile nodes to maintain information availability in localized geographic areas. This approach not only reduces backhaul dependency but also enhances data resilience in highly dynamic environments, such as smart cities and disaster recovery zones. The final sections of this chapter focused on the Convergence of Quantum Technologies and Complexity Science. Here, the principles of quantum mechanics–such as entanglement, superposition, and quantum teleportation–are leveraged to introduce unprecedented levels of security and synchronization in 6G networks. Quantum Key Distribution (QKD) and Quantum Network Slicing redefine how secure communication and real-time synchronization are achieved across globally distributed infrastructures.

The content mined from your recent participation in the conference on AI-driven Telco transformation further underscores the significance of AI-native telecommunications as the cornerstone of 6G. Concepts such as Generative AI (GenAI) enablement in Telcos represent a new paradigm where artificial intelligence is not merely a supporting technology but a foundational element of telecommunication infrastructures. The idea of AI-driven transformation highlights a shift towards autonomous, self-optimizing networks where AI orchestrates processes such as network management, anomaly detection, and resource allocation in real-time. The concept of LLMs (Large Language Models) in Telecommunications, as introduced in your slides, emphasizes the role of generative AI in transforming communication networks. LLMs, with their ability to process multimodal data (text, images, voice), are envisioned to power autonomous wireless networks, where decision-making and optimization are executed with minimal human intervention. This capability aligns perfectly with the self-organizing principles described earlier in this chapter.

Furthermore, the integration of LLM-powered AGI (Artificial General Intelligence) within telcos signals a transformative leap towards self-evolving networks, capable of dynamic configuration and real-time optimization. Concepts such as Autonomous SON (Self-Organizing Networks), AI accelerators, and GenAI-driven automation point towards a future where network operations are intelligently managed by AI agents, optimizing both performance and energy efficiency without centralized control. The technological advancements discussed in this chapter are not isolated phenomena; they interact with and are influenced by the broader dynamics of social behavior and collective interactions. In Chap. 4, there is the extension of the concepts introduced here into the realm of social complexity and digital anthropology. Specifically, the discussion shows how social behaviors, collective dynamics, and emergent phenomena can shape the very structure of communication networks.

The role of digital anthropology in understanding user behavior, fake news diffusion, misinformation spread, and emergent collective behaviors in digital environments will be central to this discussion. These phenomena not only impact the flow of information across networks but also influence the architectural design and optimization of telecommunication infrastructures. For instance, the application of LLMs in sentiment analysis, customer behavior prediction, and network-based collective intelligence paves the way for AI-native telcos that are socially aware and adaptive. This connection underscores a critical insight: the study of complex systems in telecommunications is not solely a technical endeavor–it is deeply intertwined with understanding collective social dynamics. There will be a deep understanding of this interplay between social complexity and technological evolution, setting the stage for a holistic understanding of how communication networks are both shaping and being shaped by global societal behavior.

References

1. Dorogovtsev, S.N., Mendes, J.F.: Evolution of Networks: From Biological Nets To the Internet and WWW. Oxford University Press, Oxford, UK (2013). https://doi.org/10.1093/acprof:oso/9780198515906.001.0001
2. Barrat, A., Barthelemy, M., Pastor-Satorras, R., Vespignani, A.: The architecture of complex weighted networks. Proc. Natl. Acad. Sci. U.S.A. **101**(11), 3747–3752 (2004)
3. Scatà, M., La Corte, A., Marotta, A., Graziosi, F., Cassioli, D.: A complex network and evolutionary game theory framework for 6G function placement. IEEE Open J. Commun. Soc. (2024). https://doi.org/10.1109/OJCOMS.2024.XXXXXXX
4. Amin, M., et al.: National infrastructures as complex interactive networks. In: Automation, Control, and Complexity: An Integrated Approach, vol. 3, pp. 263–286. Wiley, Hoboken, NJ, USA (2013)
5. Gu, B., Liu, J.: A systematic review of resilience in the maritime transport. Int. J. Logist. Res. Appl. **28**(3), 257–278 (2025)
6. Müller, F.: State-of-the-art in ecosystem theory. Ecol. Model. **100**(1–3), 135–161 (1997)
7. Shavandi, A., Khedmati, M.: A multi-agent deep reinforcement learning framework for algorithmic trading in financial markets. Expert Syst. Appl. **208**, 118124 (2022)
8. Lee, C.Y., Hasegawa, H., Gao, S.: Complex-valued neural networks: a comprehensive survey. IEEE/CAA J. Autom. Sin. **9**(8), 1406–1426 (2022)
9. Bento, F., Tagliabue, M., Sandaker, I.: Complex systems and social behavior: bridging social networks and behavior analysis. In: Behavior Science Perspectives on Culture and Community, pp. 67–91. Springer, Berlin (2020)
10. Hobbs, W., Burke, M.: Connective recovery in social networks after the death of a friend. Nat. Hum. Behav. **1**(92), 1–6 (2017)
11. Scatà, M., Di Stefano, A., Giacchi, E., La Corte, A., Liò, P.: The bio-inspired and social evolution of node and data in a multilayer network. In: Proceedings of the 5th International Conference Data Communication Network (DCNET), pp. 1–6 (2014)
12. Sergiou, C., Lestas, M., Antoniou, P., Liaskos, C., Pitsillides, A.: Complex systems: a communication networks perspective towards 6G. IEEE Access **8**, 89007–89030 (2020)
13. Scatà, M., Attanasio, B., La Corte, A.: Cognitive profiling of nodes in 6G through multiplex social network and evolutionary collective dynamics. Futur. Internet **13**(5), 135 (2021). https://doi.org/10.3390/fi13050135
14. Mitchell, M.: Complexity: A Guided Tour. Oxford University Press, Oxford (2009)
15. Kurtz, C.F., Snowden, D.J.: The new dynamics of strategy: sense-making in a complex and complicated world. IBM Syst. J. **42**(3), 462–483 (2003)
16. Newman, M.E.J.: Networks, 2nd edn. Oxford University Press, Oxford (2018)
17. Scatà, M., Attanasio, B., Aiosa, G.V., La Corte, A.: The dynamical interplay of collective attention, awareness and epidemics spreading in the multiplex social networks during COVID-19. IEEE Access **8**, 189203–189223 (2020)
18. Letaief, K.B., Chen, W., Shi, Y., Zhang, J., Zhang, Y.-J.A.: The roadmap to 6G: AI empowered wireless networks. IEEE Commun. Mag. **57**(8), 84–90 (2019)
19. Andrews, J.G., Buzzi, S., Choi, W., Hanly, S.V., Lozano, A., Soong, A.C., Zhang, J.C.: What will 5G be? IEEE J. Sel. Areas Commun. **32**(6), 1065–1082 (2014)
20. Boccardi, F., Heath, R.W., Lozano, A., Marzetta, T.L., Popovski, P.: Five disruptive technology directions for 5G. IEEE Commun. Mag. **52**(2), 74–80 (2014)
21. Dang, S., Amin, O., Shihada, B., Alouini, M.S.: What should 6G be? Nat. Electron. **3**(1), 20–29 (2020)
22. Giordani, M., Polese, M., Mezzavilla, M., Rangan, S., Zorzi, M.: Toward 6G networks: use cases and technologies. IEEE Commun. Mag. **58**(3), 55–61 (2020)
23. Saad, W., Bennis, M., Chen, M.: A vision of 6G wireless systems: applications, trends, technologies, and open research problems. IEEE Netw. **34**(3), 134–142 (2019)

24. You, X., Wang, C.X., Huang, J., Gao, X., Zhang, Z., Wang, Y., Li, X.: Towards 6G wireless communication networks: vision, enabling technologies, and new paradigm shifts. Sci. China Inf. Sci. **64**(1), 1–74 (2021)
25. ITU-R.: IMT vision – framework and overall objectives of the future development of IMT for 2020 and beyond. Recommendation ITU-R M.2083-0 (2015)
26. Zhang, S., Zhang, H., Di, B., Song, L.: Joint trajectory and power optimization for UAV sensing over cellular networks. IEEE Trans. Commun. **68**(1), 584–599 (2020)
27. Zhang, K., Yang, Z., Başar, T.: Multi-agent reinforcement learning: a selective overview of theories and algorithms. In: Handbook of Reinforcement Learning and Control, pp. 321–384. Springer (2021)
28. 5G Architecture Working Group.: The 6G architecture landscape: European perspective. 5GPPP, Pisa, Italy, White Paper (2022)
29. Wu, W., et al.: AI-native network slicing for 6G networks. IEEE Wirel. Commun. **29**(1), 96–103 (2022)
30. Cao, H., et al.: Resource-ability assisted service function chain embedding and scheduling for 6G networks with virtualization. IEEE Trans. Veh. Technol. **70**(4), 3846–3859 (2021)
31. Ji, B., et al.: A survey of computational intelligence for 6G: key technologies, applications and trends. IEEE Trans. Ind. Inf. **17**(10), 7145–7154 (2021)
32. Masaracchia, A., Sharma, V., Canberk, B., Dobre, O.A., Duong, T.Q.: Digital twin for 6G: Taxonomy, research challenges, and the road ahead. IEEE Open J. Commun. Soc. **3**, 2137–2150 (2022)
33. Cerquitelli, T., Meo, M., Curado, M., Skorin-Kapov, L., Tsiropoulou, E.E.: Machine learning empowered computer networks. Comput. Netw. **230**, 109807 (2023)
34. Addanki, R., Bojja Venkatakrishnan, S., Gupta, S., Mao, H., Alizadeh, M.: Learning generalizable device placement algorithms for distributed machine learning. In: Proceedings of the 33rd Advances in Neural Information Processing Systems, vol. 32 (2019)
35. Zahedi, S.R., Jamali, S., Bayat, P.: A power-efficient and performance-aware online virtual network function placement in SDN/NFV-enabled networks. Comput. Netw. **205**, 108753 (2022)
36. Gupta, R., Gupta, J.: Federated learning using game strategies: state-of-the-art and future trends. Comput. Netw. **225**, 109650 (2023)
37. Van, N.T.T., Luong, N.C., Feng, S., Nguyen, V.D., Kim, D.I.: Evolutionary games for dynamic network resource selection in RSMA-enabled 6G networks. IEEE J. Sel. Areas Commun. **41**(5), 1320–1335 (2023)
38. Iyer, S., Killingback, T.: Evolution of cooperation in social dilemmas on complex networks. PLoS Comput. Biol. **12**(2), e1004 (2016)
39. Stefano, A., Scatà, M., Attanasio, B., Corte, A., Lió, P., Das, S.K.: A novel methodology for designing policies in mobile crowdsensing systems. Pervasive Mob. Comput. **67**, 101230 (2020)
40. Bianconi, G., et al.: Complex systems in the spotlight: next steps after the 2021 Nobel prize in physics. J. Phys. Complex **4**(1), 010201 (2023)
41. Boccaletti, S., Latora, V., Moreno, Y., Chavez, M., Hwang, D.U.: Complex networks: structure and dynamics. Phys. Rep. **424**(4), 175–308 (2006)
42. Battiston, F., Nicosia, V., Latora, V.: The new challenges of multiplex networks: measures and models. Eur. Phys. J. Spec. Top. **226**(3), 401–416 (2017)
43. Iacovacci, J., Wu, Z., Bianconi, G.: Mesoscopic structures reveal the network between the layers of multiplex data sets. Phys. Rev. E **92**(4), 042806 (2015)
44. Giacchi, E., Di Stefano, A., La Corte, A., Scatà, M.: A dynamic context-aware multiple criteria decision making model in social networks. In: Proceedings of the International Conference on Information Society (i-Society), pp. 157–162 (2014)
45. Helbing, D.: Globally networked risks and how to respond. Nature **497**, 51–59 (2013)
46. Buldyrev, S.V., Parshani, R., Paul, G., Stanley, H.E., Havlin, S.: Catastrophic cascade of failures in interdependent networks. Nature **464**, 1025–1028 (2010)

47. Havlin, S., Kenett, D.Y., Ben-Jacob, E., Bunde, A., Hermann, H., Kurths, J., Kirkpatrick, S., Solomon, S., Portugali, J.: Challenges of network science: applications to infrastructures, climate, social systems and economics. Eur. J. Phys. Spec. Top. **214**, 273–293 (2012)
48. Kenett, D.Y., Gao, J., Huang, X., Shao, S., Vodenska, I., Buldyrev, S.V., Paul, G., Stanley, H.E., Havlin, S.: Network of interdependent networks: overview of theory and applications. In: Networks of Networks: The Last Frontier of Complexity, pp. 3–36. Springer, Berlin (2014)
49. Kivelä, M., Arenas, A., Barthelemy, M., Gleeson, J.P., Moreno, Y., Porter, M.A.: Multilayer networks. J. Complex Netw. **2**, 203–271 (2014)
50. Boccaletti, S., Bianconi, G., Criado, R., Del Genio, C., Gómez-Gardeñes, J., Romance, M., Sendiña-Nadal, I., Wang, Z., Zanin, M.: The structure and dynamics of multilayer networks. Phys. Rep. **544**, 1–122 (2014)
51. Gao, J., Buldyrev, S.V., Stanley, H.E., Havlin, S.: Networks formed from interdependent networks. Nat. Phys. **8**, 40–48 (2012)
52. Bashan, A., Parshani, R., Havlin, S.: Percolation in networks composed of connectivity and dependency links. Phys. Rev. E **83**, 051127 (2011)
53. Parshani, R., Buldyrev, S.V., Havlin, S.: Interdependent networks: reducing the coupling strength leads to a change from a first to second order percolation transition. Phys. Rev. Lett. **105**, 048701 (2010)
54. Bashan, A., Berezin, Y., Buldyrev, S.V., Havlin, S.: The extreme vulnerability of interdependent spatially embedded networks. Nat. Phys. **9**, 667–672 (2013)
55. Havlin, S., Stanley, H.E., Bashan, A., Gao, J., Kenett, D.Y.: Percolation of interdependent network of networks. Chaos, Solitons Fractals **72**, 4–19 (2015)
56. Kenett, D.Y., Perc, M., Boccaletti, S.: Networks of networks-an introduction. Chaos, Solitons Fractals **80**, 1–6 (2015)
57. Arenas, A., Díaz-Guilera, A., Kurths, J., Moreno, Y., Zhou, C.: Synchronization in complex networks. Phys. Rep. **469**, 93–153 (2008)
58. Perc, M., Gómez-Gardeñes, J., Szolnoki, A., Floría, L.M., Moreno, Y.: Evolutionary dynamics of group interactions on structured populations: a review. J. R. Soc. Interface **10**, 20120997 (2013)
59. Szabó, G., Fáth, G.: Evolutionary games on graphs. Phys. Rep. **446**, 97–216 (2007)
60. Wang, Z., Szolnoki, A., Perc, M.: Interdependent network reciprocity in evolutionary games. Sci. Rep. **3**, 1183 (2013)
61. Wang, B., Chen, X., Wang, L.: Probabilistic interconnection between interdependent networks promotes cooperation in the public goods game. J. Stat. Mech. **2012**, P11017 (2012)
62. Szolnoki, A., Perc, M.: Information sharing promotes prosocial behaviour. New J. Phys. **15**, 053010 (2013)
63. Wang, Z., Szolnoki, A., Perc, M.: Rewarding evolutionary fitness with links between populations promotes cooperation. J. Theor. Biol. **349**, 50–56 (2014)
64. Pacheco, J.M., Vasconcelos, V.V., Santos, F.C.: Climate change governance, cooperation and self-organization. Phys. Life Rev. **11**, 573–586 (2014)
65. D'Orsogna, M.R., Perc, M.: Statistical physics of crime: a review. Phys. Life Rev. **12**, 1–21 (2015)
66. Rothkötter, M., Weise, J., Mostaghim, S.: Towards a new model for a 6G network-of-networks. In: Proceedings of the IEEE Conference on Artificial Intelligence (CAI), pp 255–256. IEEE (2023)
67. Ozger, M., Godor, I., Nordlow, A., Heyn, T., Pandi, S., Peterson, I., Viseras, A., Holis, J., Raffelsberger, C., Kercek, A., et al.: 6G for connected sky: a vision for integrating terrestrial and non-terrestrial networks. In: Proceedings of the 2023 Joint European Conference on Networks and Communications and 6G Summit (EuCNC/6G Summit), pp. 711–716. IEEE (2023)
68. Letaief, K.B., Shi, Y., Lu, J., Lu, J.: Edge artificial intelligence for 6G: vision, enabling technologies, and applications. IEEE J. Sel. Area Commun. **40**(1), 5–36 (2021)
69. Beniiche, A., Rostami, S., Maier, M.: Society 5.0: internet as if people mattered. IEEE Wirel. Commun. **29**(6), 160–168 (2022)

70. Bernardos, C.J., Uusitalo, M.A.: European vision for the 6G network ecosystem. Zenodo, Technical Report (2021). https://zenodo.org/record/5007671
71. Gao, J., Buldyrev, S.V., Havlin, S., Stanley, H.E.: Robustness of a network of networks. Phys. Rev. Lett. **107**, 195701 (2011)
72. Majdandzic, A., Podobnik, B., Buldyrev, S.V., Kenett, D.Y., Havlin, S., Stanley, H.E.: Spontaneous recovery in dynamical networks. Nat. Phys. **10**(1), 34–38 (2014). Nature Publishing Group UK, London. https://doi.org/10.1038/nphys2819
73. Zhang, X., Podobnik, B., Kenett, D.Y., Stanley, H.E.: Systemic risk and causality dynamics of the world international shipping market. Phys. A **415**, 43–53. Elsevier (2014). https://doi.org/10.1016/j.physa.2014.07.068
74. Wang, Z., Szolnoki, A., Perc, M.: Self-organization towards optimally interdependent networks by means of coevolution. New J. Phys. **16**, 033041 (2014)
75. Zhou, D., Bashan, A., Cohen, R., Berezin, Y., Shnerb, N., Havlin, S.: Simultaneous first-and second-order percolation transitions in interdependent networks. Phys. Rev. E **90**, 012803 (2014)
76. Wang, Z., Szolnoki, A., Perc, M.: Optimal interdependence between networks for the evolution of cooperation. Sci. Rep. **3**(1), 2470 (2013). Nature Publishing Group UK, London. https://doi.org/10.1038/srep02470
77. Mao, B., Liu, J., Wu, Y., Kato, N.: Security and privacy on 6G network edge: a survey. IEEE Commun. Surv. Tutor. **25**(2), 1095–1127 (2023). https://doi.org/10.1109/COMST.2022.3145678
78. Resilient and Intelligent NextG Systems (RINGS).: Available via NSF (2021). https://www.nsf.gov/pubs/2021/nsf21581/nsf21581.pdf
79. Expanded 6G Vision, Use Cases and Societal Values.: Available via Hexa-X (2021). https://hexa-x.eu/wp-content/uploads/2021/05/Hexa-X_D1.2.pdf
80. ITU Focus Group on Technologies for Network: Network 2030: A Blueprint of Technology, Applications and Market Drivers Towards the Year 2030 and Beyond. ITU, Geneva (2019)
81. Andrews, J.G., Buzzi, S., Choi, W., Hanly, S.V., Lozano, A., Soong, A.C., Zhang, J.C.: What will 5G be? IEEE J. Sel. Areas Commun. **32**(6), 1065–1082 (2014)
82. Shafi, M., Molisch, A.F., Smith, P., Haustein, T., Zhu, P., Silva, P.J., Tufvesson, F., Benjebbour, A., Wunder, G.: 5G: a tutorial overview of standards, trials, challenges, deployment, and practice. IEEE J. Sel. Areas Commun. **35**(6), 1201–1221 (2017)
83. Huawei.: 5.5G. Available via Huawei (2020). https://www.huawei.com/en/news/2020/11/mbbf-shanghai-huawei-david-wang-5dot5g
84. Tataria, H., Shafi, M., Molisch, A.F., Dohler, M., Sjöland, H., Tufvesson, F.: 6G wireless systems: vision, requirements, challenges, insights, and opportunities. Proc. IEEE **109**(7), 1–34 (2021)
85. Maier, M., Ebrahimzadeh, A., Rostami, S., Beniiche, A.: The internet of no things: making the internet disappear and 'see the invisible'. IEEE Commun. Mag. **58**(11), 76–82 (2020)
86. Ali, S., Saad, W., Steinbach, D., Ahmad, I., Huusko, J.: White paper on machine learning in wireless communication networks. 6G Res Vis 7:1–36. Available via 6G Research Vision (2025). https://www.6gvision.com/whitepaper
87. Wang, J., Li, B., Song, W., Zhou, L., Zhang, Q.: Interplay between RIS and AI in wireless communications: fundamentals, architectures, applications, and open research problems. IEEE J. Sel. Areas Commun. **39**(8), 2271–2288 (2021)
88. Kim, H., Jiang, Y., Rana, R.B., Kannan, S., Oh, S., Viswanath, P.: Communication algorithms via deep learning. In: Proceedings of the International Conference on Learning Representations (ICLR), pp. 1–19 (2018)
89. O'Shea, T., Hoydis, J.: An introduction to deep learning for the physical layer. IEEE Trans. Cogn. Commun. Netw. **3**(4), 563–575 (2017)
90. Hoydis, J., Aoudia, F.A., Valcarce, A., Viswanathan, H.: Toward a 6G AI-native air interface. IEEE Commun. Mag. **59**(5), 76–81 (2021)
91. Yang, K., Shi, Y., Zhou, Y., et al.: Federated machine learning for intelligent IoT via reconfigurable intelligent surface. IEEE Netw. **34**(5), 16–22 (2020)

92. Shao, J., Zhang, J.: Communication-computation trade-off in resource-constrained edge inference. IEEE Commun. Mag. **58**(12), 20–26 (2021)
93. Kairouz, E.B.P., McMahan, H.B.: Advances and open problems in federated learning. Found. Trends Mach. Learn. **14**(1), 1–210 (2021)
94. McMahan, B., Moore, E., Ramage, D., Hampson, S., Arcas, B.A.Y.: Communication-efficient learning of deep networks from decentralized data. In: Proceedings of the International Conference on Artificial Intelligence and Statistics (AISTATS), pp. 1273–1282 (2017)
95. Alistarh, D., Grubic, D., Li, J., Tomioka, R., Vojnovic, M.: QSGD: communication-efficient SGD via gradient quantization and encoding. In: Proceedings of Advances in Neural Information Processing Systems (NeurIPS), pp. 1707–1718 (2017)
96. Wen W., Xu C., Wu C., Wang Y., Chen Y., Li H. (2017) TernGrad: Ternary gradients to reduce communication in distributed deep learning. In: Proceedings of Advances in Neural Information Processing Systems (NeurIPS), pp. 1508–1518
97. Bernstein, J., Wang, Y.-X., Azizzadenesheli, K., Anandkumar, A.: SignSGD: compressed optimization for non-convex problems. In: Proceedings of the International Conference on Machine Learning (ICML), pp. 560–569 (2018)
98. Du, Y., Yang, S., Huang, K.: High-dimensional stochastic gradient quantization for communication-efficient edge learning. IEEE Trans. Signal Process. **68**, 2128–2142 (2020)
99. Shlezinger, N., Chen, M., Eldar, Y.C., Poor, H.V., Cui, S.: UVeQFed: universal vector quantization for federated learning. IEEE Trans. Signal Process. **69**, 500–514 (2021)
100. Lian, X., Zhang, C., Zhang, H., Hsieh, C., Zhang, W., Liu, J.: Can decentralized algorithms outperform centralized algorithms? a case study for decentralized parallel stochastic gradient descent. In: Proceedings of Advances in Neural Information Processing Systems (NeurIPS), pp. 5330–5340 (2017)
101. Savazzi, S., Nicoli, M., Bennis, M., Kianoush, S., Barbieri, L.: Opportunities of federated learning in connected, cooperative, and automated industrial systems. IEEE Commun. Mag. **59**(2), 16–21 (2021)
102. Boyd, S., Ghosh, A., Prabhakar, B., Shah, D.: Randomized gossip algorithms. IEEE Trans. Inf. Theory **52**(6), 2508–2530 (2006)
103. Sayed, A.H., Tu, S.-Y., Chen, J., Zhao, X., Towfic, Z.J.: Diffusion strategies for adaptation and learning over networks: an examination of distributed strategies and network behavior. IEEE Signal Process. Mag. **30**(3), 155–171 (2013)
104. Xin, R., Kar, S., Khan, U.A.: Decentralized stochastic optimization and machine learning: A unified variance-reduction framework for robust performance and fast convergence. IEEE Signal Process. Mag. **37**(3), 102–113 (2020)
105. Wang, J., Joshi, G.: Cooperative SGD: a unified framework for the design and analysis of communication-efficient SGD algorithms (2018). arXiv:1808.07576
106. Koloskova, A., Loizou, N., Boreiri, S., Jaggi, M., Stich, S.: A unified theory of decentralized SGD with changing topology and local updates. In: Proceedings of the International Conference on Machine Learning (ICML), pp. 5381–5393 (2020)
107. Li M., Andersen, D.G., Smola, A.J., Yu, K.: Communication-efficient distributed machine learning with the parameter server. In: Proceedings of Advances in Neural Information Processing Systems (NeurIPS), pp. 19–27 (2014)
108. Gupta, O., Raskar, R.: Distributed learning of deep neural network over multiple agents. J. Netw. Comput. Appl. **116**, 1–8 (2018)
109. Chen, T., Zhang, K., Giannakis, G.B., Basar, T.: Communication-efficient policy gradient methods for distributed reinforcement learning. IEEE Trans. Control Netw. Syst. (2021)
110. Lowe, R., Wu, Y., Tamar, A., Harb, J., Abbeel, P., Mordatch, I.: Multi-agent actor-critic for mixed cooperative-competitive environments. In: Proceedings of Advances in Neural Information Processing Systems (NeurIPS), pp. 6382–6393 (2017)
111. Dwork, C., Roth, A.: The algorithmic foundations of differential privacy. Found. Trends Theor. Comput. Sci. **9**(3–4), 211–407 (2014)
112. Wu, Z., Ling, Q., Chen, T., Giannakis, G.B.: Federated variance-reduced stochastic gradient descent with robustness to Byzantine attacks. IEEE Trans. Signal Process. **68**, 4583–4596 (2020)

113. Yin, D., Chen, Y., Kannan, R., Bartlett, P.: Byzantine-robust distributed learning: towards optimal statistical rates. In: Proceedings of the International Conference on Machine Learning (ICML), pp. 5650–5659 (2018)
114. Kim, H., Park, J., Bennis, M., Kim, S.-L.: Blockchained on-device federated learning. IEEE Commun. Lett. **24**(6), 1279–1283 (2020)
115. Lin, X., Kundu, L., Dick, C., Obiodu, E., Mostak, T., Flaxman, M.: 6G digital twin networks: from theory to practice. IEEE Commun. Mag. **61**(11), 72–78 (2023)
116. Sheraz, M., Chuah, T.C., Lee, Y.L., Alam, M.M., Al-Habashna, A., Han, Z.: A comprehensive survey on revolutionizing connectivity through artificial intelligence-enabled digital twin network in 6G. IEEE Access **12**, 49184–49215 (2024)
117. Bhatia, M., Kumar, R.: Digital twin technology in wireless communication: a comprehensive engineering survey. IEEE Internet Things J. (2025)
118. Allen, B.D.: Digital Twins and Living Models at NASA. NASA Technical Reports Server (2021). Retrieved from https://ntrs.nasa.gov/citations/20210023699
119. Khan, L.U., Yaqoob, I., Tran, N.H., Kazmi, S.M.A., Dang, T.N., Hong, C.S.: Edge-computing-enabled smart cities: a comprehensive survey. IEEE Internet Things J. **7**(10), 10200–10232 (2020)
120. Shen, X., Gao, J., Wu, W., Li, M., Zhou, C., Zhuang, W.: Holistic network virtualization and pervasive network intelligence for 6G. IEEE Commun. Surv. Tutor. **24**(1), 1–30 (2021)
121. Tang, F., Chen, X., Rodrigues, T.K., Zhao, M., Kato, N.: Survey on digital twin edge networks (DITEN) toward 6G. IEEE Open J. Commun. Soc. **3**, 1360–1381 (2022)
122. Alcaraz, C., Lopez, J.: Digital twin: a comprehensive survey of security threats. IEEE Commun. Surv. Tutor. **24**(3), 1475–1503 (2022)
123. Barricelli, B.R., Casiraghi, E., Fogli, D.: A survey on digital twin: definitions, characteristics, applications, and design implications. IEEE Access **7**, 167653–167671 (2019)
124. Minerva, R., Lee, G.M., Crespi, N.: Digital twin in the IoT context: a survey on technical features, scenarios, and architectural models. Proc. IEEE **108**(10), 1785–1824 (2020)
125. Khan, L.U., Saad, W., Niyato, D., Han, Z., Hong, C.S.: Digital-twin-enabled 6G: vision, architectural trends, and future directions. IEEE Commun. Mag. **60**(1), 74–80 (2022)
126. Bellavista, P., Berardinelli, L., Corradi, A., Foschini, L., Perillo, M.: Application-driven network-aware digital twin management in industrial edge environments. IEEE Trans. Ind. Inform. **17**(11), 7658–7667 (2021). https://doi.org/10.1109/TII.2020.3045823
127. Girletti, L., Groshev, M., Guimarães, C., Bernardos, C.J., de la Oliva, A.: An intelligent edge-based digital twin for robotics. In: 2020 IEEE Globecom Workshops (GC Wkshps), pp. 1–6. IEEE (2020) https://doi.org/10.1109/GCWkshps50303.2020.9367549
128. Hu, C., Fan, W., Zen, E.: A digital twin-assisted real-time traffic data prediction method for 5G-enabled internet of vehicles. IEEE Trans. Ind. Inform. **17**(5), 1–1 (2021). https://doi.org/10.1109/TII.2021.3083596
129. Xie, Y., Wu, Q., Fan, P.: Digital twin vehicular edge computing network: task offloading and resource allocation (2024). arXiv:2407.11310. https://doi.org/10.48550/arXiv.2407.11310
130. Zhang, K., Cao, J., Maharjan, S., Zhang, Y.: Digital twin empowered content caching in social-aware vehicular edge networks. IEEE Trans. Ind. Inform. **17**(7), 4968–4977 (2021). https://doi.org/10.1109/TII.2020.2979871
131. Tao, F., Zhang, M., Liu, Y., Nee, A.Y.C.: Digital Twin Driven Smart Manufacturing. Academic Press (2019). https://doi.org/10.1016/B978-0-12-817630-6.00001-5
132. Dai, Y., Zhang, K., Maharjan, S., Zhang, Y.: Deep reinforcement learning for stochastic computation offloading in digital twin networks. IEEE Trans. Ind. Inform. **17**(7), 4968–4977 (2020). https://doi.org/10.1109/TII.2020.2979871
133. Van Huynh, D., Nguyen, V.D., Sharma, V., Dobre, O.A., Duong, T.Q.: Digital twin empowered ultra-reliable and low-latency communications-based edge networks in industrial IoT environment. IEEE Trans. Ind. Inform. **18**(4), 2753–2761 (2021). https://doi.org/10.1109/TII.2021.3056789
134. Khan, L.U., Saad, W., Niyato, D., Han, Z., Hong, C.S.: Digital-twin-enabled 6G: vision, architectural trends, and future directions. IEEE Commun. Mag. **60**(1), 74–80 (2022). https://doi.org/10.1109/MCOM.001.2100511

135. Jiang, W., Han, B., Habibi, M.A., Schotten, H.D.: The road towards 6G: a comprehensive survey. IEEE Open J. Commun. Soc. **2**, 334–366 (2021). https://doi.org/10.1109/OJCOMS.2021.3057679
136. Han, B., Jiang, W., Habibi, M.A., Schotten, H.D.: An abstracted survey on 6G: drivers, requirements, efforts, and enablers (2021). arXiv:2101.01062
137. Lu, Y., Maharjan, S., Zhang, Y.: Adaptive edge association for wireless digital twin networks in 6G. IEEE Internet Things J. **8**(22), 16219–16230 (2021). https://doi.org/10.1109/JIOT.2021.3064958
138. Zeb, S, Mahmood, A, Hassan, S.A., Piran, M.J., Gidlund, M., Guizani, M.: Industrial digital twins at the Nexus of NextG wireless networks and computational intelligence: a survey. J. Netw. Comput. Appl. **200**, 103309 (2021). https://doi.org/10.1016/j.jnca.2021.103309. https://api.semanticscholar.org/CorpusID:236965761
139. Habibi, M.A., Han, B., Nasimi, M., Kuruvatti, N.P., Fellan, A., Schotten, H.D.: Towards a fully virtualized, cloudified, and slicing-aware RAN for 6G mobile networks. In: 6G Mobile Wireless Networks (Computer Communications and Networks), pp. 327–358. Springer, Cham (2021)
140. Habibi, M.A., Yousaf, F.Z., Schotten, H.D.: Mapping the VNFs and VLs of a RAN slice onto intelligent PoPs in beyond 5G mobile networks. IEEE Open J. Commun. Soc. **3**, 670–704 (2022). https://doi.org/10.1109/OJCOMS.2022.3150567
141. Habibi, M.A., Han, B., Yousaf, F.Z., Schotten, H.D.: How should network slice instances be provided to multiple use cases of a single vertical industry? IEEE Commun. Stand. Mag. **4**(3), 53–61 (2020). https://doi.org/10.1109/MCOMSTD.001.2000008
142. Liu, Q., Tang, L., Wu, T., Chen, Q.: Deep reinforcement learning for resource demand prediction and virtual function network migration in digital twin network. IEEE Internet Things J. **10**(21), 19102–19116 (2023). https://api.semanticscholar.org/CorpusID
143. Sheraz, M., Ahmed, M., Hou, X., Li, Y., Jin, D., Han, Z., Jiang, T.: Artificial intelligence for wireless caching: schemes, performance, and challenges. IEEE Commun. Surv. Tutor. **23**(1), 631–661 (2021)
144. Hui, Y., Qiu, Y., Cheng, N., Yin, Z., Chen, R., Liang, K., Luan, T.H.: Digital twin enabled on-demand content delivery in HetVNets. IEEE Internet Things J. **10**(16), 14028–14041 (2023)
145. Zhang, K., Cao, J., Zhang, Y.: Adaptive digital twin and multi-agent deep reinforcement learning for vehicular edge computing and networks. IEEE Trans. Ind. Inf. **18**(2), 1405–1413 (2022)
146. Zheng, J., Luan, T.H., Gao, L., Zhang, Y., Wu, Y.: Learning based task offloading in digital twin empowered internet of vehicles (2021). arXiv:2201.09076
147. Chen, L., Gu, Q., Jiang, K., Zhao, L.: A3C-based and dependency-aware computation offloading and service caching in digital twin edge networks. IEEE Access **11**, 57564–57573 (2023)
148. Dai, Y., Zhang, Y.: Adaptive digital twin for vehicular edge computing and networks. J. Commun. Inf. Netw. **7**(1), 48–59 (2022). https://api.semanticscholar.org/CorpusID
149. Lu, Y., Huang, X., Zhang, K., Maharjan, S., Zhang, Y.: Communication-efficient federated learning and permissioned blockchain for digital twin edge networks. IEEE Internet Things J. **8**(4), 2276–2288 (2021)
150. Liu, T., Tang, L., Wang, W., Chen, Q., Zeng, X.: Digital-twin-assisted task offloading based on edge collaboration in the digital twin edge network. IEEE Internet Things J. **9**(2), 1427–1444 (2022)
151. Qu, Z., Li, Y., Liu, B., Gupta, D., Tiwari, P.: DTQFL: A digital twin-assisted quantum federated learning algorithm for intelligent diagnosis in 5G mobile network. IEEE J. Biomed. Health Inform. (2023)
152. Dembski, F., Wössner, U., Letzgus, M., Ruddat, M., Yamu, C.: Urban digital twins for smart cities and citizens: the case study of Herrenberg, Germany. Sustainability **12**(6), 2307 (2020). https://www.mdpi.com/2071-1050/12/6/2307
153. Dembski, F., Yamu, C., Wössner, U.: Digital twin, virtual reality and space syntax: Civic engagement and decision support for smart, sustainable cities. In: Proceedings of the 12th International Space Syntax Symposium, pp. 1–14 (2019). https://api.semanticscholar.org/CorpusID:197725347

154. Deckert, A., Dembski, F., Ulmer, F., Ruddat, M., Wössner, U.: Digital tools in stakeholder participation for the German energy transition. In: The Role of Public Participation in Energy Transitions. Academic, Cambridge (2020)
155. Deng, T., Zhang, K., Shen, Z.J.M.: A systematic review of a digital twin city: a new pattern of urban governance toward smart cities. J. Manag. Sci. Eng. **6**(2), 125–134 (2021). https://api.semanticscholar.org/CorpusID:233678798
156. Shirowzhan, S., Tan, W., Sepasgozar, S.M.E.: Digital twin and CyberGIS for improving connectivity and measuring the impact of infrastructure construction planning in smart cities. ISPRS Int. J. Geo.-Inf. **9**(4), 240 (2020) . https://api.semanticscholar.org/CorpusID:216266776
157. Dembski, F., Wössner, U., Letzgus, M., Ruddat, M., Yamu, C.: Urban digital twins for smart cities and citizens: the case study of Herrenberg, Germany. Sustainability **12**(6), 2307 (2020). https://api.semanticscholar.org/CorpusID:216404355
158. Salunke, A.A.: Reinforcement learning empowered digital twins: Pioneering smart cities towards optimal urban dynamics. EPRA Int. J. Res. Develop. (IJRD) **8**(7), 265–269 (2023). https://api.semanticscholar.org/CorpusID:260393398
159. Bellini, P., Colombo, C.M., Fanfani, M., Nesi, P., Pantaleo, G., Pisanu, R., Adreani, L.: (2022) Digital twin framework for smart city solutions. In: Proceedings of the International Distribution Multimedia System Conference Visualization and Visual Language, pp. 1–8. https://api.semanticscholar.org/CorpusID:251776924
160. Pervez, Z., Khan, Z., Ghafoor, A., Soomro, K.: SIGNED: smart city digital twin verifiable data framework. IEEE Access **11**, 29430–29446 (2023)
161. Pradhan, M., Suri, N., Fuchs, C., Bloebaum, T.H., Marks, M.: Toward an architecture and data model to enable interoperability between federated mission networks and IoT-enabled smart city environments. IEEE Commun. Mag. **56**(10), 163–169 (2018)
162. Mongelli, M., Scanzio, S.: Approximating optimal estimation of time offset synchronization with temperature variations. IEEE Trans. Instrum. Meas. **63**(12), 2872–2881 (2014)
163. Liu, Y. et al.: Reconfigurable intelligent surfaces: Principles and opportunities. IEEE Commun. Surv. & Tutor. **23**(3), 1546–1577 (2021)
164. Wang, H., Hu, J., Xing, C., Zhang, L.J.: Federated learning for 6g edge intelligence: concepts, challenges and solutions. In: International Conference on AI and Mobile Services, pp. 99–112. Springer, Berlin (2021)
165. Mills, J., Hu, J., Min, G.: Communication-efficient federated learning for wireless edge intelligence in IoT. IEEE Internet Things J. **7**(7), 5986–5994 (2019)
166. Ott, J., Kärkkäinen, T., Keränen, A.: Floating content: Information sharing in urban areas. In: 2011 IEEE International Conference on Pervasive Computing and Communications Workshops (PERCOM Workshops), pp. 157–162. IEEE (2011). https://doi.org/10.1109/PERCOMW.2011.5766885
167. Keränen, A., Ott, J., Kärkkäinen, T.: The ONE simulator for DTN protocol evaluation. In: Proceedings of the 2nd International Conference on Simulation Tools and Techniques, pp. 1–10. ICST (Institute for Computer Sciences, Social-Informatics and Telecommunications Engineering) (2009). https://doi.org/10.4108/ICST.SIMUTOOLS2009.5674
168. Asadi, A., Wang, Q., Mancuso, V.: A survey on device-to-device communication in cellular networks. IEEE Commun. Surv. Tutor. **16**(4), 1801–1819 (2014). https://doi.org/10.1109/COMST.2014.2319555
169. Boldrini, C., Conti, M., Passarella, A.: ContentPlace: social-aware data dissemination in opportunistic networks. Pervasive Mob. Comput. **6**(3), 342–357 (2010). https://doi.org/10.1016/j.pmcj.2009.11.003
170. Karagiannis, T., Le Boudec, J.Y., Vojnović, M.: Power law and exponential decay of inter contact times between mobile devices. IEEE Trans. Mob. Comput. **9**(10), 1377–1390 (2010). https://doi.org/10.1109/TMC.2010.99
171. Pelusi, L., Passarella, A., Conti, M.: Opportunistic networking: data forwarding in disconnected mobile ad hoc networks. IEEE Commun. Mag. **44**(11), 134–141 (2006). https://doi.org/10.1109/MCOM.2006.248176

172. Barabási, A.L.: Linked: the new science of networks. Perseus Publishing, Cambridge (2002). 9780738206677
173. Watts, D.J., Strogatz, S.H.: Collective dynamics of 'small-world' networks. Nature **393**(6684), 440–442 (1998). https://doi.org/10.1038/30918
174. Bonabeau, E., Dorigo, M., Theraulaz, G.: Swarm Intelligence: From Natural to Artificial Systems (No. 1). Oxford university press (1999)
175. Ning, W., Zhu, Y., Song, C., Li, H., Zhu, L., Xie, J., Chen, T., Xu, T., Xu, X., Gao, J.: Blockchain-based federated learning: a survey and new perspectives. Appl. Sci. **14**(20), 2076–3417 (2024)
176. Mao, B., Shen, H., Lu, K., et al.: Security and privacy on 6G network edge: a survey. IEEE Commun. Surv. Tutor. 1–1 (2023)
177. Yu, S., Wang, C., Ren, K., Lou, W.: Attribute-based encryption for fine-grained access control of encrypted data. IEEE Trans. Inf. Forensics Secur. **5**(4), 768–778 (2010). https://doi.org/10.1109/TIFS.2010.2064828
178. Gentry, C.: Fully homomorphic encryption using ideal lattices. In: Proceedings of the 41st Annual ACM Symposium on Theory of Computing, pp. 169–178. ACM (2009). https://doi.org/10.1145/1536414.1536440
179. Feng, Y., Guo, Y., Hou, Y., Wu, Y., Lao, M., Yu, T., Liu, G.: A survey of security threats in federated learning. Complex Intell. Syst. **11**(2), 1–26 (2025). Springer. https://doi.org/10.1007/s40747-025-00234-1
180. Dorri, A., Kanhere, S.S., Jurdak, R., Gauravaram, P.: Blockchain for IoT security and privacy: the case study of a smart home. In: 2017 IEEE International Conference on Pervasive Computing and Communications Workshops (PerCom Workshops), pp. 618–623. IEEE (2017). https://doi.org/10.1109/PERCOMW.2017.7917634
181. Grove, N., Baumann, O.: Complexity in the telecommunications industry: When integrating infrastructure and services backfires. Telecommun. Policy **36**(1), 40–50 (2012). https://doi.org/10.1016/j.telpol.2011.11.019
182. Martin, V., Brito, J.P., Escribano, C., et al.: Quantum technologies in the telecommunications industry. EPJ Quantum Technol. **8**(19) (2021). https://doi.org/10.1140/epjqt/s40507-021-00108-9
183. Wehner, S., Elkouss, D., Hanson, R.: Quantum internet: a vision for the road ahead. Science **362**(6412), eaam9288 (2018). https://doi.org/10.1126/science.aam9288
184. Meter, R.: Quantum Networking. Wiley (2019)
185. Grover, L.K.: A fast quantum mechanical algorithm for database search. In: Proceedings of the 28th Annual ACM Symposium on Theory of Computing, pp. 212–219 (1996). https://doi.org/10.1145/237814.237866
186. Chen, L., Chen, L., Jordan, S., Liu, Y.-K., Moody, D., Peralta, R., Perlner, R., Smith-Tone, D.: Report on post-quantum cryptography. NISTIR 8105, National Institute of Standards and Technology (2016). https://doi.org/10.6028/NIST.IR.8105
187. Bennett, C.H., Brassard, G.: Quantum cryptography: Public key distribution and coin tossing. In: Proceedings of IEEE International Conference on Computers, Systems and Signal Processing, pp. 175–179. IEEE, Bangalore, India (1984)
188. Foukas, X., Patounas, G., Elmokashfi, A., Marina, M.K.: Network slicing in 5G: survey and challenges. IEEE Commun. Mag. **55**(5), 94–100 (2017)
189. Chaoub, A., Mämmelä, A., Martinez-Julia, P., Chaparadza, R., Elkotob, M., Ong, L., Krishnaswamy, D., Anttonen, A., Dutta, A.: Hybrid self-organizing networks: evolution, standardization trends, and a 6G architecture vision. IEEE Commun. Stand. Mag. 7(1):14–22 (2023). https://doi.org/10.1109/MCOMSTD.2023.3267164
190. Qureshi, K.N., Newe, T.: Artificial internet of things: a new paradigm of connected networks. In: Artificial Intelligence of Things (AIoT), pp. 3–20. CRC Press (2024)
191. Cai, L., Pan, J., Yang, W., Ren, X., Shen, X.: Self-evolving and transformative protocol architecture for 6G. IEEE Wirel. Commun. **30**(4), 178–186 (2022). https://doi.org/10.1109/MWC.2022.9845324

192. Prehofer, C., Bettstetter, C.: Self-organization in communication networks: principles and design paradigms. IEEE Commun. Mag. **43**(7), 78–85 (2005). https://doi.org/10.1109/MCOM.2005.1470822
193. Kalantari, S., Nazemi, E., Masoumi, B.: (2020) Emergence phenomena in self-organizing systems: a systematic literature review of concepts, researches, and future prospects. J. Organ. Comput. Electron. Commer. **30**(3), 224–265. https://doi.org/10.1080/10919392.2020.1761896
194. Nikolić, J., Jubatyrov, N., Pournaras, E.: Self-healing dilemmas in distributed systems: fault correction versus fault tolerance. IEEE Trans. Netw. Serv. Manag. **18**(3), 2728–2741 (2021). https://doi.org/10.1109/TNSM.2021.3072837
195. Mahmood, M.R., Alizadeh, M., Esmaili, F., Kumar, A.: A comprehensive review on artificial intelligence/machine learning algorithms for empowering the future IoT toward 6G era. IEEE Access **10**, 87535–87562 (2022)
196. Zeydan, E., Singh, V., Ng, D.W.K., Imran, M.A.: Quantum technologies for beyond 5G and 6G networks: applications, opportunities, and challenges (2025). arXiv:2504.17133
197. Yu, P., Zhu, L., Song, C., Li, H., Zhu, L., Xie, J., Chen, T., Xu, T., Xu, X., Gao, J.: Digital twin driven service self-healing with graph neural networks in 6G edge networks. IEEE J. Sel. Areas Commun. **41**(11), 3607–3623 (2023). https://doi.org/10.1109/JSAC.2023.3316249
198. Gelenbe, E., Lent, R., Nunez, A.: Self-aware networks and QoS. Proc. IEEE **92**(9), 1478–1489 (2004). https://doi.org/10.1109/JPROC.2004.832980
199. Papidas, A.G., Polyzos, G.C.: Self-organizing networks for 5G and beyond: a view from the top. Futur. Internet **14**(3), 95 (2022). https://doi.org/10.3390/fi14030095
200. Liao, J., He, B., Wang, J., Wang, J., Qi, Q.: Network Intelligentization. In: Key Technologies for On-Demand 6G Network Services, pp. 25–61. Springer, Berlin (2024)

Chapter 4
Digital Anthropology and Communication Networks: Understanding Social Impact in 6G

Abstract This chapter introduces the concept of *Digital Anthropology* as a critical framework for understanding and designing 6G communication networks. Rooted in complex systems theory, Digital Anthropology analyzes how human behaviors, social interactions, and emergent dynamics influence and are shaped by digital infrastructures. The chapter explores the propagation of information and misinformation, the co-evolution of network structures and social behaviors, and the impact of collective dynamics on communication resilience and trust. By integrating insights from social contagion models, higher-order network theory, and adaptive intelligence, the chapter presents Digital Anthropology as a practical approach for the development of human-centric, resilient, and ethically aware 6G systems. It concludes by aligning these insights with the broader goals of Society 5.0, emphasizing the transformative role of complexity science in enabling inclusive, adaptive, and socially responsive communication infrastructures.

4.1 Digital Anthropology and Communication Networks: Understanding Social Impact in 6G

The advent of 6G communication networks introduces not only technological advancements but also significant transformations in how social behaviors are analyzed and understood. Within this evolving landscape, the concept of *Digital Anthropology* emerges as a novel framework to explore the interplay between digital interactions and communication infrastructures. Digital Anthropology, as proposed in this chapter, represents the study of human behaviors, social contagion, and collective dynamics within digital ecosystems, with a particular focus on how these phenomena influence and reshape telecommunication designs.

As this chapter introduces several emerging and interdisciplinary concepts, Table 4.1 provides a concise glossary of key terms. This reference is intended to support the reader in navigating the analytical and theoretical discussions on Digital Anthropology and its role in shaping 6G communication networks.

M. Scatà, *The Power of Complex Systems*, Studies in Systems, Decision and Control 630,
https://doi.org/10.1007/978-3-032-04243-9_4

Table 4.1 Key concepts introduced in this chapter

Concept	Definition
Digital anthropology	The study of human behavior and collective dynamics in digital ecosystems, focusing on how interactions in communication networks shape and are shaped by societal patterns
Social contagion	The process by which behaviors, attitudes, or information spread through social networks, analogously to infectious diseases
Emergence	A phenomenon in which collective behaviors arise from the interactions of individual agents, producing outcomes not predictable from individual actions alone
Self-organization	A decentralized process through which patterns and structures emerge in a system without external control, often observed in digital communities and network traffic
Multiplex network	A network model where nodes are connected through multiple types of relationships (layers), such as social media, messaging, and physical proximity
Complex contagion	A spreading process requiring reinforcement from multiple neighbors before a behavior or idea is adopted, typical in digital activism or misinformation acceptance
Digital twin	A virtual replica of physical or social environments used for simulating and optimizing behaviors in smart cities or network infrastructures
Contextual feedback loops	Dynamic interactions where user behavior influences the system and vice versa, shaping communication flow and content exposure in real-time

Traditionally, *social anthropology* has focused on understanding human societies and cultures by examining various aspects of human life, including social structures, cultural norms, rituals, beliefs, and practices. Social anthropologists employ methods such as participant observation, interviews, and ethnographic research to study different societies and comprehend the complexities of human behavior within cultural contexts. This classical approach provides deep insights into how communities form, evolve, and sustain themselves through shared cultural practices.

In the context of advanced communication networks and pervasive digitalization, a new branch of anthropology could be defined, **Digital Anthropology**. This term extends the anthropological lens to digital spaces, where human interactions are increasingly mediated by technology, for example, in terms of digital communications and social networks and so on. Digital Anthropology in the field of communication networks is defined as the study of human behavior, social needs, and collective dynamics as they manifest within digital ecosystems created by modern communication technologies. This includes analyzing how individuals and communities engage, disseminate information, form collective behaviors, and influence network infrastructures in virtual environments.

Unlike traditional anthropology, Digital Anthropology could leverages large-scale data collection through communication networks, data science, and artificial intelligence (AI) to study human interactions at unprecedented scales. It seeks to understand not only the social and cultural implications of digital interactions but also how these emergent behaviors impact and, conversely, are shaped by communication infrastructures [1, 2]. In this sense, Digital Anthropology provides a dual perspective: it both reflects societal behaviors through the lens of digital communication and can have the power to change the design of communication networks to be more resilient, adaptive, and socially aware.

Digital Anthropology is inherently linked to the principles of complex systems. In traditional communication networks, interactions are often modeled as isolated events. However, in hyper-connected 6G ecosystems, individual behaviors influence and are influenced by the collective dynamics of the network. This bidirectional influence aligns with the theory of complex systems, where local interactions among nodes generate emergent global behaviors without centralized control [3]. For instance, the rapid spread of information, or misinformation, can be seen as a contagion process, analogous to epidemic spreading in multiplex networks [4]. Multiplex modeling, which captures the layered nature of social interactions across different platforms, allows for a more granular understanding of how digital behaviors propagate and evolve [5]. This modeling is particularly relevant for 6G networks, where the integration of digital twins, edge intelligence, and IoT devices creates multi-dimensional communication channels that must be synchronized and resilient to perturbations [2]. Through these models, Digital Anthropology not only explains how digital interactions manifest but also predicts network behaviors, enhancing the design and optimization of telecommunication infrastructures.

The primary motivation behind integrating Digital Anthropology into 6G design lies in its ability to predict and manage emergent behaviors [3]. Social contagion models illustrate how behaviors, ideas, and misinformation spread across interconnected layers of communication, often with exponential acceleration [6, 7]. Understanding these propagation mechanisms is crucial for optimizing information dissemination, mitigating the spread of fake news, and enhancing digital trust [8]. Furthermore, the principles of Digital Anthropology support the design of self-organizing communication networks capable of adaptive response to dynamic social behaviors. By leveraging real-time data from digital interactions, 6G networks can optimize resource allocation, enhance latency management, and prevent cascading failures resulting from misinformation or cyber-attacks [4]. This framework also informs the development of resilient network architectures that adapt to social shifts and collective dynamics, reflecting real-world human behavior in virtual spaces.

The study of Digital Anthropology in the context of 6G paves the way for novel research directions, including:

- **Enhanced Social Sensing Mechanisms**: Developing algorithms for real-time detection of collective attention shifts and social contagion events in communication networks [9].
- **Resilience to Digital Misinformation**: Employing multiplex network models to trace and mitigate the spread of fake news and coordinated disinformation campaigns [1].

- **Integration with Digital Twins**: Synchronizing social behavior analytics with digital twin environments to enable predictive social modeling and resilient communication infrastructures [10].
- **Complex Adaptive Security Mechanisms**: Utilizing evolutionary game theory to model competitive and cooperative behaviors within digital ecosystems, enhancing resilience against cyber threats [11].

This introduction could be essential for exploring how Digital Anthropology not only describes but actively impacts on the design of resilient, adaptive, and socially aware 6G networks.

4.2 Social Behavior and Network Dynamics

The evolution of communication networks towards 6G introduces new paradigms in how social behavior is mediated, influenced, and propagated. Unlike previous generations, 6G networks are designed not only to connect individuals but also to integrate and reflect social dynamics at unprecedented scales and resolutions. This capability emerges from the integration of advanced communication technologies, massive IoT deployments, and artificial intelligence, enabling real-time sensing, analysis, and prediction of collective social behaviors.

Social behavior, in the context of digital communication, is defined as the collective actions and interactions of users within networked environments [6, 7]. These behaviors are increasingly influenced by digital platforms, social media, and pervasive communication technologies that allow instantaneous sharing of information and collective decision-making [1]. Digital Anthropology, as introduced earlier, could provides a framework for analyzing these behaviors by extending traditional anthropological methods to virtual environments where interactions are mediated by digital infrastructures.

In hyper-connected 6G ecosystems, social behaviors are not isolated phenomena; they are networked interactions that propagate across layers of digital communication. For example, the dissemination of information in social media platforms often follows patterns similar to epidemic spreading, where information cascades through communities based on proximity and influence [2]. These interactions can lead to phenomena such as:

- **Information Cascades**: When a piece of information, whether true or false, rapidly spreads across a network due to its perceived relevance or endorsement by influential nodes [1].
- **Collective Attention Shifts**: Sudden, large-scale focus on specific events or topics driven by social cues, media coverage, or viral sharing [9].
- **Echo Chambers and Polarization**: Network topologies that reinforce specific ideologies or beliefs, limiting exposure to diverse perspectives and contributing to social polarization [8].

These dynamics are not merely byproducts of connectivity; they are influenced by the architecture of the communication networks themselves. Multiplex and multilayer networks capture these social phenomena by modeling different types of interactions—such as friendship ties, information flows, and trust relationships—across overlapping layers [5]. This structural complexity allows for more accurate modeling of how behaviors emerge, spread, and solidify in digital spaces.

The propagation of behaviors, information, and even misinformation in digital networks can be understood through the lens of *spreading processes*. These processes are governed by the topology of the network, the nature of the content, and the behavioral characteristics of users [4]. Classical models such as the Susceptible-Infected-Recovered (SIR) and Susceptible-Infected-Susceptible (SIS) frameworks have been adapted to model how information spreads similarly to biological contagions [2].

However, in the context of multilayer social networks, these models evolve to capture the complexity of concurrent spreading processes:

- **Interacting Contagions**: In multiplex networks, multiple types of contagions (e.g., information and misinformation) can spread simultaneously, influencing each other. For instance, the spread of a rumor may be countered by the simultaneous dissemination of fact-checking information [5].
- **Threshold Models and Influence Cascades**: Social influence often requires a critical mass before behaviors or beliefs spread. Threshold models illustrate how individuals adopt behaviors only when a certain percentage of their neighbors do the same [4].
- **Co-evolution of Network and Behavior**: In many social platforms, the network topology evolves as behaviors change. For example, during political events or crises, users may form new connections or strengthen existing ones based on shared beliefs, thereby altering the propagation paths [2].

The analysis of these processes is crucial for understanding phenomena like viral misinformation, digital activism, and collective behavior during critical events. For instance, during the COVID-19 pandemic, the spread of both health information and misinformation exhibited contagion dynamics that were deeply intertwined with social network structures [2].

One of the hallmarks of complex systems, including digital social networks, is *emergence*—where collective behaviors arise from the interactions of individual agents, often without centralized control. In 6G networks, emergence is anticipated to play a critical role in shaping information flows, public opinion, and even coordinated action [4].

For example:

- **Flash Mobs and Digital Protests**: Collective organization through social media platforms can rapidly mobilize thousands of individuals based on shared goals or grievances [8].

- **Smart City Responses**: In urban IoT environments, collective behavior driven by social sensors (e.g., traffic congestion reporting, pollution monitoring) can trigger automated responses in smart infrastructure [5].
- **Cognitive Collectives in Virtual Environments**: Digital twins and virtual communities foster collective problem-solving by simulating real-world scenarios and facilitating collaborative decision-making [10, 21].

These emergent behaviors are not only reflections of social dynamics but are also drivers of network adaptation and optimization. As users interact and share information, the underlying communication infrastructure must dynamically adjust to accommodate fluctuating demand, shifting interests, and emergent social phenomena.

The study of social behavior in digital ecosystems offers valuable insights for the design and optimization of 6G communication networks:

- **Resilient Information Dissemination**: Understanding social contagion mechanisms enables the design of networks that are more resilient to misinformation and capable of promoting verified content [12–14].
- **Dynamic Resource Allocation**: Collective behavior analytics can inform resource allocation strategies, optimizing bandwidth and latency during large-scale digital events [9, 15].
- **Behaviorally-Informed Network Slicing**: In 6G, network slices could be dynamically configured based on collective behaviors, ensuring optimal service delivery during high-demand periods [5].
- **Adaptive Edge Intelligence**: Real-time analysis of social dynamics allows edge-based AI to preemptively allocate computational and communication resources where emergent behaviors are detected [20].

By leveraging Digital Anthropology as a framework, 6G architectures can be designed not only to respond to social behaviors but to anticipate them, optimizing network resilience, information dissemination, and adaptive capacity in real-time.

4.2.1 Fake News, Misinformation, and Digital Trust

The proliferation of digital communication platforms has transformed the mechanisms by which information is produced, disseminated, and consumed. While this digital shift has democratized access to information, it has also facilitated the widespread distribution of fake news, misinformation, and disinformation. These phenomena have critical implications for public opinion, political stability, and societal cohesion, particularly in hyper-connected environments like those anticipated for 6G networks [1]. In this context, the principles of *Digital Anthropology* and *Complex Systems Theory* provide robust analytical frameworks to understand the propagation dynamics of misinformation and its impact on digital trust.

Fake news, misinformation, and disinformation are distinct yet interrelated concepts that disrupt the integrity of information in digital ecosystems:

- **Fake News**: Fabricated information that mimics news media content but lacks the editorial norms and processes to ensure accuracy. It is deliberately crafted to mislead or manipulate public perception [1].
- **Misinformation**: Incorrect or misleading information shared without harmful intent. This often includes rumors, misconceptions, and accidental falsehoods spread through social interactions [16].
- **Disinformation**: Deliberately false or manipulated information created with the intent to deceive and influence public opinion or obscure the truth [1]. This often includes strategic propaganda aimed at destabilizing political systems or manipulating social dynamics.

Empirical findings highlight that fake news dissemination is not uniformly distributed among users. In fact, only a small subset of individuals—often older users or those embedded in ideologically homogeneous clusters—account for the majority of misleading content sharing [12]. Furthermore, polarization within online communities creates echo chambers that amplify emotionally charged and misleading content. Early-warning signals, such as the rapid formation of polarized clusters, have been proposed as indicators for potential misinformation outbreaks [13]. These insights align with the complex systems perspective, in which nonlinear thresholds, feedback mechanisms, and clustered structures shape diffusion outcomes.

The surge in misinformation and disinformation is particularly problematic in crisis situations such as pandemics, natural disasters, and political elections. During the COVID-19 pandemic, for instance, false information regarding health protocols and vaccine efficacy spread rapidly through social media platforms, complicating public health responses and undermining collective trust in institutions [16]. In highly interconnected 6G environments, where communication is decentralized and real-time, the threat of misinformation is expected to grow exponentially, necessitating resilient strategies to protect information integrity. Recent studies have proposed that the diffusion of fake news behaves more like a complex contagion than a simple one, requiring multiple exposures and social reinforcement before adoption [14]. In this view, misinformation propagation depends not only on network topology but also on affective resonance, ideological alignment, and confirmation bias. Such characteristics suggest that traditional countermeasures based solely on fact-checking may be insufficient without also addressing the structural and cognitive dynamics that sustain belief formation and viral spread.

The dissemination of fake news and misinformation in digital ecosystems follows complex propagation dynamics that can be analyzed through the lens of *social contagion theory* [7]. This theory posits that information spreads similarly to infectious diseases, permeating social networks through repeated interactions and influence-based mechanisms. The spread is particularly accelerated in *multiplex networks*, where individuals are connected across multiple communication platforms such as Twitter, Facebook, WhatsApp, and Telegram, enabling rapid cross-platform diffusion [2, 5].

Several key factors influence this dissemination:

- **Network Topology**: The structure of social networks plays a critical role in the spread of fake news. In *scale-free networks*, a small number of highly connected hubs serve as primary dissemination points, dramatically amplifying the reach of misinformation [5]. These hubs often correspond to influential social media users or widely followed digital news outlets.
- **Echo Chambers and Filter Bubbles**: Social media algorithms optimize user engagement by prioritizing content that aligns with individual beliefs and preferences. This creates isolated echo chambers where misinformation is continuously reinforced without exposure to corrective information [8].
- **Information Cascades**: When users observe others sharing content, they are more likely to replicate the behavior, regardless of the information's veracity. This phenomenon, known as an *information cascade*, is particularly potent in highly interconnected networks where multiple exposures amplify perceived credibility [6].
- **Interconnected Layers in Multiplex Networks**: Misinformation can propagate simultaneously across multiple communication channels, reinforcing its credibility and extending its reach. For example, a piece of fake news shared on Twitter can be further disseminated on WhatsApp and Facebook, increasing its visibility and perceived legitimacy [2, 5].
- **Psychological and Social Reinforcement**: Studies have shown that misinformation spreads faster than corrections due to cognitive biases such as confirmation bias and the illusory truth effect, where repeated exposure increases believability [16].

The concept of *complex contagions* is particularly relevant, where misinformation often requires multiple exposures across different network layers before individuals accept and propagate it [2]. This is in contrast to simple contagions, like viral marketing, which can spread with a single exposure. In 6G-enabled ecosystems, where communication is pervasive and instantaneous, these contagion dynamics are likely to become even more pronounced, challenging traditional information control mechanisms.

In response to these challenges, research in computational social science and AI has begun to develop sophisticated detection mechanisms for misinformation. These approaches leverage deep learning, graph-based analysis, and user-behavioral signals to identify deceptive content at scale [17]. However, technical tools alone are insufficient. Studies in cognitive psychology highlight that misinformation persists even after correction, due to biases such as the continued influence effect and resistance to debiasing [18]. Therefore, addressing the spread of fake news requires a combination of algorithmic interventions, psychological insights, and trust-aware communication design.

Moreover, in decentralized 6G environments, the verification of content authenticity must occur without relying on centralized authorities. In this regard, blockchain-based infrastructures offer a promising solution by enabling distributed provenance tracking and tamper-evident records [19]. These mechanisms can enhance digital trust by providing cryptographically verifiable information trails, thereby reducing vulnerability to manipulation.

Digital trust is foundational to the functioning of modern communication networks. It encompasses user confidence in the authenticity, reliability, and integrity of online platforms and the information they disseminate. However, the pervasive spread of misinformation significantly undermines this trust, leading to skepticism, polarization, and reduced societal cohesion [8]. Several mechanisms drive the erosion of digital trust in networked environments:

- **Algorithmic Amplification**: Social media algorithms prioritize sensational and polarizing content to maximize engagement, often amplifying fake news and disinformation [8].
- **Deepfakes and Synthetic Media**: Advances in AI have enabled the creation of hyper-realistic synthetic media, further complicating the verification of authentic information [1].
- **Decentralized Communication**: While decentralized platforms enhance privacy, they also limit the capacity for effective moderation, allowing false information to propagate unchecked [7].
- **Fragmentation in Digital Anthropology**: From a Digital Anthropology perspective, the fragmentation of information across virtual communities contributes to isolated echo chambers, reinforcing biases and diminishing collective awareness [2].
- **Pandemic and Crisis Scenarios**: During crises like the COVID-19 pandemic, the rapid dissemination of false information exacerbated public confusion, undermined trust in health institutions, and led to non-compliance with safety measures [16].

The erosion of digital trust is particularly critical in hyper-connected 6G environments, where communication is near-instantaneous and decentralized. Rebuilding trust requires robust verification mechanisms and resilient information infrastructures capable of identifying and mitigating misinformation at its source.

To address the risks posed by misinformation, 6G networks must incorporate advanced technological and social countermeasures:

- **Blockchain for Provenance Verification**: Distributed Ledger Technologies (DLT) such as blockchain can provide immutable records of information provenance, enhancing content verification [8].
- **Edge-Based Content Verification**: In decentralized 6G environments, edge nodes can perform real-time fact-checking and verification of data before dissemination [5].
- **Digital Twin Monitoring**: Digital Twins of social platforms can simulate the spread of misinformation, enabling predictive analytics and real-time intervention [2].

- **Reputation-based Trust Models**: Integrating reputation-based scoring in communication protocols to filter out unreliable sources during information propagation [1].
- **Quantum Cryptography for Data Integrity**: Quantum Key Distribution (QKD) ensures secure communication channels, safeguarding data authenticity and integrity [8].

These measures are essential to building trust-enhanced communication infrastructures in the 6G era.

4.2.2 Emergent Behaviors in Digital Ecosystems

Digital ecosystems are characterized by the dynamic interplay of human, technological, and social entities that continuously interact, adapt, and evolve within complex networked environments. These ecosystems, particularly in the context of 6G and beyond, embody complex adaptive systems where emergent behaviors arise from the interactions among autonomous agents without any central coordination. The concept of emergence is central to understanding how local actions and decisions collectively shape the global dynamics of these ecosystems. From the perspective of digital anthropology, emergent behaviors in digital ecosystems offer a profound lens to analyze how collective social dynamics influence communication networks, shaping their design and operational paradigms. Emergence refers to the spontaneous formation of complex patterns or behaviors from simpler interactions among system components. In digital ecosystems, emergent behaviors often manifest as collective phenomena that cannot be easily inferred from the properties of individual agents, as deeply explained in Chap. 2. These behaviors arise from a combination of network effects, interaction rules, and adaptive responses to environmental changes. The study of emergence in digital ecosystems is intrinsically linked to complex systems theory, which emphasizes how local interactions give rise to global order. This principle is particularly evident in social networks, where human behaviors such as information diffusion, opinion formation, and collective decision-making evolve in unexpected ways [6]. In the context of 6G networks, emergent behaviors significantly impact communication paradigms, from the propagation of information through mobile ad hoc networks (MANETs) to the self-organization of edge nodes in decentralized architectures [3].

Traditional network models often fail to capture the complexity of real-world social interactions, which frequently involve group dynamics rather than simple pairwise connections. Higher-order networks, represented as hypergraphs, multiplex networks, or simplicial complexes, provide a more nuanced framework for studying emergent phenomena in social systems [22]. In these topologies, interactions among groups (hyperedges) can significantly alter the diffusion of behaviors and information. Recent studies have shown that the overlap between hyperedges can drive explosive transitions in systems with higher-order interactions, leading to abrupt changes in collective states [22]. This phenomenon is particularly relevant for understanding how coordinated behavior or rapid consensus emerges in digital ecosystems, such as during viral content dissemination or coordinated social movements. In social dilemmas, cooperation among agents is often influenced by the network structure. Higher-order interactions, where multiple individuals interact simultaneously, have been shown to promote cooperative behavior under specific conditions. For instance, the existence of overlapping groups and interlinked communities can enhance mutual support and collective resilience, especially in decentralized digital ecosystems [23]. The evolutionary dynamics within higher-order social networks reveal that cooperative strategies can stabilize through mechanisms of group selection and network clustering. Such insights are crucial for designing resilient digital communication systems that rely on distributed consensus and cooperative data sharing, as seen in federated learning scenarios within 6G edge networks [24]. Social dynamics in digital ecosystems are inherently complex due to the interplay of human behavior, technological interventions, and the adaptive nature of networked systems. Emergent phenomena such as collective attention, opinion polarization, and the spontaneous spread of misinformation are fundamental aspects that influence the integrity and functionality of communication networks [1, 8].

Collective attention during shocking events or crises can lead to sudden shifts in network dynamics, causing bursts of information dissemination and the rapid formation of opinion clusters. For instance, during the COVID-19 pandemic, the dissemination of health-related information and misinformation demonstrated how collective attention can influence public behavior and network robustness [2, 16]. The role of multiplex networks in capturing these multilayered social processes is crucial, as it highlights the interaction between different communication channels and their combined effect on the propagation dynamics [5].

Social contagion theory posits that behaviors, beliefs, or attitudes can spread through social networks analogously to infectious diseases [7]. In digital ecosystems, misinformation and fake news can proliferate rapidly through social contagion mechanisms, especially when higher-order interactions are involved. Studies have shown that misinformation can significantly compromise the stability of social systems, leading to polarized communities and disrupted information flows [1, 8]. Complex contagion occurs when the adoption of behaviors requires reinforcement from multiple neighbors, contrasting with simple contagion, where a single exposure suffices. In digital ecosystems, complex contagion often governs the spread of socially risky behaviors or controversial opinions. Such dynamics are particularly prominent in online communities where consensus emerges only after sufficient peer validation [25].

Understanding emergent behaviors in digital ecosystems is essential for designing resilient and adaptive 6G communication networks. Integrating complexity-inspired models into network architecture can help predict and mitigate the impact of collective phenomena, such as misinformation cascades or coordinated cyberattacks. The inherent unpredictability of emergent behaviors necessitates communication frameworks that are inherently robust and self-organizing. Leveraging higher-order topology analysis allows network designers to anticipate critical transition points and preemptively implement control measures. For instance, integrating hypergraph-based algorithms can enhance the identification of pivotal nodes in social contagion processes, allowing for targeted interventions to curb misinformation spread [22, 23]. Incorporating adaptive edge intelligence into 6G architectures enables real-time monitoring of social dynamics, particularly in urban environments where human mobility and digital interactions are densely intertwined. By utilizing graph neural networks (GNNs) and federated learning models, edge nodes can autonomously detect emerging patterns and adapt resource allocation strategies accordingly [21].

Emergent behaviors are inherently non-linear and stochastic, posing challenges for classical analytical methods. The integration of quantum computing in 6G networks can enhance the detection and analysis of emergent patterns by leveraging quantum algorithms for real-time pattern recognition and anomaly detection [26]. This approach is especially beneficial for detecting abrupt transitions in higher-order network states, which classical methods may overlook. Emergent behaviors in digital ecosystems challenge traditional communication paradigms, especially within the context of 6G and beyond. By adopting complexity science perspectives, network designers can better understand how local interactions lead to global patterns, enabling more resilient and adaptive infrastructures. Higher-order networks, in particular, offer promising frameworks for modeling complex contagion and cooperative behavior, essential for maintaining stability in digital ecosystems. As digital anthropology continues to evolve, incorporating insights from social dynamics and emergent behavior into communication design will be crucial for developing socially robust and technologically advanced networks.

4.2.3 Digital Anthropology as a Framework for 6G Design

The concept of Digital Anthropology, as introduced and defined in this chapter, represents a paradigm shift in understanding the interplay between human behaviors, social interactions, and telecommunication infrastructures. Traditionally, anthropology has explored human societies through cultural, social, and behavioral lenses. However, with the rise of digital technologies, there is an urgent need to extend this exploration into the digital realm, where social behaviors are increasingly mediated and influenced by technology. In the context of 6G and beyond, Digital Anthropology serves as a theoretical and practical framework to analyze how digital interactions evolve, spread, and influence communication networks, shaping not only the way information flows but also how network infrastructures are designed and optimized.

The Foundations of Digital Anthropology in Communication Networks Digital Anthropology in the context of telecommunications is rooted in the understanding that digital interactions are not isolated occurrences but rather collective phenomena that mirror the principles of complex systems. Just as social behaviors in physical communities emerge from individual interactions, digital behaviors propagate through networks, creating collective patterns of communication, misinformation, collaboration, and trust. These patterns are not merely reflective of human interaction—they actively shape the architecture and dynamics of communication networks.

In Digital Anthropology, three core principles derived from complex systems theory are fundamental:

- Emergence: Digital interactions among individuals give rise to collective behaviors such as viral information spreading, digital activism, and collective response during crises. These emergent behaviors cannot be inferred from the properties of individual nodes (users) but arise from the interactions within the digital ecosystem.
- Self-Organization: Digital networks often self-regulate without centralized control. Hashtags trend, communities form, and information cascades autonomously through decentralized mechanisms, reflecting the principles of self-organizing systems observed in natural and social ecosystems.
- Adaptation and Co-evolution: Digital communication platforms adapt to user behaviors while simultaneously influencing them. For example, algorithms adjust information feeds based on interaction patterns, which in turn shape user engagement and communication habits.

These principles not only inform the analysis of digital interactions but also serve as a blueprint for the design of 6G networks that must anticipate and adapt to emergent communication patterns. Unlike 5G, which primarily focuses on connectivity and speed, 6G aims to enhance contextual awareness, predictive maintenance, and self-healing capabilities, all of which require an understanding of how digital communities evolve and respond to network changes. Digital Anthropology in the Design of Next-Generation Communication Networks The vision for 6G networks extends far beyond enhanced bandwidth and latency; it encompasses a seamless integration of the physical, digital, and social layers into a cohesive communication fabric. This vision demands that network architectures not only support ultra-reliable low-latency communication (URLLC) and massive machine-type communication (mMTC) but also reflect the socio-digital interactions that characterize modern communication landscapes.

Digital Anthropology emphasizes the importance of human-centric design in telecommunications. In practice, this means that network optimization must consider not just technical metrics like throughput and latency but also social behaviors, digital footprints, and collective dynamics. For instance, during periods of collective attention (e.g., global events, emergencies), communication networks should dynamically reconfigure to support surges in connectivity and data sharing. The integration of Digital Twins with Digital Anthropology principles can enhance real-time context-awareness in communication networks. Digital Twins serve as virtual replicas of physical and social environments, providing a platform for simulating and predicting network behaviors in response to human interactions. This capability is particularly critical for applications such as smart cities, emergency response systems, and urban mobility networks, where real-time adjustments are necessary for efficiency and safety. Understanding the spread of information, both truthful and deceptive, is crucial for maintaining network resilience. As discussed in previous sections, misinformation and fake news propagate through digital networks in ways that can destabilize information ecosystems. By leveraging Digital Anthropology, 6G networks can be designed with built-in resilience mechanisms that detect and mitigate the rapid spread of harmful information, ensuring that critical communication channels remain reliable and secure.

Artificial Intelligence (AI), when integrated with Digital Anthropology, can predict emergent behaviors in real-time, allowing for proactive network management. For example, predictive analytics powered by AI can forecast spikes in network usage during mass social events or crises, enabling preemptive scaling of resources to maintain service quality. This anticipatory capacity is a direct application of complex systems theory, where understanding local interactions enables the prediction of global dynamics. The application of Digital Anthropology as a framework for 6G design is not merely theoretical—it aligns with global trends toward more interconnected, intelligent, and adaptive communication networks.

In this vision, 6G represents more than a mere technological advancement—it marks a shift toward networks that inherently integrate social awareness, cognitive intelligence, contextual adaptability, and structural resilience. These networks will be capable of understanding collective human behaviors to optimize data dissemination, leveraging artificial intelligence to anticipate and manage social dynamics, dynamically adjusting to real-time interactions, and embedding self-healing mechanisms along with strategies to counter misinformation. The implications for network design are profound: future communication infrastructures will need to function not only as technological backbones but also as social platforms, reflecting and responding to the patterns, needs, and collective behaviors of the societies they are intended to serve.

4.3 Society 5.0 and the Future of Communication

4.3.1 Toward a Human-Centric Communication Infrastructure

The emergence of Society 5.0, a concept introduced by the Japanese government, marks a paradigmatic transition in the evolution of human societies—from an information society (Society 4.0) toward a human-centered, super-smart society. Society 5.0 envisions a world in which physical and digital spaces are seamlessly integrated to enhance human well-being and sustainability through the application of technologies such as artificial intelligence, the Internet of Things, robotics, and high-performance communication networks. Within this vision, communication infrastructure is not simply a supporting element but a constitutive layer of societal evolution [27].

As highlighted by Beniiche et al. [27], the infrastructure required for Society 5.0 is inherently socio-technical and complex. It must accommodate dynamic feedback loops between human behavior, machine learning models, and evolving societal needs. Therefore, designing 6G systems that align with this trajectory necessitates a profound rethinking of network architectures—one that embraces the tenets of complexity science, ethical AI, and context-sensitive communication.

4.3.2 Complex Systems as the Backbone of Society 5.0

Complex systems theory provides an essential framework to model the rich interdependencies and emergent properties of Society 5.0. In particular, it offers tools for understanding how individual components—humans, sensors, devices, algorithms—interact across spatial and temporal scales to generate collective behaviors that shape public policy, healthcare, mobility, and information diffusion [28].

These interactions are far from linear. In cities, for instance, the deployment of Digital Twins—a key enabler of Society 5.0—relies on multilayered data inputs from environmental sensors, user devices, social media, and transportation systems. The synchronization and feedback among these layers are managed through distributed, self-adaptive protocols inspired by complex networks, enabling real-time policy testing, disaster response planning, and scenario simulations [28].

In this regard, complexity becomes more than a descriptive attribute—it becomes a design principle. Multilayer network models, for example, have been used to assess vulnerability in digital health ecosystems [29] and identify centrality patterns in urban infrastructures that correlate with social accessibility and spatial equity [30]. These analytical approaches provide actionable insights for engineers and policymakers alike, enabling the integration of fairness, resilience, and adaptability into next-generation communication platforms.

4.3.3 Digital Health, Urban Resilience, and Social Security in 6G

The implementation of 6G communication technologies in Society 5.0 extends beyond connectivity. It entails rethinking healthcare delivery, urban resilience, and cybersecurity through the lens of socio-technical entanglement.

In the field of digital health, network analysis methods have been used to characterize how innovation ecosystems emerge, evolve, and distribute resources among stakeholders such as hospitals, insurers, regulators, and tech companies [31]. The adoption of Federated Learning (FL) and edge intelligence in urban health applications, as proposed by Zhang et al. [32], underscores the need for architectures that support privacy-preserving analytics and decentralized optimization—core tenets of both complexity science and digital anthropology.

From a security perspective, the design of 6G infrastructures must account for vulnerabilities arising from human behavior and social engineering. Wang et al. [33] and He et al. [34] argue that adversaries often exploit not just technical flaws but also cognitive biases and trust relationships among users. This necessitates a comprehensive strategy that integrates behavioral modeling, anomaly detection, and multi-agent learning into the very fabric of communication networks.

Moreover, disaster-resilient communication systems must support situational awareness and rapid information dissemination. Zhang et al. [35] developed a multi-layer model to simulate information dissemination in post-disaster scenarios, emphasizing the role of cross-layer interference management and delay-tolerant networking. Their work confirms the importance of designing for complexity, where heterogeneity, redundancy, and adaptability enable system survival under extreme stress.

4.3.4 Urban Intelligence and Communication as a Civic Resource

Society 5.0 also implies a shift in how we perceive communication infrastructures—not merely as technological assets but as civic resources. In this context, telecommunication networks become part of urban metabolism, facilitating collective intelligence, public participation, and decentralized governance.

For instance, the combination of crowdsensing and federated analytics enables participatory urban planning. Citizens, through smartphones and wearables, become active contributors to data-driven decisions about mobility, pollution control, and public safety [32]. This democratization of sensing and intelligence illustrates how socio-technical ecosystems can balance individual agency and collective coordination, especially when communication networks are designed as public goods.

4.3.5 Complexity, Society 5.0, and the Road to 6G

In summary, Society 5.0 is not merely a technological ambition but a call for socio-technical harmony. The road to 6G must align with this ethos by incorporating principles of fairness, adaptability, and human-centered design. Complexity science offers the methodological substrate for this alignment. It equips us with a vocabulary and toolkit to capture the nonlinearity, feedback, heterogeneity, and emergent properties of interconnected systems that define the societies of tomorrow.

To enable Society 5.0 through 6G, future research should focus on:

- Integrating multiplex and multilayer network analysis into the design of digital health, mobility, and education infrastructures.
- Embedding distributed intelligence and ethical AI in urban communication protocols.
- Ensuring robustness and inclusivity through complex systems-inspired design, leveraging redundancy, decentralization, and dynamic adaptation.
- Enhancing public trust and transparency by making digital infrastructures intelligible, participatory, and responsive to human needs.

Ultimately, as we move toward a more integrated, intelligent, and humane society, the intersection of complexity science, digital anthropology, and 6G technologies will be instrumental in building infrastructures that not only connect devices—but empower people.

4.4 Conclusion

Digital Anthropology reframes communication networks as arenas of human interaction, cultural complexity, and adaptive behavior. This chapter has argued that embedding this perspective in 6G design is not optional—it is essential.

By leveraging the principles of complex systems—emergence, self-organization, and adaptation—Digital Anthropology enables us to understand, model, and even anticipate collective social dynamics. It becomes a powerful lens for confronting contemporary challenges such as misinformation, digital trust erosion, and the unpredictable spread of behaviors in highly connected ecosystems.

Key insights presented include:

- The integration of Digital Twins and contextual analytics to synchronize social and environmental layers in real time.
- The design of adaptive, decentralized architectures that respond to emergent behaviors rather than impose fixed control.
- The strategic role of complexity-informed security and trust frameworks in safeguarding critical infrastructures.

As 6G networks evolve, we envision a shift from purely technical optimization toward infrastructures that are:

- **Socially aware**: Capable of interpreting and reacting to collective behavior.
- **Cognitively enhanced**: Able to anticipate usage and societal needs.
- **Contextually adaptive**: Reconfigurable in response to dynamic social patterns.
- **Resilient by design**: With embedded mechanisms for misinformation mitigation and trust preservation.

Future directions should explore:

- Quantum communication and Digital Anthropology for privacy and synchronization.
- Federated AI models that respect social plurality while enabling prediction.
- Ethical protocols for behavioral monitoring in socio-technical networks.

Digital Anthropology is not just a theoretical framework—it is a design paradigm for building human-centered, ethically robust, and socially responsive 6G communication ecosystems.

To encapsulate the conceptual contributions of this chapter, Fig. 4.1 provides a visual framework that synthesizes the layered relationship between human behavior, digital anthropology, and the design of resilient communication systems. At its core, the framework identifies human social interaction—encompassing users, autonomous agents, and societies—as the fundamental layer upon which Digital Anthropology operates. This interdisciplinary domain draws upon complexity theory to analyze phenomena such as social contagion, emergence, digital trust, and collective dynamics. These insights, in turn, inform a set of network functions designed to enhance adaptability, resilience, and ethical intelligence in future 6G architectures.

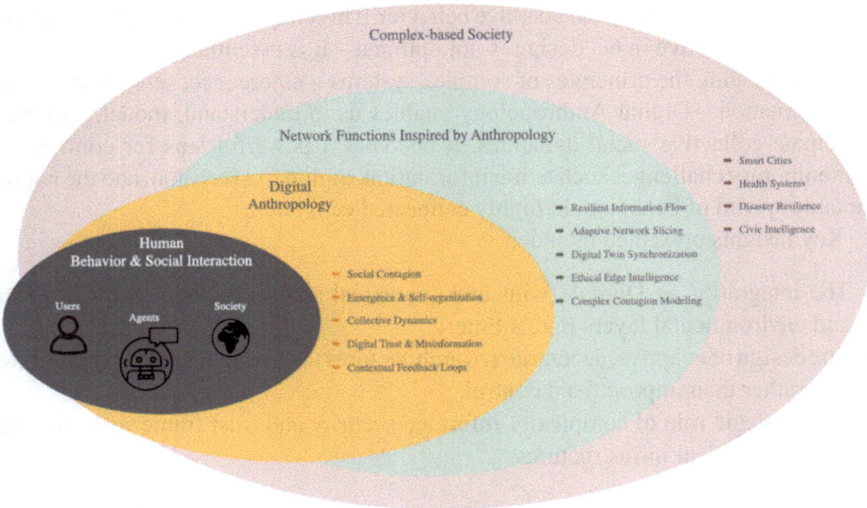

Fig. 4.1 Venn diagram illustrating the relationship between "Human Behavior & Social Interaction," "Digital Anthropology," and "Complex-based Society." The innermost circle, labeled "Human Behavior & Social Interaction," includes elements like users, agents, and society. The middle section, "Digital Anthropology," covers topics such as social contagion, emergence, self-organization, collective dynamics, digital trust, misinformation, and contextual feedback loops. The outermost section, "Complex-based Society," encompasses network functions inspired by anthropology, including resilient information flow, adaptive network slicing, digital twin synchronization, ethical edge intelligence, complex contagion modeling, smart cities, health systems, disaster resilience, and civic intelligence

Surrounding these technological implications is the broader vision of a complex-based society, where communication infrastructures are not merely functional but socially attuned, ethically grounded, and contextually responsive. By explicitly linking behavioral patterns to network design strategies, the framework underscores the pivotal role of Digital Anthropology in bridging the technical and human dimensions of next-generation communication systems.

References

1. Aïmeur, E., Amri, S., Brassard, G.: Fake news, disinformation and misinformation in social media: a review. Soc. Netw. Anal. Min. **13**(1), 30 (2023). https://doi.org/10.1007/s13278-023-00999-5
2. Scatá, M., Di Stefano, A., La Corte, A., Liò, P.: A multiplex social contagion dynamics model to shape and discriminate d2d content dissemination. IEEE Trans. Cogn. Commun. Netw. **6**(4), 1065–1075 (2020). https://doi.org/10.1109/TCCN.2020.2991424
3. Scatà, M., Attanasio, B., La Corte, A.: Cognitive profiling of nodes in 6G through multiplex social network and evolutionary collective dynamics. Future Internet **13**(5), 135 (2021). https://doi.org/10.3390/fi13050135
4. De Domenico, M., Granell, C., Porter, M.A., Arenas, A.: The physics of spreading processes in multilayer networks. Nat. Phys. **12**(10), 901–906 (2016). https://doi.org/10.1038/nphys3865
5. Bródka, P., Musial, K., Jankowski, J.: Interacting spreading processes in multilayer networks: a systematic review. IEEE Access **8**, 10316–10341 (2020). https://doi.org/10.1109/ACCESS.2020.2965084
6. Scatà, M., Di Stefano, A., La Corte, A., Liò, P.: Quantifying the propagation of distress and mental disorders in social networks. Sci. Rep. **8**(1), 5005 (2018)
7. Christakis, N.A., Fowler, J.H.: Social contagion theory: examining dynamic social networks and human behavior. Stat. Med. **32**(4), 556–577 (2013). https://doi.org/10.1002/sim.5408
8. Wang, J., et al.: Fake news in virtual community, virtual society, and metaverse: a survey. IEEE Trans. Comput. Soc. Syst. (2023). https://doi.org/10.1109/TCSS.2023.3274568
9. He, X., Lin, Y.-R.: Measuring and monitoring collective attention during shocking events. EPJ Data Sci. **6**(1), 30 (2017). https://doi.org/10.1140/epjds/s13688-017-0120-2
10. Masaracchia, A., Sharma, V., Canberk, B., Dobre, O.A., Duong, T.Q.: Digital twin for 6G: taxonomy, research challenges, and the road ahead. IEEE Open J. Commun. Soc. **3**, 2137–2150 (2022)
11. Scatà, M., Attanasio, B., Corte, A.: Cognitive multiplex modeling for trust-aware D2D communications. IEEE Trans. Cogn. Commun. Netw. **6**(4), 1012–1025 (2020)
12. Guess, A., Nagler, J., Tucker, J.: Less than you think: prevalence and predictors of fake news dissemination on Facebook. Sci. Adv. **5**(1), eaau4586 (2019)
13. Vicario, M., Quattrociocchi, W., Scala, A., Zollo, F.: Polarization and fake news: early warning of potential misinformation targets. ACM Trans. Web (TWEB) **13**(2), 10 (2019)
14. Törnberg, P.: Echo chambers and viral misinformation: modeling fake news as complex contagion. PLOS ONE **13**(9), e02039
15. Scatà, M., La Corte, A., Marotta, A., Graziosi, F., Cassioli, D.: A complex network and evolutionary game theory framework for 6G function placement. IEEE Open J. Commun. Soc. (2024). https://doi.org/10.1109/OJCOMS.2024
16. Prandi, L., Primiero, G.: Effects of misinformation diffusion during a pandemic. Appl. Netw. Sci. **5**, 1–20 (2020). https://doi.org/10.1007/s41109-020-00286-z
17. Zhou, X., Zafarani, R.: A survey of fake news: fundamental theories, detection methods, and opportunities. ACM Comput. Surv. **53**(5), 1–40 (2020)

18. Lewandowsky, S., Ecker, U.K.H., Seifert, C.M., Schwarz, N., Cook, J.: Misinformation and its correction: continued influence and successful debiasing. Psychol. Sci. Public Interes. **13**(3), 106–131 (2012)
19. Zheng, Z., Xie, S., Dai, H., Chen, X., Wang, H.: Blockchain challenges and opportunities: a survey. Int. J. Web Grid Serv. **14**(4), 352–375 (2018)
20. Letaief, K.B., Shi, Y., Lu, J., Lu, J.: Edge artificial intelligence for 6G: vision, enabling technologies, and applications. IEEE J. Sel. Area Commun. **40**(1), 5–36 (2021)
21. Yu, P., Zhu, L., Song, C., Li, H., Zhu, L., Xie, J., Chen, T., Xu, T., Xu, X., Gao, J.: Digital twin driven service self-healing with graph neural networks in 6G edge networks. IEEE J. Sel. Areas Commun. **41**(11), 3607–3623 (2023). https://doi.org/10.1109/JSAC.2023.3316249
22. Malizia, F., Ginestra, B., Squartini, T., Radicchi, F.: Hyperedge overlap drives explosive transitions in systems with higher-order interactions. Nat. Commun. **16**, 555 (2025). https://doi.org/10.1038/s41467-025-01234-5
23. Sadekar, O., Johnson, R., Schaub, M.T., Perra, N.: Drivers of cooperation in social dilemmas on higher-order networks (2025). arXiv:2502.09446. https://arxiv.org/abs/2502.09446
24. Alvarez-Rodriguez, U., Battiston, F., de Arruda, G.F., Moreno, Y.: Evolutionary dynamics of higher-order interactions in social networks. Nat. Hum. Behav. **5**, 586–595 (2021). https://doi.org/10.1038/s41562-021-01128-8
25. Guilbeault, D., Becker, J., Centola, D.: Complex contagions: a decade in review. In: Complex Spreading Phenomena in Social Systems, pp. 3–25. Springer, Cham (2018)
26. Wehner, S., Elkouss, D., Hanson, R.: Quantum internet: a vision for the road ahead. Science **362**(6412), eaam9288 (2018). https://doi.org/10.1126/science.aam9288
27. Beniiche, A., Rostami, S., Maier, M.: Society 5.0: internet as if people mattered. IEEE Wirel. Commun. **29**(6), 160–168 (2022)
28. Caldarelli, G., Nicola, R., Petri, G., Tagarelli, A., Vespignani, A.: The role of complexity for digital twins of cities. Nat. Comput. Sci. **3**(5), 374–381 (2023)
29. Moro Visconti, R.: Multilayer networks and public digital health scalability. Int. J. Comput. Sci. Mob. Appl. **12**(2), 56–83 (2024)
30. Porta, S., Crucitti, P., Latora, V.: Multiple centrality assessment in Parma: a network analysis of paths and open spaces. Urban Des. Int. **13**(1), 41–50 (2008)
31. Park, Y., Park, S., Lee, M.: Digital health care industry ecosystem: network analysis. J. Med. Internet Res. **24**(8), e3762 (2022)
32. Zhang, W., Xu, L., Wang, W., Yang, F.: Federated Learning Meets Urban Opportunistic Crowdsensing in 6G Networks: Opportunities. Challenges and Optimization Potentials. IEEE Network (2024)
33. Wang, Z., Sun, L., Zhu, H.: Defining social engineering in cybersecurity. IEEE Access **8**, 85094–85115 (2020)
34. He, W., Dong, X., Li, J., et al.: Secure control of multiagent systems against malicious attacks: a brief survey. IEEE Trans. Industr. Inf. **18**(6), 3595–3608 (2021)
35. Zhang, Y., Zhang, W., Xu, X., Qiu, T.: A multi-layer information dissemination model and interference optimization strategy for communication networks in disaster areas. IEEE Trans. Veh. Technol. **73**(1), 1239–1252 (2023)

Chapter 5
Comparative Insights on R and Python for Complex Systems and 6G Network Analysis

Abstract This chapter provides a comparative and integrative overview of R and Python as essential tools for modeling, simulating, and analyzing complex systems and 6G communication networks. It examines the distinct strengths and limitations of each language across various dimensions, including statistical modeling, machine learning, network analysis, and deployment scalability. Emphasis is placed on hybrid workflows that integrate both environments, facilitated by packages, to support advanced research in digital twins, network slicing, federated learning, and edge intelligence. Through pseudocode examples and interoperability strategies, the chapter offers actionable insights for leveraging the complementary capabilities of R and Python in interdisciplinary and future infrastructures.

5.1 Introduction

In the field of complex systems and next-generation communication networks, which is attracting growing interdisciplinary interest, the role of computational tools for data analysis, simulation, and modeling is increasingly central. Two dominant languages in this landscape, **R** and **Python**, have emerged as indispensable assets for researchers and engineers alike. Their robust ecosystems, active user communities, and wide-ranging libraries make them powerful tools for exploring the dynamics of multilayered systems, performing statistical inference, building agent-based models, and designing intelligent telecommunications infrastructures such as digital twins and network slicing mechanisms.

The analysis of communication networks through the lens of complexity science often demands an integrated approach to modeling structure and dynamics. Tasks such as community detection, link prediction, epidemic modeling, game-theoretic simulation, and topological analysis require tools that are both computationally efficient and statistically expressive. Python and R meet these needs with complementary strengths. Python, with its general-purpose nature and seamless integration with machine learning and artificial intelligence libraries, is particularly well-suited to scalable simulation and deep learning applications. R, on the other hand, excels

M. Scatà, *The Power of Complex Systems*, Studies in Systems, Decision and Control 630,
https://doi.org/10.1007/978-3-032-04243-9_5

in statistical analysis, visualization, and the manipulation of complex tabular data-functions essential for interpreting simulation outputs and constructing reproducible, data-driven models.

For example, in the context of 6G, the use of Python and R has extended to domains such as:

- **Digital Twin Modeling**: Both languages support simulation frameworks for real-time digital representations of networked systems, with Python gaining traction in edge computing and AI, and R offering powerful data integration and visualization.
- **Network Slicing**: Python's integration with SDN controllers and AI-based orchestration tools makes it suitable for the dynamic allocation of network resources. R, in turn, provides statistical reliability estimation and stochastic process modeling for evaluating slice performance.
- **Agent-Based and Higher-Order Network Modeling**: Packages like `HyperNetX` (Python) and `SEMgraph` (R) support the design of models that go beyond pairwise interactions, which are increasingly relevant in modeling socio-technical communication systems.

The **reticulate** package in R has further closed the gap between these ecosystems by enabling seamless integration of Python scripts within R projects, thus allowing researchers to utilize the statistical strengths of R in tandem with Python's AI and network science capabilities [1]. This hybrid use is particularly effective in fields where both statistical rigor and algorithmic scalability are necessary, such as federated learning in urban sensing networks [2] or mental health propagation analysis in social networks [3, 4].

Moreover, the choice between R and Python is often based on the lifecycle stage of a project:

- **Exploratory and Statistical Analysis**: R's packages (`ggplot2`, `lme4`, `SEMgraph`) offer advanced statistical workflows and high-quality visualizations.
- **Algorithm Prototyping and Simulation**: Python, especially with libraries like `NetworkX`, `Scikit-learn`, `PyTorch`, and `HyperNetX`, excels in building complex and scalable simulation environments.
- **Deployment and Automation**: Python's broad integration capabilities make it suitable for deployment in production-level network automation and edge intelligence platforms.

In this chapter, we discuss a comparative and integrative exploration of R and Python in the context of complex systems and 6G networks. We address the theoretical underpinnings, practical applications, interoperability strategies, and emerging trends. Through code examples, pseudo-algorithms, and performance insights, the aim is to give a starting guide in selecting and combining these tools for optimal research and development outcomes in the hyper-connected, self-organizing infrastructures of tomorrow.

5.2 Background and Evolution of Analytical Tools in Network Science and Complex Systems

Over the last two decades, the fields of complex systems and network science have undergone significant methodological and technological transformation, driven largely by the evolution of statistical computing tools and open-source programming languages. In the context of communications engineering, and more recently in the design of 6G infrastructures, the ability to model multilayered networks, simulate emergent behaviors, and conduct massive-scale data analytics is indispensable. Two dominant tools have emerged in this landscape: **R** and **Python**.

Initially, R was adopted within academic and statistical communities due to its robust suite of libraries for mathematical modeling, hypothesis testing, and data visualization [5]. Its development environment, enhanced by RStudio, provided a powerful IDE with support for markdown documentation, reproducible research, and advanced graphics through packages like *ggplot2*, *igraph*, and *SEMgraph* [6]. These capabilities made R a natural fit for theoretical modeling in social systems, agent-based simulations, and structural equation modeling.

The foundational role of R in education and practical statistics is well-documented. As highlighted by Field et al. [7], R supports a transparent and interactive style of statistical computing, allowing users to both understand and apply complex models–from t-tests to multilevel modeling–through hands-on simulation and visualization. This pedagogical value makes R an ideal entry point for learners and researchers seeking to bridge theory and application, especially within the increasingly interdisciplinary domain of network science and complex systems.

At the same time, Python gained momentum for its flexibility, readability, and integration with external APIs and data pipelines. It expanded beyond general-purpose programming into specialized domains like machine learning (ML), complex network modeling, and telecommunications simulation, thanks to libraries such as *NetworkX*, *Scikit-learn*, *TensorFlow*, and *HyperNetX* [8]. Python's applicability for both prototyping and production environments made it especially appealing in industrial and multidisciplinary research contexts. With the emergence of digital twins, edge computing, and AI-native network slicing in 6G, both R and Python have become pivotal tools in modeling communication networks and their dynamics. Python's modularity and ML capabilities allow it to simulate end-to-end 6G network behavior, train models for resource optimization, and integrate with quantum or edge-based architectures [1]. Meanwhile, R has maintained its role in statistical validation of simulated topologies, reliability analysis, and higher-order network modeling. In the study of complex systems, particularly those involving multilayer and higher-order topologies, R packages such as *igraph*, *Statnet*, and *SEMgraph* allow for algebraic and probabilistic representations of cascading failures, social contagion, and collective dynamics [9]. At the same time, Python-based libraries like *HyperNetX* and *EoN* (Epidemics on Networks) are used to simulate dynamic spreading processes across complex networks, often with superior performance and scalability [8]. As these languages matured, researchers have increasingly pursued the idea of combining their strengths. Using packages like *reticulate*, R users can embed Python code within R scripts and notebooks, allowing for hybrid workflows that utilize the best tools for each task [1]. This has proven especially effective in edge-AI scenarios for 6G, where Python handles ML training while R performs model interpretation and statistical diagnostics. Despite their strengths, both languages have notable limitations. According to a recent study comparing Python and R bug patterns, R presents more issues in intermediate data flow management, whereas Python's flexibility can lead to a higher incidence of bugs during preprocessing [10]. Furthermore, Python's verbosity and overhead in some statistical operations make it less efficient than R for compact, high-throughput computations like ANOVA and regression diagnostics [11].

As the analytical demands of 6G networks evolve–particularly in light of increasing data volume, time sensitivity, and system interconnectivity–both languages continue to adapt. Their evolution reflects broader shifts in computational science, from closed, monolithic software to open, modular, and interdisciplinary ecosystems. In summary, R and Python have co-evolved with the needs of complex systems research and modern telecommunications. Each provides a distinct philosophy–statistical rigor versus flexible integration–and both are increasingly used in tandem. Understanding their origins and development sets the stage for a comparative evaluation of their capabilities, which follows in the next sections.

5.3 Comparative Strengths and Limitations of R and Python in Complex Systems and 6G Network Analysis

As data-driven methodologies become central to 6G research and complex systems analysis, understanding the comparative advantages and limitations of R and Python is critical for researchers, engineers, and decision-makers. This section provides an extended evaluation of each language across a range of dimensions, including usability, scalability, domain-specific capabilities, visualization features, and integration with modern AI ecosystems.

Strengths of R

R has matured into an indispensable tool for researchers in statistical science and network theory, particularly in academic and exploratory phases of complex systems modeling:

- **Statistical Modeling**: R was designed with statistical analysis at its core. Core libraries like `stats`, `lme4`, and `survey` enable the execution of advanced techniques such as generalized linear models (GLMs), Bayesian inference, mixed-effect models, and post-hoc hypothesis testing. These are particularly valuable in modeling stochastic processes and performance evaluation of 6G systems.

- **Data Visualization**: With powerful packages like `ggplot2`, `plotly`, and `shiny`, R offers unparalleled capabilities for crafting interactive, publication-grade visualizations. These features are essential in interpreting simulation results and designing real-time dashboards for network monitoring.
- **Network Modeling**: R provides mature support for network analysis through libraries such as `igraph`, `statnet`, and `SEMgraph`. These tools support multilayer networks, community detection, contagion simulations, and structural equation models–key features in the analysis of sociotechnical systems and higher-order topologies.
- **Reproducible Research and Academic Ecosystem**: R's documentation standards, Markdown integration, and CRAN repository ensure reproducibility and foster a robust academic community. This makes R especially suitable for long-term research projects requiring clarity, validation, and traceability.

In their comprehensive treatment, Kolaczyk and Csárdi [12] emphasize how R enables rigorous statistical analysis of network data, especially in applications involving uncertainty quantification, model selection, and latent variable modeling. The integration of probabilistic graphical models and Bayesian techniques allows researchers to represent hidden structures, test dependencies, and analyze cascading dynamics within multilayer or temporal networks–capabilities critical to studying resilience, diffusion, and evolution in complex adaptive systems.

Strengths of Python

Python's dominance in software development and machine learning has made it a staple in both academic and industrial applications:

- **Machine Learning and AI**: Python offers extensive ML and AI capabilities through libraries like `scikit-learn`, `TensorFlow`, `PyTorch`, `XGBoost`, and `LightGBM`. These tools are instrumental in implementing predictive analytics, reinforcement learning, and digital twin synchronization in future 6G networks.
- **System Integration and Deployment**: Python's compatibility with REST APIs, microservices (Flask, FastAPI), and containerized environments (Docker, Kubernetes) makes it suitable for deployment scenarios, such as real-time traffic control or edge-intelligence models.
- **Topological and Higher-Order Modeling**: Python supports advanced topological data analysis through libraries such as `HyperNetX`, `Simplicial`, and `PyG`. These are vital for investigating non-pairwise interactions, such as in federated and decentralized learning.
- **Parallelism and Cloud Scalability**: Python's ability to scale across distributed architectures via `Dask`, `Ray`, or native multiprocessing modules offers a clear edge in large-scale network simulations and real-time inference pipelines.

According to Zinoviev [13], Python's approach to network science follows a sequential five-step framework–Recognize, Construct, Visualize, Analyze, Interpret–that enables both researchers and engineers to build meaningful abstractions of real-world systems. This workflow not only facilitates exploratory data analysis but also aligns with the modular logic required for scalable applications in 6G infrastructure and socio-technical simulations.

Pseudocode Comparison: Network Analysis Workflow

Below we present pseudocode comparing a basic network analysis and visualization workflow in R and Python. These examples illustrate how both languages handle network generation, degree analysis, and shortest path computation.

```
// Pseudocode in R using igraph
library(igraph)
g <- erdos.renyi.game(100, p=0.05, directed=FALSE)
plot(g, vertex.size=5, main="ER Network")
mean(degree(g))
shortest.paths(g)
```

```
// Pseudocode in Python using NetworkX
import networkx as nx
import matplotlib.pyplot as plt
G = nx.erdos\_renyi\_graph(100, 0.05)
nx.draw(G, node\_size=50)
avg\_degree = sum(dict(G.degree()).values()) / G.number\_of\_nodes()
sp\_lengths = dict(nx.shortest\_path\_length(G))
```

This example highlights that while Python is syntactically more verbose, it provides extensibility across the software stack, making it particularly valuable in cloud environments, production pipelines, or multi-agent simulations.

Limitations of R

- **Production Readiness**: R lacks seamless integration with APIs, web interfaces, and containerized services, which limits its usage in large-scale production environments.
- **Machine Learning Ecosystem**: Though packages like `caret`, `mlr3`, and `xgboost` exist, the overall depth, breadth, and speed of ML libraries in R lag behind Python's ecosystem.

- **Concurrency and Distributed Computing**: R's support for parallel and distributed execution, while improving (e.g., via `future`, `parallel`, or `BiocParallel`), remains less intuitive and flexible than Python's `multiprocessing`, `Joblib`, or `Dask`.

Limitations of Python

- **Statistical Precision**: While Python offers many statistical modules, they often provide fewer diagnostic options and model assumptions compared to R. For rigorous experimental design or hierarchical models, R is typically preferred [11].
- **Data Wrangling Complexity**: Despite the capabilities of `pandas`, operations involving chained transformations or grouped summaries can be more verbose and less readable compared to R's `dplyr` or `data.table`.
- **Graph Aesthetics and Reporting:** Visualization in Python requires more code and customization to match the default graphical polish provided by `ggplot2` or `shiny` in R.

Interoperability: Reticulate and Beyond

The emergence of `reticulate` and hybrid notebooks has transformed research pipelines. In a typical 6G simulation scenario, Python may be used to train neural networks for traffic prediction, while R handles post-hoc statistical validation, uncertainty estimation, and exploratory visual analytics. This cross-language approach bridges theoretical modeling with scalable, deployable AI components, enabling more efficient workflows and reproducibility in interdisciplinary research.

R and Python are complementary, not competitive, languages in the evolving landscape of complex systems and 6G network analysis. R's expressiveness in statistical modeling and academic rigor makes it ideal for early-stage theory development and inferential analysis. Python's software engineering maturity and machine learning prowess provide unparalleled tools for implementation and deployment. In the future of communication systems–marked by adaptivity, edge learning, and sociotechnical dynamics–a hybrid use of R and Python stands out as the most pragmatic and powerful solution.

To summarize the comparative analysis between R and Python, Table 5.1 presents a consolidated view across the most relevant dimensions for complex systems modeling and 6G network design.

Table 5.1 Comparative strengths of R and python in complex systems and 6G network analysis

Dimension	R	Python
Statistical modeling	Excellent (GLM, mixed models, Bayesian analysis via `lme4`, `brms`)	Moderate (`statsmodels`, `scipy.stats`)
Machine learning & AI	Good (`caret`, `mlr3`, limited deep learning support)	Excellent (`scikit-learn`, `TensorFlow`, `PyTorch`)
Network analysis	Strong (`igraph`, `SEMgraph`, `statnet`)	Strong (`NetworkX`, `HyperNetX`, `PyG`)
Visualization	Outstanding (`ggplot2`, `shiny`, `plotly`)	Good (`matplotlib`, `seaborn`, `plotly`)
Production deployment	Weak (limited API and container support)	Excellent (native API integration, microservices, containers)
Integration flexibility	Moderate (via `reticulate`, `plumber`)	Outstanding (used in cloud-native ML and orchestration)
Scalability	Moderate (`parallel`, `future`)	Excellent (`Ray`, `Dask`, multiprocessing)
Use in 6G systems	Preprocessing, statistical diagnostics, visual analytics	Modeling, ML inference, deployment, federated learning

5.4 Integration Scenarios: Combining R and Python for Advanced Analytics

The integration of R and Python represents a powerful paradigm for advanced analytics in complex systems modeling and 6G communication network design. Each language brings its unique strengths: R excels in statistical modeling and data visualization, while Python offers robust support for machine learning, large-scale simulations, and automation workflows. The increasing complexity and interdisciplinarity of current research problems demand hybrid solutions that maximize the complementary capabilities of both ecosystems.

The Reticulate Framework

At the core of this integration lies the `reticulate` package in R, which provides a seamless interface to Python. With `reticulate`, R users can:

- Import Python modules directly into R sessions.
- Call Python functions and classes from R scripts.
- Use Python for tasks such as machine learning while relying on R for preprocessing and statistical interpretation.

A typical workflow using `reticulate` might involve pre-processing network metrics in R (e.g., using `SEMgraph` or `igraph`), exporting these to Python for predictive modeling using `scikit-learn` or `networkx`, and finally returning results to R for visualization with `ggplot2`.

```
# Pseudo-code using reticulate

library(reticulate)
py_run_string("import networkx as nx")
py_run_string("G = nx.erdos_renyi_graph(1000, 0.02)")
py_run_string("centrality = nx.betweenness_centrality(G)")
centrality_scores <- py$centrality
```

Joint Use Cases in 6G Systems and Complex Networks

Joint R-Python environments are particularly useful in the context of:

- **Digital Twins and Edge AI**: Using R for statistical modeling of user behaviors, while Python drives federated learning algorithms at the edge.
- **Network Slicing**: Employing R to analyze topological patterns and clustering, and Python to apply reinforcement learning to dynamically reallocate slices.
- **Epidemic and Information Diffusion Modeling**: Using R's compartmental modeling (`EpiModel`) alongside Python's agent-based libraries (`Mesa`, `NetLogo`, or `HyperNetX`) for hybrid simulation.

Architectures for Mixed Environments

Modern workflows increasingly rely on containerization and orchestration tools (e.g., Docker, JupyterHub) to manage hybrid R-Python environments. Solutions often adopt microservice designs:

- R handles the statistical modeling and diagnostics layer.
- Python manages scalable simulation or deep learning inference in the backend.

This separation of concerns allows interdisciplinary teams–statisticians, data scientists, and engineers–to collaborate effectively in scalable, reproducible, and performant research pipelines.

The ability to combine Python's ML power and R's statistical expressiveness not only enhances modeling performance but also strengthens reproducibility. Hybrid workflows are particularly aligned with the needs of complex adaptive systems and 6G infrastructures, where networked behaviors must be modeled, simulated, and adapted dynamically. Moreover, as both languages evolve toward better interoperability and shared cloud infrastructures (e.g., RStudio Workbench, Jupyter with R kernel), their combined application is likely to become a staple in next-generation telecom and systems science research.

5.5 Integration Strategies: Combining R and Python for Hybrid Workflows

While both R and Python offer powerful capabilities for complex systems and network science, the growing demands of modern telecommunications, particularly in 6G, Digital Twin Networks (DTNs), and edge intelligence, necessitate hybrid approaches that exploit the best of both environments. This Section provides an in-depth discussion on the integration strategies that allow researchers to leverage R's statistical robustness and Python's advanced machine learning and graph analytics in unified workflows.

Motivation for Integration

The motivation for combining R and Python stems from the complementary nature of their ecosystems:

- **R** excels in statistical inference, visualization, and complex regression modeling through packages like `ggplot2`, `lme4`, and `SEMgraph` [6].
- **Python** dominates in scalable machine learning, agent-based modeling, and hypergraph analytics via `scikit-learn`, `HyperNetX`, and `NetworkX` [8].

Integrating the two languages allows for seamless transitions between statistical preprocessing in R and simulation or deployment pipelines in Python, particularly in fields like 6G and AI-native networking.

The `reticulate` Package

The `reticulate` package in R serves as a primary bridge for executing Python code from within R scripts and environments. It supports:

- Direct calling of Python functions from R.
- Conversion of R objects into Python objects and vice versa.
- Jupyter Notebook interoperability.

Example Pseudocode: Calling Python from R with `reticulate`

```
library(reticulate)
py_run_string("
import numpy as np
def simulate_network(n):
    return np.random.rand(n, n)
")
network_matrix <- py$simulate_network(10)
```

This code block shows how an R script can call a Python function that generates a random network matrix using NumPy, capturing the result in an R object.

Use Cases in Complex Systems and 6G Networks

- **Federated Learning Analysis**: Use R for exploratory data analysis and preprocessing (e.g., PCA), then train models in Python using TensorFlow Federated [1].
- **Digital Twin Data Calibration**: Apply R for structural equation modeling of DT data, and pass the calibrated parameters to Python for reinforcement learning-based control strategies.
- **Epidemic Spread in Multiplex Networks**: Simulate epidemic models in Python using agent-based modeling, while using R for visualization and regression-based evaluation of parameter sensitivity [4, 14, 15].

5.5.1 Challenges and Best Practices

- **Version Management**: Tools like `renv` in R and `virtualenv` in Python should be used to maintain consistent environments.
- **Data Exchange Format**: Use `feather`, `parquet`, or JSON formats to preserve data schema across platforms.
- **Performance Monitoring**: Leverage profiling tools in both languages to identify computational bottlenecks, and selectively offload tasks to the most efficient environment.

Integrating R and Python in a unified pipeline unlocks substantial analytical power, essential for the design, simulation, and real-time adaptation of complex communication infrastructures. This hybrid workflow reflects the very nature of modern socio-technical systems–heterogeneous, distributed, and adaptive.

5.6 Conclusion

The integration of R and Python for analyzing complex systems and 6G networks is not only a reflection of current research practices but also a pointer toward future technological and methodological advancements. Section 5.6 highlights emerging trends, projected developments, and actionable recommendations that researchers and practitioners should consider when adopting these languages in the evolving landscape of next-generation communication systems. As quantum communication and computation emerge in 6G, the analytical tools will need to adapt. Python's growing ecosystem around quantum programming–especially through libraries such as `Qiskit`, `Cirq`, and `PennyLane`–positions it as a front-runner for developing quantum-ready simulations. While R currently lags in native quantum computing support, efforts to bridge this gap through interfaces and quantum-aware statistical models are expected to rise.

Develop hybrid workflows where Python handles quantum simulation and R complements with statistical verification of outcomes, especially for quantum-enhanced machine learning or security protocols in network slicing. Reinforcement Learning (RL) is gaining momentum as a framework for autonomous resource allocation and control in self-organizing 6G architectures. Python leads this space with libraries like `Stable-Baselines3`, `Ray RLlib`, and `OpenAI Gym`. Meanwhile, R has started integrating reinforcement learning frameworks through `ReinforcementLearning` and `rTorch`. Use R for interpretability and pre-modeling tasks, such as environment encoding, while deploying RL agents in Python for high-efficiency training and simulation of adaptive 6G network functions.

The black-box nature of many AI models in communications calls for greater transparency and causal interpretability. R's deep tradition in causal modeling (e.g., `dagitty`, `SEMgraph`) complements Python's emerging XAI toolkits such as `SHAP`, `LIME`, and `Captum`. These tools are vital in validating the fairness, trust, and robustness of decision-making in digital twin networks and AI-native communication infrastructures. Adopt a dual-stage approach–using Python to train deep learning models for 6G analytics and R to analyze post-hoc causal relationships or audit model behavior with statistical transparency.

Emerging frameworks like digital anthropology and socio-technical design challenge computational models to be culturally aware and behaviorally grounded. Python offers scalability and diverse APIs to capture behavioral data, while R is ideal for clustering, semantic analysis, and hypothesis testing in social systems. Combine both languages in frameworks analyzing collective attention, fake news, and socio-technological resilience in 6G environments by embedding simulation agents (Python) with ethnographic behavior models (R).

Federated learning (FL) is critical for edge intelligence and privacy-aware modeling. Python's tools (`TensorFlow Federated`, `Flower`) support scalable FL simulations, whereas R's strengths in statistical diagnostics and visualization can support model inspection and explainability. Use Python to implement and orchestrate FL pipelines, while leveraging R for anomaly detection, differential privacy analysis, and federated model comparison across edge clients.

R and Python will continue to serve as complementary instruments in the modeling and analysis of complex systems, edge intelligence, and network evolution. Researchers should embrace their interoperability through frameworks like `reticulate`, containerized reproducible environments (e.g., Docker, Binder), and cloud-native notebooks. As network infrastructures become more intelligent, resilient, and socio-cognitive, the joint use of R and Python will be central to exploring new paradigms in the co-design of artificial and social network systems.

References

1. García Navarro, A.L., et al.: A comprehensive guide to combining r and python code for data science, machine learning and reinforcement learning (2024). arXiv:2407.14695
2. Zhang, W., Liu, X., Zhu, C., Varjonen, S.K.L., Wang, F., Tarkoma, S.A.O.: Federated learning meets urban opportunistic crowdsensing in 6g networks: opportunities, challenges, and optimization potentials. IEEE Netw. **39**(2), 36–43 (2025). https://doi.org/10.1109/mnet.2024.3520552
3. Scatà, M., Di Stefano, A., La Corte, A., Liò, P.: Quantifying the propagation of distress and mental disorders in social networks. Sci. Rep. **8**(1), 5005 (2018)
4. Scatà, M., Attanasio, B., Aiosa, G.V., La Corte, A.: The dynamical interplay of collective attention, awareness and epidemics spreading in the multiplex social networks during COVID-19. IEEE Access **8**, 189203–189223 (2020). https://doi.org/10.1109/ACCESS.2020.3029261
5. Team, R.S.: RStudio: Integrated Development Environment for R. RStudio, PBC, Boston (2020). http://www.rstudio.com/
6. Palluzzi, F., Grassi, M.: SEMgraph: an R package for causal network analysis of high-throughput data with structural equation models (2021). arXiv:2103.08332
7. Field, A., Field, Z., Miles, J.: Discovering Statistics Using R. Sage (2012)
8. Praggastis, B., et al.: HyperNetX: a python package for modeling complex network data as hypergraphs (2023). arXiv:2310.11626
9. Antonov, M., et al.: igraph enables fast and robust network analysis across programming languages (2023). arXiv:2311.10260
10. Ahmed, S., Ren, J., Jiang, Z., Xu, B.: Characterizing bugs in python and R data analytics programs (2023). arXiv:2306.08632
11. Hill, C., Kresta, J., Wu, J., Li, Q.: Comparing programming languages for data analytics: accuracy of estimation in Python and R. Wiley Interdiscip. Rev.: Data Min. Knowl. Discov. **14**(3), e1531 (2024)
12. Kolaczyk, E.D., Csárdi, G.: Statistical Analysis of Network Data with R, vol. 65. Springer, Berlin (2014)
13. Zinoviev, D.: Complex Network Analysis in Python: Recognize-Construct-Visualize-Analyze-Interpret. Springer, Berlin (2018)
14. Scatà, M., Di Stefano, A., La Corte, A., Liò, P.: A multiplex social contagion dynamics model to shape and discriminate D2D content dissemination. IEEE Trans. Cogn. Commun. Netw. (2020)
15. Alvarez-Rodriguez, U., et al.: Evolutionary dynamics of higher-order interactions in social networks. Nat. Hum. Behav. **5**(5), 586–595 (2021)

MIX
Papier aus verantwortungsvollen Quellen
Paper from responsible sources
FSC® C105338

FSC
www.fsc.org

If you have any concerns about our products,
you can contact us on
ProductSafety@springernature.com

In case Publisher is established outside the EU,
the EU authorized representative is:
Springer Nature Customer Service Center GmbH
Europaplatz 3, 69115 Heidelberg, Germany

Printed by Libri Plureos GmbH
in Hamburg, Germany